Mutant

'Money is making biology mutate. Capital nowadays reaches ever deeper into organisms to reformat their genes, metabolisms and more. This book is a lucid and provocative guide to this brave new world.'

—Stefan Helmreich, Elting E. Morison Professor of Anthropology, Massachusetts Institute of Technology

'An ambitious and timely critique of biopolitical economy that traces the long history of genetics through to the high-tech biofoundries that serve as capital's hidden abode of genomic production. The battle against capital requires a struggle over the means of genomic production, and *Mutant Ecologies* provides an essential, historically and theoretically rigorous assessment of the terrain.'

—Jesse Goldstein, Associate Professor of Sociology, Virginia Commonwealth University

'*Mutant Ecologies* is an incandescent illumination of capital's own molecular revolution. With deep research and smart theory, Borg and Policante take us into the planet factory's latest abodes of production, where genomic tools manufacture life-forms tailor-made for accumulation on a scorching planet; a must-read.'

—Nick Dyer-Witheford, author of *Cyber-Marx* and *Cyber-Proletariat*

'Capitalism is becoming genomic, Erica Borg and Amedeo Policante contend. Reinventing the critique of political economy in this new conjuncture of capital accumulation is the task they successfully pursue in this book. *Mutant Ecologies* is a major work, a must-read both for scholars of capitalism and for anybody interested in the history and present of genomics.'

—Sandro Mezzadra, Professor of Political Philosophy, University of Bologna

Mutant Ecologies

Manufacturing Life in the Age of Genomic Capital

Erica Borg and Amedeo Policante

Illustrations by Asbjørn Skou

First published 2022 by Pluto Press
New Wing, Somerset House, Strand, London WC2R 1LA
and Pluto Press Inc.
1930 Village Center Circle, Ste. 3-384, Las Vegas, NV 89134

www.plutobooks.com

British Library Cataloguing in Publication Data
A catalogue record for this book is available from the British Library

ISBN 978 0 7453 4453 9 Hardback
ISBN 978 0 7453 4452 2 Paperback
ISBN 978 0 7453 4456 0 PDF
ISBN 978 0 7453 4454 6 EPUB

This book is printed on paper suitable for recycling and made from fully managed and sustained forest sources. Logging, pulping and manufacturing processes are expected to conform to the environmental standards of the country of origin.

Typeset by Stanford DTP Services, Northampton, England

Simultaneously printed in the United Kingdom and United States of America

Contents

Fragments of Tomorrow

In the biotech Utopia conjured by venture capitalists, entrepreneurial scientists and techno-plutocrats, monoculture plantations populated by genetically modified crops cover most of the Earth. Row after row of perfectly identical corn-stalks, producing pesticide in their cellular machinery. Fast-growing sugarcane, shooting to full height in half the time of its unmodified kin. Herbicide-resistant vegetables, prospering in the midst of ever-increasing amounts of poison in the air and soil. Spring is silent; no insects dwell in the fields – a cocktail of bio-engineered pesticide-producing plants and genetic extinction technologies have wiped them out. A few lonely genetically modified honeybees, engineered to withstand the pesticides dumped on never-ending fields, buzz among the rows of identical plants.

In a nearby factory farm, genetically engineered cattle grow bodybuilder-type, hyper-muscled bodies. They no longer have pesky horns that might damage workers or profits – these have been genetically removed. The space is overcrowded, but the veterinary doctors are confident that the last round of genetic modification may contain the next epidemic. Up in the hills, where salmon used to swim in rivers before they were all dammed, sit massive tanks in which the fish grow to maturity in half the natural time.

In the rapidly shrinking forests, bioprospectors are busy mining a disappearing biodiversity, hoping to sequence their genomes for the latest proprietary pharmaceutical commodity. Using DNA barcode scanners, they create a digital copy of the forest in server farms located thousands of kilometres away. Timber companies are busy chopping down poplar trees disposed in neat geometric patterns. The gene-edited trees grow fast, their accelerated life cycle boosting the expected rate of profit. Yet, no animal will munch their bark. The grey squirrels who used to eat into loggers' profits have been genetically sterilised and wiped out.

At sea, genetic life-support keeps corals breathing in a warming ocean. Further beyond, the global dead zone spreads. There, floating farms grow genetically modified algae in water that is too oxygen-poor to host marine wildlife. Oil rigs in the distance keep pumping, but ships are already releasing gene-edited microbes to lap up the oil spills. Beneath the waves, in the deep time of the Hadal Zone, unmanned vehicles hunt extremophilic microbes by hydrothermal vents, sequencing their genomes in search of potentially valuable genetic traits to patent.

In crowded cities around the world, conscious urbanites have a choice between a variety of bio-engineered foods. 'Become the envy of your friends and followers with this highly sought-after delicacy! Pinkglow® will look phenomenal on whatever social media platform is en vogue by the time you read this.' Biofortified foods produced by billionaire philanthropies are offered to the many who have been forcefully removed from the land. In tiny rental studios, miniaturised pigs keep company to the housebound workforce.

Capital's internal contradictions appear temporarily suspended; no need for radical social and ecological change.

Introduction

Fast-growing salmon, herbicide-resistant plants, hyper-muscled cattle, synthetic bacteria, Trojan mosquitoes, humanised mice, autocidal rats, pigs growing human organs – these are but a few of the dazzling new life-forms that have recently emerged from corporate and university laboratories around the world, all promising to lubricate the circuits of capital accumulation in distinct ways. Each of these living organisms is a metabolic being, which systematically transforms raw materials into energy, molecules and waste by-products. Each is both a destructive agent (breaking down molecular compounds in processes of digestion and respiration) and a productive agent (building up new molecular compounds by synthesising the proteins, carbohydrates, lipids and nucleic acids they need in order to persist, grow, act and reproduce). Each is both a catabolic and anabolic force. Through their metabolic processes of reproduction, all living organisms participate in the shaping of the global environment. Indeed, we could say that the biosphere is the ever-changing result of the multiple metabolic exchanges set in play by living organisms as they go about their existence.

Mutant Ecologies traces the spinning of new synthetic threads into the web of life. Genetic engineering seizes the living vectors that make and unmake the world. It is a set of techniques, elaborated in the last half a century, which offers new ways of manufacturing commodities by taking possession of living bodies, redesigning metabolic pathways and thereby affecting global ecological processes at ever-larger scales. The emergence of new genomic tools has opened up new horizons of industrial production, enabling the manufacture of life-forms adapted to the peculiar requirements of capital accumulation. Capital no longer contents itself with simply appropriating the living bodies of plants and animals. It purposefully engineers their internal metabolism, thereby reshaping the countless living vectors that constitute the global biosphere. Genomic science and genome editing are increasingly central to business operations in a number of sectors – including agriculture, aquaculture, livestock breeding, pharmaceutical production, the chemical industry, the textile industry, and many more. In all of these branches of production, the deliberate induction of genetic mutations enables a multiplicity of metabolic shifts, whose impact on the socio-ecological (re)construction of the biosphere remains uncertain. Capital is driving a biological revolution, which ripples through the ecosphere.

The introduction of genome editing technologies has rapidly transformed the ways in which life is imagined, manipulated and exploited for profit. It has stimulated the elaboration of novel technological solutions to many of the most pressing economic, social and ecological problems; a new understanding of the relation between nature and culture, ontology and technology, society and the self; and a new vision of the future that negates the necessity for radical political change. CRISPR/Cas9, in particular, has been variously celebrated as a 'breakthrough technology' that may 'control evolution', 'improve lives', 'protect the Earth's biodiversity', 'change society', 'edit the human race' and even 'save the world'.[1] According to the Royal Swedish Academy, which awarded the 2020 Nobel Prize in Chemistry to Emmanuelle Charpentier and Jennifer Doudna, 'these genetic scissors have already taken the life sciences into a new epoch and, in many ways, are bringing the greatest benefit to humankind'.[2]

In writing this book, we aim to develop a critique of this sort of contemporary biotech utopianism by investigating the complex relationship between capital accumulation and the genome-editing industry. As the Nobel Committee put it in their award statement, CRISPR may certainly 'take the life sciences into a new epoch', 'rewrite the code of life' and make 'us gaze out onto a vast horizon of unimagined potential'.[3] And yet, this technological innovation continues to operate within a socio-historical context defined by capitalist competition and exploitation. In the following pages, we investigate some of the structural political-economical tendencies that are shaping, and will continue to shape for the foreseeable future, the development of these genetic biotechnologies. Genomic science, and its biotechnological applications, is drastically transforming the constitutive relationship between what Charles Darwin called 'the economy of nature' and political economy; between biological and social life; between living beings and capital. In what ways has the logic of capital shaped the development of these biotechnologies? And, on the other hand, how is genomics reshaping the political economy of capital accumulation on a global scale?

MOLECULAR MARX

At the cusp of the Second Agricultural Revolution, Karl Marx could still write that 'it is impossible to deliver a five-year-old animal before the lapse of five years'.[4] At least in the short term, Marx suggested, capitalists must take the form of plants and animals as a given: there is a given time when the calf must grow to maturity, 'abandoned to the sway of natural processes'. In any industry which uses living means of production, this maturation period is time that capital is *dormant*, unable to valorise itself in the market. In its endless quest to reduce turnover time and accelerate production, capital faces a series of biological barriers. Capital has long strived to compress biological time. In the same way that it aims to increase the extraction of relative surplus value by continually accelerating the pace of work, it strives to accelerate the bodily metabolism of bacteria, plants and animals.

Marx already noticed that nineteenth-century breeders were developing new ways to get 'animals ready for their destination in less time by changing the way of treating them'. Reducing the 'maturation time' – in which no human labour is applied and thus no surplus value is produced – provided a powerful way to reduce turnover time and increase profits.[5] Biotechnologies provide new means to erode these temporal constrictions – however imperfectly and with countless unintended effects. Today, a single corporation can conceive in the abstract how best to modify its living means of production to speed up the labour process and accelerate capital turnover. The time and money necessary to implement those modifications is rapidly declining.

The process of real subsumption has reached the genetic threshold. Capital not only transforms the labour process, and the environment in which it takes place; it develops artificial life-forms, conceived and designed to support, accelerate and extend the valorisation process. The twentieth century has been profoundly shaped by technological processes driving the real subsumption of labour under capital: in the midst of political struggles, labour processes have been re-organised, rationalised and automated. The twenty-first century is increasingly shaped by genomic interventions tending towards the real subsumption of life under capital: the metabolic processes of countless living beings are being re-designed to facilitate, extend and secure industrial production.

Through bioengineering, capital creates its own, adequate genetic foundation: not only 'a world in its own image', but also a new life and a new biology. Constant revolutionising of the living means of production, uninterrupted disturbance of all ecological relations, everlasting uncertainty and agitation distinguish the genomic age from all earlier ones. In the endless quest for relative surplus value, all life-forms become targets of genetic improvements and mutations. All new-formed ones become antiquated before they can ossify. Yesterday's innovative recombinant life-form is rapidly turned into a discarded technological fossil. In writing this book we aim to investigate the historical relationship between capital accumulation, the life sciences and the gene-editing industry. We try to face, with sober senses, our mutated conditions of life and our ecological relations with other kinds of life.

We do this by bringing Marxist methods to the molecular scale, asking: What can the emergence of proprietary life-forms reveal not only about the development of molecular biology, but of capitalist development in general? Who owns the means of molecular production? What world-views are embodied in genetically modified life-forms? In patented living artefacts? Marx suggested that one of the key features of the capitalist mode of production is a perpetual revolutionary transformation in technology. What does this analysis entail when the technology that is perpetually improved, revolutionised and transformed is the living body and the living cell? We begin to unravel these questions by studying how genetic biotechnologies have evolved as part of an emergent biopolitical economy: a strategy of accumulation and a mode of government, which increasingly relies on the development of a technoscientific knowledge of biological processes and new means of manipulating them.[6]

Marx's scattered notes on methods have greatly influenced our approach to this research. We read *Capital* as a work of natural history, insofar as it offers an analysis of how the internal functioning of capital becomes a life-shaping force which revolutionises not just human societies, but their ecological conditions. Marx argues that 'technology reveals the active relation' between humans and the rest of nature, insofar as technology forms the 'organs that are the material basis of every particular organisation of society'.[7] In our reading, from the vantage point of the so-called Anthropocene, a critical history of technology captures the dialectical relationship between 'natural' and 'social' history: retracing a historical trajectory that is inherently *socionatural*. As a work of historical-geographical materialism, this book aims to tease out the many messy determinations that facilitated the historical emergence of molecular biology and its application in genomic biotechnologies.

We pursue, in other words, a critique of biopolitical economy that draws upon, while being distinct from, existing research in 'bioethics' and 'biopolitics'. Foucault's reflection on biopolitics – while explicitly haunted by the emergence of genetic engineering in his time – interrogates the historical trajectory by which society started to pose itself the problem of how to foster life *in general*, of how to make it multiply. But what is left out of Foucault's thinking is the more granular question of *what kind* of social processes and forces participate in the social construction of *particular forms of life* through genetic engineering? Since the 1970s, Foucault's reflections on biopolitics have prompted many to focus on the emergence of a political rationality that takes the administration of living populations as its fundamental object. Today, critical theory offers a multitude of reflections on the contradictions of a biopower geared 'to ensure, sustain, and multiply life, to put this life in order'.[8] These reflections, however, seldom consider the constituent relation between this emergent political rationality and capital's fundamental drive to manipulate the living means of production it mobilises for its self-valorisation.

More recently, a vast literature on 'bioethics' has grown up around the doubtful assumption that genetic biotechnologies are a neutral tool that simply needs to be handled correctly, providing employment to a whole army of moral philosophers tasked to determine the ethical way forward. This literature often recognises the dangers and risks associated with contemporary uses of genetic biotechnologies, but it relies on the assumption that other uses may be promoted by appealing to the morality of the reader and/or to the enlightened wisdom of public regulators. This obscures the fact that structural tendencies are at play, which profoundly shape the research paths chosen by scientists; the funding strategies pursued by governmental agencies, public universities and private corporations; the regulations crafted by bureaucrats and jurists. To paraphrase Marx's famous incipit to *The Eighteenth Brumaire of Louis Bonaparte*, we could say that human beings may hold the power to shape their own mutant ecologies, but they do not make them as they please; they do not make them under self-selected circumstances, but under circumstances existing already, given and transmitted from the past.[9]

ECOLOGICAL RIFTS, METABOLIC SHIFTS AND THE GENOMIC FIX

In March 1997, *Business Week* released an enthusiastic special issue titled *The Biotech Century.* On the front cover, a young sheep stands still on the grass, gazing into the reader's eyes. It is the first published picture of Dolly, the first mammal to be produced through cloning technologies. The issue opens with an interview with the managing director of PPL Therapeutics – Dolly's mother-company – discussing the experiment as part of a larger commercial strategy. 'Our aim', he says, 'is to make drugs, and to make money. The fact that we use animals to do it is interesting, but it's not what it's all about.'[10] In the following pages, Bloomberg's most iconic magazine paints the silhouette of a dazzling future, profoundly revolutionised by the application of genomic knowledge to the production process: a future in which PPL Therapeutics will go on breeding animal bodies genetically designed to produce cheap proteins, useful drugs, and even human organs; a future that promises hyper-productive agriculture, restored biodiversity and, ultimately, an epochal transcendence of human and environmental limits. *Business Week* presents to its readers the dream of a fully revitalised global economy – one in which hunger, disease and environmental catastrophe are reduced to simple biotechnical issues, to be solved by the systematic overcoming of the limits to production inscribed in the DNA of plant, animal and human life.

Technotriumphalism reigned supreme in the final decade of the twentieth century: genetically modified crops were touted as spelling the end of poverty and hunger, while cloning technologies inspired fantasies about immortality. Several social and ecological crises later, this optimism seems hopelessly naïve. Dolly herself died a premature death, and three decades later cloning technologies remain fraught with deadly inconsistencies. The GM crops that were supposed to end hunger rather exacerbated unequal access to food, expanded colonial land relations, and generated unpredicted ecological problems. The recent recognition that the Earth has entered into a new geological era, characterised by an ongoing 'Great Acceleration' in industrial production and pollution, infuses new meaning to attempts to engineer life in the biosphere.[11] Earth System scientists have christened this new epoch the Anthropocene – the geological age of *anthropos*, the human.[12] The history of modernity, which has long been narrated as a steady march towards mastery over nature, suddenly appears as a blind race towards ecological collapse.

The existential threat represented by multiple ongoing shifts in the biogeochemistry of the planet is increasingly recognised. Many are now turning to a desperate search for technical fixes designed to contain the rapidly proliferating socio-ecological crises that ravage the biosphere. Such fixes tinker with surface dynamics, while leaving the root of the problem unaddressed: they foster change only so that things can stay the same. They aim to adapt the Earth's biogeochemical cycles to the rhythms of capital accumulation. In the context of current debates concerning the Anthropocene, we interrogate the intersection between ecological crisis and emergent modalities of *genetic bioengineering* and *genetic control*: What kind of eco-social politics arise from the recognition that twentieth century industrialisation has profoundly undermined the conditions for social reproduction? What visions of a bioengineered planet are being conjured, promoted, and funded? What role do genomic technologies play in this barely secularised salvation drama?

The desperate quest for new techno-fixes is the social context in which the hyper-genomic age germinates and takes root. In an article published in *Foreign Affairs* in 2020, William Henry 'Bill' Gates III revisits the bio-futurism of *Business Week*, injecting it with an apocalyptic spin. In 'Gene Editing for Good', the co-founder of Microsoft stresses that 'gene editing could help humanity overcome some of the biggest and most persistent challenges in global health and development' thanks to its inherent 'potential to save millions of lives and empower millions of people to lift themselves out of poverty'. Gates also sounds a warning, insisting that there are no feasible alternatives to the biotech revolution and that, therefore, 'it would be a tragedy to pass up the opportunity'.[13] Gates urges the public to face an epochal choice between genome engineering and planetary tragedy. Meanwhile, Cascade Investment – one of Gates's main assets firms – is investing heavily in the biotechnology industry, acquiring stocks in a number of biotech companies including Ginkgo Bioworks (which aspires to 'replace technology with biology'), Memphis Meat (which specialises in the production of lab-grown meat produced from animal stem cells) and Vir Biotechnology (whose website promises 'a world without infectious diseases'). Bill Gates's complex affair with genetic biotechnologies – at once an object of his political writings, a recipient of generous donations from his philanthropic foundation, and a frontier of investments for his capital funds – is symptomatic of the increasingly ubiquitous, and ambiguous, role of these technologies in addressing social and ecological issues.

The recent introduction of a new generation of genome editing technologies informed by research on CRISPR-Cas9 has stimulated a new vision of the future that displaces and negates the necessity for radical political change. CRISPR has been celebrated as a 'genomic fix', capable of addressing issues in areas as diverse as agriculture, livestock farming, biodiversity conservation, climate change and human health. Jennifer Doudna, one of the developers of CRISPR, has been prominent in marketing the molecular tool as a form of 'genetic command and control', which will enable unprecedented mastery over biological processes and new ways of governing the Anthropocene. In *A Crack in Creation*, she gives a daring account of how CRISPR delivers us into 'a new age in genetic engineering and biological mastery – a revolutionary era in which the possibilities are limited only by our collective imagination'. The 'supreme mastery' exercised 'over genetic material inside living cells', Doudna assures her readers, offers new ways of directing the natural history of our own species, and possibly of the entire biosphere. Natural history is neatly cut in two epochs: before and after CRISPR. 'For billions of years, life progressed according to Darwin's theory of evolution'; but now 'scientists have succeeded in bringing this primordial process fully under human control'.[14]

Genome engineering transmogrifies the genome into an adjustable lever, whose manipulation promises to steer natural history, direct the evolution of species, and thereby shape local ecosystems and the planetary biosphere. In this way it reflects the logic underlying its sibling discipline, Earth system science. When atmospheric chemist Paul Crutzen first proposed the concept of the Anthropocene, he concluded with a bold vision that has since taken root in social discourse: the human species must become an active steward of the Earth by monitoring its ongoing transformations, anticipating critical thresholds and endlessly intervening to re-establish ecological balance. New monitoring technologies will constantly track the shifting parameters of Earth's multiple subsystems, rendering the planet into a cybernetic system, which can be centrally governed and purposefully engineered.[15] Bioengineering represents a central element in this emerging political vision, offering novel means to redesign the biosphere and terraform the Earth. Since each species shapes its own ecological niche, through its metabolic interactions with the surrounding environment, genetic engineering constitutes an indirect way of stewarding the planet through the industrial production of living-artefacts. The metabolic processes taking place at an intracellular scale within each organism shape local and planetary metabolic assemblages and biogeochemical cycles. Who controls the gene controls the body, who controls the body controls the species, and who controls the species controls the world. Or at least, this is the vision conjured by today's biogeoengineers.

Yet, there is a troubling tension at the heart of molecular biology between the growing realisation that even the most simple organisms remain too complex to be properly understood at the molecular level, and the hyper-modernist attempt to modify those very organisms to make them increasingly legible, predictable and industrially exploitable. Rather paradoxically, as molecular biology makes visible a world of irreducible biological complexity, genome engineering is embarking on increasingly ambitious programmes to rationalise that living complexity. This hyper-modernist attitude reflects a form of scientific reductionism in which living organisms are increasingly conceived as reprogrammable 'molecular factories', whose metabolism can be functionally tweaked and engineered.

MUTANT ECOLOGIES

The book's title conveys multiple meanings. As biologists often quip, mutations have always been 'the raw material of evolution'.[16] They are the unpredictable molecular alterations that generate 'the genetic variation on which the evolutionary process depends'.[17] It is easy to understand how a mutation that allows an organism to feed, grow or reproduce more effectively could cause the mutant variant to become more abundant over time. What starts off as a mutant form that deviates from the genomic norm can slowly become the new normality. Mutations are the 'genetic events', which constantly redirect natural history: the Epicurean swerves that deflect evolutionary patterns from following a straight monotonous line.[18]

Since the early twentieth century, molecular biologists have dreamed of establishing control over these processes of mutation, using X-rays, nuclear radiations and atomic power. By the 1970s new technologies based on recombinant DNA provided the first tools to induce genetic mutations in predictable ways. Recombinant DNA transformed mutagenesis into a form of labour: mutant forms of life were planned in the abstract, their genome designed on paper, and tentatively realised in practice.[19] Today, genome engineering and synthetic biology provide increasingly scalable tools, which are turning these artisanal laboratory practices into an industrial process. While mutations continue to be 'the raw material of evolution', induced mutations have become a prominent strategy of capital accumulation.

Industrial production is increasingly mobilising forms of life whose genome is purposefully manipulated through a variety of mutagenic techniques: X-ray mutagenesis, recombinant DNA and novel methods of genomic engineering. Mutant life-forms are the final products of the biotech industry, which are patented and sold as commodities on global markets. They are also the living means of production that, once out of the biotech laboratories that generate them, are transforming labour processes in many sectors of the global economy. Contemporary farms increasingly rely on genetically modified crops. The livestock industry is developing fast-growing cattle and disease-resistant pigs. Aquaculture companies are already marketing genetically engineered salmon. The pharmaceutical industry employs humanised mice, bacterial bioreactors, viral vectors and pharma-goats. Even the conservation industry is developing bespoke organisms to carve out its 'rewilded' ecologies: pseudo-mammoths, specicidal rats and Friendly™ Mosquitoes are only a selected few of these mutant 'ecosystem engineers'. Mutant ecologies are no longer exclusively confined in laboratory spaces, nor within the fenced pastures controlled by agribusiness.[20] Multiple processes of manufacturing life are spawning mutant ecologies at every corner.

If we say that capital increasingly relies on 'manufacturing life', this has a dual meaning. The term refers firstly to the fact that genome editing technologies are currently directed towards the industrial production of living organisms, whose metabolism is purposefully engineered through targeted-mutagenesis. Genomic biotechnologies promise to accelerate capital accumulation, while fixing a number of structural contradictions that endanger socioecological reproduction: from climate change to biodiversity loss, from malnutrition to viral pandemics. In the twenty-first century, *manufacturing lives* – using tools and knowledges deriving from molecular and synthetic biology – has become an accumulation strategy *and* a form of depoliticising technoscientific governance. The claim that particular life-forms are *products of manufacture* is central to corporate strategies of appropriation, legitimising patents on countless genetically engineered organisms. But there is also another way in which the concept of 'manufacturing lives' may be read. Many of the gene-edited organisms that populate today's mutant ecologies are conceived and designed as molecular factories and living means of production. If the first industrial revolution was built on 'manufacturing machines' like the Spinning Jenny, the contemporary bioindustrial revolution is built on *manufacturing lives*: life-forms engineered to produce a growing variety of valuable commodities.

This book interrogates the intersection of these two historical processes. It asks: How is life industrially engineered in the twenty-first century? What kind of lives emerge from contemporary biofoundries and genomic assembly lines? How does the claim to 'manufacturing life' legitimise processes of enclosure and appropriation? But also: How are these gene-edited organisms subsequently deployed as living means of production in the chemical industry, in agriculture, in medicine and pharmaceutics, etc.? How are they transforming the way people work and live in agricultural lands, in industrial facilities, in barnyards and slaughterhouses, in biotech labs and medical clinics?

* * *

This book places the emergence of 'mutant ecologies' – populated by life-forms that are at once 'manufactured' and 'manufacturing' – in the context of the increasingly important role played by genomic science and genomic biotechnologies in contemporary neoliberal societies. We use the term 'genomic capital' to indicate branches of industry that employ genetic matter as raw material and genomic biotechnologies as means of production. The term reveals a common development across a number of different economic sectors: from agriculture to health, to industry and forestry. Capitalism is becoming genomic insofar as capital accumulation increasingly relies on: the accumulation of genomic material (Chapter 2), its transformation in valuable digital data (Chapter 3), and its purposeful manipulation (Chapters 4 to 8).

The book proceeds in two parts. The first half dissects the ways in which the dynamics of capital accumulation have shaped the emergence of genomics and molecular biology. Chapter 1 traces the history of genetics from the mid-1800s, unearthing its roots in the persistent quest for a theory of life that might provide control over living processes. We investigate the webs of power/knowledge that have shaped genetic biotechnologies, paying particular attention to the role of capitalist competition and ideological hegemony. Chapter 2 follows the transformation of genomic science into an industry dedicated to the systematic production of rent-yielding, patentable bio-objects. We consider the creation of a neoliberal framework for the accumulation of genetic property in the 1980s and 1990s, and the subsequent expansion of the global bio-prospecting industry. Gene-grabbing has supported the creation of novel value-chains, while sparking a new genomic politics. Multiple biopolitical struggles have emerged surrounding the contested ownership and control over isolated genetic sequences and genetically engineered organisms.

Chapter 3 analyses the emergence of a novel technoscientific paradigm based on genomic extractivism, underpinned by the conversion of life-forms into information-systems. We interrogate how the Human Genome Project spurred the construction of a critical infrastructure: a 'molecular railroad', for the genomic industry. We further analyse the rise of personal genomics as an industry based on the transformation of DNA molecules into genomic data and of genomic data into capital. Finally, we consider how genomic data from non-human life is being heralded not only as a precious raw material, but also as the basis for new ways of governing the ecological crisis. Chapter 4 surveys new 'genome editing' techniques and charts their potential for accelerating the production of mutant lives adapted to capital. It traces ongoing battle for proprietary control over CRISPR and sketches the emergence of a new 'synthetic kingdom' of life.

In the second half, we survey how genomics and molecular biology are transforming practices of production and opening new avenues for capital accumulation. We focus on concrete applications of novel genomic biotechnologies to diverse areas of industry. Chapter 5 analyses how genome engineering is transforming capitalist agriculture. It follows the dialectic of metabolic crises and chemical fixes of the twentieth century before stepping into the 'molecular factory farms' being constructed with genome engineering. We situate the genomic subsumption of the metabolism of plants and animals as a continuation of the Green and the Gene Revolution, and in the context of long-standing struggles over access to the means of life. Chapter 6 surveys the shifting topographies of biogeoengineering and genetic control. Contemporary projects of 'resurrecting' woolly mammoths, exterminating 'alien species' from islands, or winning the 'war on the mosquito' reflect a growing ambition to craft a bespoke biodiversity for the Anthropocene.

Chapter 7 is concerned with the production of custom-made mutants in medicine and the pharmaceutical industry. It follows the technonatural history of the laboratory mouse as a living commodity and as an experimental body. Today, genome engineering is delivering a new generation of mutant mice, specifically designed to more closely resemble *Homo sapiens*. Humanised mice represent both *living assets*, as patented organisms, and *living commodities*, as bodies sold on the world market. Big Pharma is redesigning animal bodies into increasingly specialised experimental platforms. It is also turning them into bio-foundries in which human proteins, skin and organs are grown and harvested for commercialisation in the world market. Pharmaceutical production is increasingly molecularised, taking assembly lines out of the factory and into the body of genetically modified microbes, plants and animals.

Chapter 8 depicts the penetration of genome editing under the human skin. The first section analyses the emergence of recombinant and genetic vaccines as commodities and instruments of security and government. It highlights how genetic vaccines molecularise the production of antigen proteins from the factory to the human cell. It then considers practices of genome engineering in the human species, focusing first on existing programmes of gene-therapy to then turn to the thorny issue of germline editing. We recount how the encounter of neoliberal thought and genome engineering technologies has birthed an increasingly vocal movement explicitly advocating for a 'liberal eugenics' relying on the deregulation of germline editing.

Throughout the book, we maintain that the advent of genomic technoscience is contributing to the emergence of a new phase of capitalist accumulation characterised by the increasing subsumption of life under capital.

1

Life's Inner Workings

Cracking Codes, Mutant Flies and Recombinant Lives

The science of genetics is so central to twenty-first-century societies that it can be difficult to remember what it stands for: a research programme focusing on the patterns and mechanisms of heredity, aimed at understanding and potentially controlling that process. Genetic science, and the genetic biotechnologies it engenders, is transforming how living beings are understood, exploited and governed. The genome represents not only an object of scientific study, but also a field of power. It is not only a biological entity shaped by natural history, but also a technoscientific terrain shaped by economic, social and political forces. It is no longer possible to think of the molecular realm of the genome as being isolated from socio-economic tendencies, technological interventions and political struggles. One may argue that this is nothing radically new. The history of the twentieth century, after all, was already traversed by eugenic fantasies and practices. This chapter recounts how the dream of establishing control over hereditary processes has energised the development of molecular genetics from its inception, while stressing some of the scientific advancements and transformations that resulted from this long-standing quest.

In the early twentieth century, when geneticists celebrated new techniques of artificial mutagenesis as the first step in a technoscientific revolution that would 'place the process of evolution in our hands', their Promethean vision appeared to be the stuff of science fiction. This fantasy, nevertheless, would resurface repeatedly in the writings of molecular biologists in the first half of the twentieth century. It was regularly and ambiguously evoked either as an aspiration, or as a looming threat. By the end of the 1970s, these abstract dreams began materialising in flesh as mutant commodities swept the world market. Early biotech companies transformed genetic engineering into an industrial strategy, which promised to have profound social, political, ecological and economic effects.

Looking back at the relatively short history of molecular biology, we do not intend to offer a complete account of this scientific discipline. Rather, we point out some of the main socio-economic tendencies that have led to an increasing investment – both in terms of labour power, capital and social imagination – in the collective search for technological means that would allow the manipulation of hereditary patterns, and thus the conscious stewardship of evolutionary history. This is the history of a collective drift. It is a chronicle of how one of the technoscientific dreams that define our present emerged in modern times as a result of a whole set of social, political and economic contradictions; and how this imaginary was gradually reified in scientific theories, technological apparatuses and the flesh of the living. The history of genetics, as a theory of life, and the development of genetic biotechnologies, as a form of power over life, were constantly intertwined. We trace some of the multiple trajectories that led to the present moment that we inhabit together with mutant flies, molecular chimaeras and recombinant bacteria.

We review this historical trajectory by focusing on three distinct moments. First, we show how the science of genetics introduced a new conception of life. Second, we consider how and why this particular theory of life – which in the early 1950s was still highly heterodox even among molecular biologists – achieved growing financial support, academic power and intellectual influence. Third, we reflect upon the rise of genetic engineering as a form of power over life enabled and legitimised by the rise of molecular biology, focusing particularly on the first wave of genetic engineering techniques.

MUTANT FLIES OVER ATOMIC GARDENS

'Evolution in Our Hands'

In the middle of the nineteenth century, an economically motivated interest in hereditary processes gradually led to the establishment of a new scientific discipline. School books often portays the colourful life of Gregor Mendel as a sort of idyllic origin myth. Working in the gardens of the monastery of Brnö, so the story goes, the monk noted that some of the visible characteristics of pea plants – such as seed shape and flower colour – changed, generation after generation, in rather predictable ways. On this basis, he postulated the existence of invisible hereditary 'elements', which would determine the visible characteristics of each and every individual plant according to precise rules that could be determined experimentally. While forgotten for the rest of the nineteenth century, the 'Mendelian laws of inheritance' would fundamentally influence early genetic science by providing a set of theoretical questions that would guide scientific research for decades.

Mendel's peas play a symbolic role in the history of genetics, comparable only to the one of Newton's apples in the history of physics.[1] Yet, this origin-myth tends to place Mendel's studies in a monastic social vacuum: his laws of inheritance appear as a mystic vision, which could have taken place at any time in history and at any point in space. Once contextualised in the political and economic milieu in which they first emerged, however, the driving logic behind these studies emerges from the mist. Mendel was part of a political, economic and scientific community that profoundly influenced the direction taken by his scientific work. In the nineteenth century, Brnö was the pulsing core of a thriving textile industry and one of the most important economic centres of the Austro-Hungarian Empire. The 'Austrian Manchester', as it was known at the time, directly challenged the hegemony of northern England in wool production. It constituted another stage on which the first steps of the Industrial Revolution were tentatively trodden. We may then ask: what difference does it make to look at the origins of capitalist modernity from a monastery in Brnö, rather than from a pin factory in Manchester?

There were elementary reasons why Brnö was known as the Austrian Manchester: it was, after all, an early industrial town and the site of a prosperous, market-oriented textile industry. There were also differences. For instance, while technical advances in mechanics are often placed right at the centre of most historical accounts of nineteenth-century Manchester, advances in breeding technologies and biological science were arguably one of the forces that drove economic modernisation in Brnö. The city's Agricultural Society was, then, one of the most thriving scientific institutions in central Europe. It regularly hosted talks by agriculturalists and natural scientists, investigating the potential offered by different practices of hybridisation. Its Sheep Breeders' Society regularly published reports exploring how 'climate' and 'seed' influenced wool production. It hosted frequent debates on the advantages offered by various breeding methods. It financially supported studies that could promote a more elaborate understanding of heredity, and some form of control over its dynamics. It envisioned a new science in the service of industry, 'seeking ways to regulate the process of heredity'.[2]

The origin of the Mendelian laws of inheritance was directly influenced by these breeders' dreams. Cyril Napp, the abbot of the Augustinian monastery of St Thomas, was a prominent member of the association. The church owned large tracts of agricultural land, and Napp was interested in promoting the study of new breeding practices. In his opening address to the 1837 meeting of the Society, he argued that a new approach was necessary to improve wool and textile production: 'What we should have been dealing with is not the theory and process of breeding. Rather, the question should be: what is inherited and how?'[3] In this way, Napp defined the fundamental challenge that faced the new science of heredity. But Napp did more than simply pose an important question. In the following years, he organised a cohort of dedicated researchers to search for an answer.

It was among them that Gregor Mendel developed his systematic research on pea plants. He grew convinced that inheritance was based on invisible hereditary elements, which would combine to determine the visible properties assumed by a living organism. The visible form of each plant was little more than an epiphenomenon, an appearance resulting from the specific combination of hereditary factors passed from parents to offspring. According to his theory, these factors came in pairs, one from each parent, and when the two factors clashed, only one of them would dominate and express itself visibly.[4] Although it offered key insights into the mechanism of inheritance, Mendel's hypothesis did not answer the question posed by his abbot. What was inherited and how remained a mystery to be solved. Yet, his studies helped set the science of heredity on the particular track it would follow for much of the following century. Twentieth century biologists continued to question the nature of the hypothetical hereditary elements: Did they really exist? What were they made of? Where were they located? How did they shape the development of living organisms? And, finally, could they be observed and manipulated?

In 1909, Wilhelm Johannsen introduced the term 'gene' to indicate the hypothetical hereditary elements imagined by Mendel.[5] The study of heredity increasingly embraced a mechanistic conception of life, seeking to grasp vital processes by closely analysing the multiplicity of minute biological building blocks that participate in their unfolding. The gene was perceived as holding the key to their geometries and dispositions. No other scientist embodied the mechanist-reductionist ideal quite like Jacques Loeb. In *The Mechanistic Conception of Life*, Loeb addressed the ambitious question of whether 'life, i.e., the sum of all life phenomena, can be unequivocally explained in physico-chemical terms'.[6] If this is indeed the case, Loeb continued, 'our social and ethical life will have to be put on a scientific basis and our rules of conduct must be brought into harmony with the results of scientific biology'.[7] Loeb's engineering ideal for biology, though controversial, profoundly influenced the development of early molecular biology and the conception of heredity emerging from it. Inspired by Mendelian genetics and spurred by Loeb's clarion call, scientists of various backgrounds began to systematise not only a mechanistic *conception* of life but a molecular method of study which could support it.

From the 1910s and throughout the 1930s, Loeb's long-time associate Thomas Hunt Morgan and his colleagues at Columbia University attempted to answer these classic questions by developing an experimental and psychochemical approach to the study of vital processes. They observed generations after generations of *Drosophila melanogaster* – a variety of vulgar fruit fly which breeds quickly and is easy to maintain in a laboratory environment – tracking how traits like eye colour and wing shape mutated and varied.[8] In 1927, Hermann Muller, one of Morgan's closest collaborators, discovered that irradiating the flies with X-rays greatly boosted the rate of random mutations. This indicated not only that 'the gene was the basis for life', but that it was something material, which could be manipulated like any other physical molecule.[9] Scientific efforts could then rapidly shift from simply studying hereditary processes to manipulating them.

'The attention of those working along classical genetic lines', Muller wrote in a 1927 paper titled *Artificial Transmutation of the Gene*, 'may be drawn to the opportunity, afforded by the use of X rays, of creating in the chosen model organisms a series of artificial races'. In a subsequent Nobel Prize Lecture, he insisted that the 'production of mutations by radiation is a method', which may soon 'prove of increasing practical use in plant and animal improvement, in the service of man'. Radiation, in other words, could be turned into a means of influencing evolutionary patterns, and thus the physiology of living organisms. He was already thinking of genes as 'controllable levers' that could provide some control over 'those forces, far-reaching, orderly, but elusive, that make and unmake our living worlds'. Speaking in front of the Nobel Committee and the media, Muller predicted that genetic research would soon 'reach down into the secret places of the great universe', giving humanity the tools to refashion itself into 'an increasingly sublime creation – a being beside which the mythical divinities of the past will seem more and more ridiculous'. Artificial mutagenesis, Muller concluded, might 'place the process of evolution in our hands'.[10]

Many enthusiastically embraced Muller's futurist vision. An article published in May 1928 by the *New York Times* captures the cultural climate of the day, reporting that Muller 'succeeded in obtaining at least 100 different varieties that had never before seen, varieties that bred true, thereby showing that the relationship of the genes had been permanently changed. Some of his flies had short wings, some notched wings, some no wings at all. This is one of the outstanding successes in forcibly directing evolution into new channels. [...] If, even with our present meager knowledge and skill, a Muller can predict that with a given dosage of X-rays a wingless race of fruit flies shall be born, surely we are on the road toward controlling human evolution.'[11] Eight years later, the *New York Times* looked back to Muller's experiments as an historical moment to be celebrated and remembered: 'When some future historian contrasts our barbaric twentieth century with his own happy era, he will not stint himself of praising Muller. To his monstrous fruit flies we trace the first deliberate, successful scientific interference with the process of heredity by external agencies'.[12] A few months later, George Gray – then one of the most prominent science writers and a publicist of the Rockefeller Foundation – stressed that 'radiation assuredly provides the energy to drive our wheel of life; demonstratedly it has provided a probe with which to reach into the living cell and alter and test the mechanism of life [...]'.[13]

The discovery that radiation could be turned into a technological means for altering genetic matter – thereby indirectly influencing the physiology of individual living organisms – represented the first step in an historical quest for control over hereditary processes. Already in the late 1950s, several programmes of so-called 'radiation breeding' were introduced around the world. By exposing seeds and young plants to radiation, scientists endeavoured to induce random genetic mutations in the hope of stumbling into a few useful ones. They could control the intensity of the radiation and thus the extent of the genetic scrambling, but not the outcome. To know the repercussions, they had to plant the seed, let it grow and examine the results. Most often the experiments killed the seed, or resulted in a debilitated organism. But in a few instances, the process produced mutants that could make a profit: a peanut with a tougher hull proved more resistant to the hardships imposed by mechanical harvesting, some mutant oats and wheat promised better yields, and an irradiated blackcurrant could grow three times the size.[14]

Encouraged by these early success stories, the United States government promoted the method as part of its 1953 *Atoms for Peace* programme. Politicians and scientists could now rebrand nuclear energy as a technological force that could be harnessed to shape life, fight famine and produce wealth for the masses.[15] Giant gamma gardens were established in the US, Europe and Japan. The programme killed innumerable plants, but it did bring to life the occasional valuable mutant such as Calrose76 (a semi-dwarf mutant variety of rice that currently accounts for about half the rice grown in California) and Star Ruby (a grapefruit with a particularly brilliant colour, which continues to be a popular cultivar).[16] Meanwhile, the Atomic Gardening Society strived to bring atomic energy and radiation breeding into the lives of ordinary citizens. 'Atomic Mutation Experimenters', wrote the president of the association, 'work as one body to help scientists produce more food more quickly for more people, and progress horticultural mutation for the masses'.[17]

Similar claims would accompany the advent of genetic engineering experimentation in the 1970s, and gene editing technologies in our own present. New forms of genetic manipulation have taken the place that radiation breeding once occupied in the collective imagination. Yet, far from being simply an eerie fossil from the past, atomic gardening continues to thrive in our own present. Gene-editing may be more precise, but radioactive stuff is cheap and abundant. Currently, the Mutant Variety Database logs over 3,000 plant varieties created through radiation breeding.[18] 'Spontaneous mutations are the motor of evolution', explains Pierre Lagoda, the current head of plant breeding and genetics at the International Atomic Energy Agency. 'We are mimicking nature in this. We're concentrating time and space for the breeder so he can do the job in his lifetime. We concentrate how often mutants appear –

going through 10,000 to one million – to select just the right one.'[19] The technique did not realise Muller's dreams of placing the process of evolution under human control. However, radiation breeding offered the first demonstration that genetics could be turned into a profitable venture; that genetic mutations could be artificially induced; and that new crops and mutant ecologies could be hurled into the present. Radiation not only offered new insights into the material basis of heredity, it also offered new living means of production to agribusiness, while endlessly rekindling the modern dream that life might one day become as malleable as clay.

CELLULAR FACTORIES AND LIVING ASSEMBLY LINES

'The Economy of the Cell'

The recognition that atomic radiation could accelerate the rate of genetic mutations within living cells had immediate applications. Artificial mutagenesis was adopted as both a means of destruction *and* as a promising means of production. While nuclear fallout imposed lethal mutations on living targets, atomic gardens demonstrated that artificial mutagenesis could be deployed as a life-shaping force. The increasing use of radiation as a means of artificial mutagenesis also reflected the fact that the field of physics and biology had become inextricably intertwined. By the second half of the 1930s, the life sciences were transformed by a steady influx of physicists and chemists, which furthered the search for a biological equivalent of the atom: a basic building block for the manipulation of living processes.[20]

Physicist Max Delbrück embodied the spirit of the new wave of genetic research in the 1930s and 1940s. Joining Morgan's team at Caltech, he described his main objective in plain terms: 'to grasp more clearly how the same matter – which in physics and chemistry displays orderly and reproducible and relative simple properties – arranges itself in the most astounding fashions as soon as it is drawn into the orbit of the living organism.'[21] He assumed that vital processes common to all living organisms – birth, growth, reproduction, mutation, and death – might be clarified by studying the simplest biological organisms. In the words of one of his students at Caltech, he 'was looking for *the* simplest system with which the fundamental problem of life could be elucidated.'[22] Delbrück thus embraced the study of bacteria and bacteriophages – simple viruses, which replicate by infecting bacteria and hijacking their cellular machinery – to construct a knowledge that, far from applying only to these lowly beings, would reveal regularities and mechanisms that define the very fact of living. As the French biologist Jacques Monod would famously quip, this method assumed that 'anything found to be true of *E. coli* must also be true of elephants'.[23]

Emphasising the originality of this approach, Warren Weaver – then director of the natural sciences section of the Rockefeller Foundation – introduced the neologism 'molecular biology'. This was, according to Weaver, a new hybrid science that would 'investigate ever more minute details of certain life processes' by analysing and manipulating molecules at the submicroscopic scale between 10^{-6} and 10^{-7} cm.[24] In the following years, the Foundation played a central role in funding and supporting the growth of molecular biology as a discipline, pouring millions of dollars into scientific laboratories and technological infrastructure. Some questions immediately arise: Why did private capital decide to bankroll this particular type of biological research? Why were biologists encouraged to focus on life processes taking place in the narrow strip of reality between ten micrometres and one nanometre? How did this research fit in the wider social objectives of a philanthropic organisation such as the Rockefeller Foundation?

Uncovering an answer to these questions has been one of the main objectives of historians of science such as Lily Kay, Michael Morange and Eric Vettel.[25] What these studies show is that the Foundation's support for molecular biology was an integral part of its larger social, political and ideological strategy: directing human behaviour, depoliticising social issues and supporting technological solutions to emerging political problems. In its early days, the Foundation heavily funded eugenic studies with the goal of finding a scientific basis to organise society, legitimise inequality and direct human evolution. As eugenics became increasingly associated with the growth of Fascist movements in Europe, the Foundation was on the lookout for alternative methods of pursuing its founding objectives. As convincingly shown by Lily Kay, the 'concerted physicochemical attack on the gene' was launched in a historical moment when eugenics had lost much of its scientific credibility, but the idea of rationalising human reproduction continued to exercise a strong appeal.[26]

The coming of the Great Depression further pushed the perceived necessity of new means of social control. Rising poverty and growing unemployment were rapidly eroding the authority of established institutions. Banking and finance were profoundly delegitimised. Industrial corporations were increasingly contested. Science was gradually losing its neutral clout as technologies increased labour productivity and profits, while millions of people were pushed into unemployment and vagrancy. In a public statement, the Foundation openly lamented 'a disarmament of the forces of law and order and social control at the very height of the engagement against crime and social deterioration'.[27] It is in this social climate that the Foundation decided to launch the 'Science of Man' project.

Warren Weaver unveiled the new course at the Foundation's 1933 planning meeting. 'Science', suggested Weaver, 'has made significant progress in the analysis of inanimate forces, but science has not made equal advances in the more delicate, more difficult and more important problem of the analysis and control of animate forces. [...] The challenge of this situation is obvious. Can man gain an intelligent control of his own power? Can we develop so sound and extensive a genetics that we can hope to breed, in the future, superior men? Can we obtain enough knowledge of the physiology and psychobiology of sex so that man can bring this pervasive, highly important and dangerous aspect of life under rational control?' In short, the Foundation would strive to obtain 'enough knowledge of [...] vital processes so that we can hope to control animate forces and rationalise human behaviour'.[28] Weaver believed that if science could grasp hereditary processes, new technological means to control, direct and manipulate living processes would soon emerge.

The apparently narrow focus of molecular biology was deceptive from the start, masking the profoundly political dimension of the new discipline. Weaver was always clear about the ultimate goals of Rockefeller's philanthropic donations: 'We are attempting to sponsor the application of experimental procedures to the study of the organization and reactions of living matter. We have chosen this activity because of a conviction that such studies will in time lay the (only?) sure foundation for the understanding and rationalization of human behavior'.[29] The patronage of wealthy donors was motivated by the search for new technologies to direct the winding paths taken by natural history, and thereby indirectly shape social and political history. Molecular biology promised exactly that: the development of new ways of controlling the human animal as a breeding living organism.

In order to pursue this ambitious programme, biology demanded unprecedented capital investments. As noted by Lily Kay, 'the reification of the molecular level as the essential locus of life, with the attendant reorientation of laboratory practice, altered the epistemological foundations of biological research, making the representation of life contingent upon technological intervention.'[30] Whereas only a century before, Mendel had inferred his laws from the observation of a few varieties of pea plants, molecular geneticists required sanitised laboratories replenished with a complex mechanical architecture made of electron microscopes, ultracentrifuges, spectrophotometers, scintillation counters and X-ray diffractometers. They required the collaboration of highly specialised actors, including legal and bureaucratic expertise to obtain compulsory authorisations and deal with legal controversies. They necessitated generous funding and institutional support, which was most likely obtained when the proposed endeavour mirrored the worldview of established institutions and affluent patrons.

Molecular biology was, from the very beginning, an extremely capital-intensive endeavour; a scientific field that would not allow easy entry to marginal actors cut off from the sources of wealth and power. It was, in other words, constituted by – as well as constitutive of – the political-economic context in which it emerged. This is not to say that the molecular conception of life championed by the Rockefeller Foundation is somehow *incorrect*, or an ideological distortion of reality. Rather, what this history shows is that the affirmation of our contemporary understanding of life cannot be explained solely on the basis of the history of science. To understand the emergence of molecular biology as a discipline, we cannot simply look at the level of abstract ideas: we must also investigate how scientific knowledge-production was materially organised and the role it played into the global political economy of the time.

As noted by Boris Hansen, Isaac Newton's *Principia Mathematica* were not just the product of individual reflection. They were also a response to specific technical problems that had become increasingly central in early capitalist societies: the problem of hydrostatics in navigation; the questions posed by simple machines in industry; the calculation of the trajectory of cannonballs in the military. The development of 'the productive forces in the age of merchant capital presented science with a number of practical tasks and urgently demanded their solution.'[31] Similarly, molecular biology was profoundly shaped by a set of research questions and priorities which were never purely scientific nor purely profit-driven, but rather part of an emergent biopolitical economy: a strategy of accumulation and a mode of government that increasingly relied on the development of a scientific knowledge of biological processes and new ways of manipulating them. The focus on genetics, the search for hereditary particles, the quest for technological means of inducing artificial mutations mirrored a growing interest in finding new ways of engineering the living means of production employed in agriculture, in industry, in medicine. It also promised new ways of tackling social problems and steering human evolution. If this line of research prospered both financially and institutionally, achieving growing scientific prestige and social influence, it was also because its proponents could convincingly refer to its social, economical and political potential.[32]

The early 1940s marked a critical point in the establishment of molecular biology as an independent discipline. On the one hand, Max Delbrück and Salvador Luria, two of the European physicists that the Foundation had helped into the United States, established one of the most influential research collectives of the early 1940s: the 'Phage Group', which gathered scientists such as Seymour Benzer, Alfred Hershey, James Watson and Renato Dulbecco. On the other hand, the publication of Erwin Schrödinger's *What is Life?* boosted the hype of the new discipline, providing considerable authority to its methodologies and aspirations. The book championed the idea that biological processes could be explained on the basis of the fundamental laws of physics, and that the principles of quantum mechanics might shed light on mechanisms of heredity and mutation. Moreover, it launched the hypothesis that chromosomes might 'contain, in some kind of code-script, the entire pattern of the individual's future development and of its functioning in the mature state'. This was a bold statement, which promoted a determinist vision of genes as detailed blueprints directing the life of all living organisms. 'In calling the structure of the chromosome fibres a code-script', Schrödinger argued, 'we mean that the all-penetrating mind [...] could tell from their structure whether the egg would develop, under suitable conditions into a black cock or a speckled hen, into a fly or a maize plant, a rhododendron, a beetle, a mouse, a woman'. Schrödinger presented genes as being at once 'law-code and executive power', 'architect's plan and builder's craft'.[33]

The long path by which molecular biologists came to accept DNA as the material base for the 'code-script' described by Schrödinger has a meandering and well-documented history.[34] In the 1940s evidence started to indicate that deoxyribonucleic acid (DNA) might be the best candidate for the role. This shift of perspective started in 1944 when Oswald Avery, Colin MacLeod and Maclyn McCarty provided strong experimental evidence supporting the view that DNA was 'functionally active in determining the biochemical activities and specific characteristics' of bacterial cells.[35]

It continued with the seminal research conducted by André Boivin, who experimentally established the possibility of 'directing the process of mutation' by manipulating 'the genes which serve as a substratum for the characters of the species'. This was taken as a confirmation of three fundamental hypotheses, which would continue to inform genetics into the twenty-first century. First, it suggested that DNA was the hereditary element originally hypothesised by Mendel. Second, it supported the idea – popularised by Schrödinger – that the difference between living species ultimately depended on variations 'within the molecular structure of one single fundamental chemical substance'. Third, and most importantly, it demonstrated that it was possible to artificially manipulate that genetic 'blueprint' to purposefully engineer living organisms.[36] Deciphering the molecular structure of DNA thus promised to further illuminate the nature of heredity, and expose new ways in which the latter may be controlled and directed. A number of questions remained outstanding. How can DNA direct the production of proteins? How does it replicate? And how do mutations occur?

In 1953, these puzzles found an answer with the publication of two groundbreaking articles in *Nature* signed by James Watson and Francis Crick. The first described the double-helical structure of DNA, and the replication mechanism suggested by that particular molecular geometry.[37] The second suggested that 'the precise sequence of the bases is the *code* which carries the genetical information'.[38] No one had ever talked of 'genetical information' before that day. It was a radical turning point. The novel concept fitted so well with the intellectual climate of the 1950s – dominated by information theory and early digital technologies – that it spread like wildfire in the scientific community. It offered a new language in which scientists, as well as the general public, could speak and think about genes and, more generally, about the fact of living. Genes are 'switched on and off'; messenger RNA is 'processed' and 'edited'; gene sequences contain 'signals' to the cellular 'machinery'; and 'operative information' is read out from the DNA 'code'. The metaphors are informational; the language, without which one can scarcely think or talk about life these days, are drawn from computing, cryptography and cybernetics. This informatic conception of life, firmly rooted in the theories and praxis of molecular biology, has become the operating framework within which most, but not all, biologists think and work.

This emergent episteme, which framed life in terms of coding and informatics, was shaped by major developments in both computing technologies and information theory. Initially employed in ballistics, cryptography and anti-aircraft artillery, computers soon started to occupy a fundamental role in the production systems of the 1950s and 1960s. These machines relied on a structured set of instructions for their operations and gradually, as noted by Edward Yoxen, 'this idea of a self-activating, self-controlling set of instructions found its way into biology'.[39] The new conception of life emerged as part of a more general turn towards the study of complex systems. The new science of cybernetics, in particular, offered a theoretical framework that erased the distinction between the natural and the artificial insofar as living cells and computers, human bodies and factories, missile systems and wolf packs were now conceptualised as being simply different types of 'complex systems'. The title of Norbert Wiener's most popular book, *Cybernetics: Or, Control and Communication in the Animal and the Machine* (1948) sums up well how cybernetics challenged conventional distinctions, presenting a global ecology composed of an infinite variety of continuously interacting informatic systems.[40] In this and other works, Wiener suggested – anticipating Watson and Crick's conclusions – that information was the essence of life, that heredity represented a system for the communication of valuable information, and that genes constituted a kind of 'memory' that could be transmitted.[41]

By the 1950s, this cybernetic conception of life had conquered a central place in both scientific and popular culture. Organisms had become complex informatic systems and genes had become coded texts, which could be transmitted, translated, read and – potentially – edited. According to Crick's subsequent research, the genetic script had one 'main function': 'to control (not necessarily directly) the synthesis of proteins'. Life now appeared, at least according to the geneticists, as a molecular one-way process: DNA directs the production of RNA, RNA directs the production of proteins, and proteins ultimately combine to create the entire organism. This was, according to Crick, 'the central dogma of molecular genetics'.[42] This apparently simple hypothesis, nevertheless, raised a myriad of puzzling questions. If DNA was indeed a code that contained detailed instructions for the assemblage of protein molecules, how could the code be decrypted? What was the 'grammar of life'?

The race to 'crack the genetic code' was officially open. Leading the race was a heterogeneous group of scientists, which had joined together to form the farcically named 'RNA Tie Club'. The president of this most curious intellectual milieu – a Russian-born cosmologist named George Gamow – stimulated the international debate on decoding by proposing an endless stream of mathematical models, which were, one by one, considered and rejected. These debates further expanded the scope and import of the new bioinformatic vision of life. *Information Transfer in a Living Cell* (1955), for instance, offers a preview of many of the metaphors that continue to dominate the biotech-imaginary. 'Comparing a living cell with a factory,' Gamow writes, 'we can consider its nucleus as the manager's office and the chromosomes as the filing cabinets where all the production plans and blueprints are stored. The main body of the cell, its cytoplasm, corresponds to the factory area where workers are manufacturing the specified product from incoming raw materials. The workers in a living cell are known as enzymes. They extract energy from the incoming food, break down the food molecules and assemble the separated units into various complex compounds needed for the growth and well-being of the organism.'[43]

Gamow's visionary description of living organisms would quickly become the commanding metaphor of the emerging biotech industry. His writings epitomised a Fordist imaginary that expanded and updated the classic machinist metaphors of early Enlightenment thinkers like René Descartes, Thomas Hobbes and Julien de La Mettrie. For these classic thinkers, the living body was a machine; for Gamow, it was a whole industrial landscape, composed of countless cellular factories – each employing a complex assemblage of molecular machines, enzymatic workers, genetic blueprints and raw materials – churning out proteins in the same way that Ford's factories produced cars.[44] Far from exceptional, Gamow's writings were symptomatic of an increasingly instrumental understanding of life, which anticipated and informed a new technoscientific aspiration: turning the living cell into a *programmable* site of production, whose factory floor could be re-organised, improved and rationalised.

In 1961, Marshall Nirenberg finally cracked the genetic 'code', elucidating how DNA and RNA direct what he would later call 'the economy of the cell'.[45] Assisted by Heinrich Matthaei, he built a synthetic RNA molecule in the form of a long strand of uracil nucleobases (U-U-U …). This was then inserted into a cell-free extract of *Escherichia coli*, which rapidly induced the bacteria's cellular machinery to begin producing phenylalanine. Nirenberg could consequently infer that the RNA code for phenylalanine was U-U-U. He had decrypted the first word of the code.[46] The news of Nirenberg's poly-U experiment fed the genetic craze. In December 1961, the *New York Times* opened with the news that 'the science of biology has reached a new frontier', leading to 'a revolution far greater in its potential significance that the atomic or hydrogen bomb'.[47] The *New York Herald Tribune* affirmed that once 'biologists know the whole hereditary code and learn how to change it at will, they may be able to control heredity by chemical means. They could raise plants and animals of almost any desired character.'[48] A few months later, the *San Francisco Chronicle* described the significance of Nirenberg's method to induce bacterial ribosomes to produce phenylalanine – and potentially other amino acids – as a step forward in the creation of automated biological factories, something akin to 'discovering that IBM cards punched out by a French car factory can direct an automatic assembly line in Detroit to make cars very much like Renaults and Citroens'.[49]

After reading Darwin's *On The Origin of Species*, Marx observed in a letter to Engels in 1862: 'It is remarkable how Darwin rediscovers, among the beasts and plants, the society of England with its division of labour, competition, opening up of new markets, "inventions" and Malthusian "struggle for existence".'[50] A century later, biologists were similarly projecting onto the cellular world a mirror-image of their own time, turning the living cell into a Fordist factory incessantly assembling the molecules described in its genetic 'blueprints'. Molecular biology was increasingly described in terms borrowed from political economy and business studies, using imagery from the industrial landscapes dominating post-war USA and Europe. The most pressing questions now became: Would it be possible to rationalise these submicroscopic factory floors? Could protein synthesis be functionally directed? Might cellular assembly lines be coaxed into producing marketable proteins on demand? Could genetics provide the scientific foundation for a molecular Fordism to come?

The scientific community was not immune to the anticipation sparked by Nirenberg's findings. At the 1961 meeting of the American Association for the Advancement of Science, Arthur Steinberg affirmed that genetic research 'might well lead in the foreseeable future to a means of directing mutations and changing genes at will'.[51] A few months later, Arne Tiselius – then chairman of the Nobel Foundation – ambiguously warned that molecular genetics might soon 'lead to methods of tampering with life, of creating new diseases, of controlling minds, of influencing heredity, even perhaps in certain desired directions'.[52] On the other side of the Atlantic, fourteen European countries – with the decisive financial support of the Volkswagen Foundation – backed the formation of the European Molecular Biology Organization: an institution committed to compete with American research in the field. Public figures such as Julian Huxley, a prominent member of the British Eugenics Society and the first secretary of UNESCO, came out as strong supporters of the new organisation. 'For any major advance in national and international efficiency', writes Huxley in *Eugenics in Evolutionary Perspective*, 'we cannot depend on haphazard tinkering with social or political symptoms [...] but must rely increasingly on raising the genetic level of man's intellectual and practical abilities'.[53] Molecular genetics was hailed as a source of new technologies, an engine for economic growth, and an effective alternative to social and political reform.

Meanwhile, Nirenberg was busy synthetising new RNA molecules and testing them, a research that would lead to the identification of at least one genetic sequence for each of the 20 standard, proteinogenic amino acids.[54] Just a few months before receiving the 1968 Nobel Prize for his 'interpretation of the genetic code and its function in protein synthesis', Nirenberg offered some reflections on the social and political significance of his research. The article, published in *Science* under the ominous title 'Will Society Be Prepared?' boldly stressed that 'the genetic language now is known, and it seems clear that most, if not all, forms of life on this planet use the same language, with minor variations. Simple genetic messages now can be synthesised chemically. Genetic surgery, applied to microorganisms, is a reality. Genes can be prepared from one strain of bacteria and inserted into another, which is then changed genetically'. It affirmed that, in all likelihood, 'cells will be programmed with synthetic messages within 25 years'. Finally, it sounded a warning, since 'man may be able to program his own cells with synthetic information long before he will be able to assess adequately the long-term consequences of such alterations, long before he will be able to formulate goals, and long before he can resolve the ethical and moral problems which will be raised'.[55]

As Francis Crick noted, this was 'the end of an era in molecular biology'.[56] A major revision of the fundamental conception of biological processes had been pushed through. Scientists increasingly thought of living organisms and their physiological activity in terms of genetic information, code, structure, sequence, transcription, translation and feedback. Now that the gene was no longer a mysterious entity – no matter how sketched and imprecise these early descriptions of it really were – the age of early genetic explorations seemed to approach an abrupt end. To be replaced by what? Knowing which DNA sequence corresponds to which amino-acid did not directly offer new ways to manipulate the cellular metabolism; knowing the 'grammar of life' was still not sufficient to manipulate 'genetic information'. Despite widespread hallucinatory visions of highly disciplined molecular factories, the practice of genetic manipulation had not moved very far from Muller's early experimentation with X-ray mutagenesis. The dream of placing 'the process of evolution in our hands' remained exactly that: a far-fetched vision from the future, at once stimulating and haunting molecular biologists, corporate sponsors, public officials and the general public.

At the end of the 1960s, Nirenberg's suggestion that genetic engineering would soon pose a number of rather troubling political dilemmas went largely unnoticed. Yet within five years, the revolution he had predicted was in full swing. Corporations would soon acquire the capability to target and alter specific sequences of nucleotides, thereby controlling – however imperfectly – the metabolism of the living cell. In turn, the efficacy of such genetic manipulations would further magnify the social authority of molecular genetics, persuading both biologists and the public that genes indeed held the key to the mastery of life itself.

MOLECULAR CHIMAERAS AND RECOMBINANT BACTERIA

'We Played a Cruel Trick on Mother Nature'

A new era in molecular biology suddenly opened in the 1970s, following a series of unexpected findings in an obscure corner of basic microbiology. A small number of scientists had worked for decades on bacterial immune systems: the set of tools by which bacteria ward off viral infections, protecting their cellular machinery from being hijacked into producing copies upon copies of the invading virus. It was eventually understood that bacterial immune systems rely on special enzymes that identify and fragment the DNA of invading viral agents. The phenomenon was studied in detail by Werner Arber in the 1960s, leading to the identification of a group of restriction enzymes which, rather than simply cutting up the viral genome in a blind fury, appeared to cleave it only at a very specific point in the base sequence. Since then, thousands of these 'Type 2 Restriction Enzymes' have been isolated in bacteria, each recognising a particular base sequence and breaking the DNA at the same or related point. By selecting the correct enzyme, it became possible to operate a cut in a precise location on the DNA strands.[57]

The second step on the path to genetic engineering was the identification by Martin Gellert and Bob Lehman of another type of enzyme called ligase, which can join two DNA fragments together. Combining restriction and ligase enzymes, Paul Berg was finally able to create the first molecule of 'recombinant DNA' by splicing together two DNA fragments from different viral genomes.[58] Yet, it was not immediately clear how the new technology could be effectively deployed in living cells. The collaboration between Stanley Cohen of Stanford University and Herbert Boyer of the University of California answered this outstanding question.[59] Using the restriction enzyme EcoRI, they generated two fragments from separate DNA molecules – one conferring resistance to tetracycline antibiotics, the other to kanamycin antibiotics – and managed to join them using DNA ligase. They had thus created a recombinant DNA molecule. Next, they placed the plasmids into a suspension of cold calcium chloride populated by *E. coli* bacteria. The temperature was rapidly raised and lowered, creating a heat-shock that induced the bacteria to take up the DNA. Finally, the bacteria were exposed to a combination of tetracycline and kanamycin. Only bacteria resistant to both antibiotics could survive. And some did survive. Boyer recalls examining the critical gels: 'I can remember tears coming into my eyes. It was so nice.'[60] He was looking at the first genetically modified organisms.

In subsequent publications, Boyer and Cohen described a replicable laboratory method to recombine DNA: sewing together the genetic fabric of unrelated organisms to create new genomic patterns.[61] The moment could be seen as the culmination of a century-long search for methods to manipulate heredity. Boyer certainly saw his contribution in this way. Interviewed in 1975, he remarked: 'Things just came together at that time: the study of small plasmids, transformation of *E. coli* with DNA, the restriction enzyme business; it was all coming to fruition at the same time [...] It was straightforward. There was not much in the way of struggle. The first experiments more or less worked as intended.'[62] A year later, the two applied the same technique to insert a gene from a frog (*Xenopus laevis*) into a bacterium, proving that it was possible to propagate eukaryotic genes in simple prokaryotic systems. Bacteria would not only accept eukaryotic DNA, they would replicate it as their own. Finally, they speculated – still without definite proof, but anticipating further experiments – that bacteria might be able to express any genetic information derived from higher organisms and could thus be tricked into constructing useful proteins.[63]

The first gene-splicing experiments were immediately perceived and portrayed as a revolutionary turning point.[64] A report presented to the first British committee formed to assess the implications of recombinant DNA sums up some of the questions posed by this early method of genetic engineering. 'It cannot be argued', writes Sidney Brenner (author of the report and future Nobel Laureate), 'that this is simply another, perhaps easier way to do what we have been doing for a long time with less direct methods. For the first time, there is now available a method which allows us to cross very large evolutionary barriers and to move genes between organisms which have never had genetic contact.'[65] Recombinant DNA technology allowed scientists to break down interspecies barriers and open up alternative venues for genetic circulation. The cultural perception of bacteria, moulds, yeast, plants, farm animals and human beings radically shifted. Nature no longer appeared as it did in the eighteenth century, as a grand canvas of fixed and well-defined organic forms: a hierarchical chain of beings filled with unmodifiable divine creations. Even the Darwinian description of a genealogical tree of life – populated by discrete species, slowly evolving through the combined effect of random mutations and natural selection – suddenly appeared outdated. It was now possible to deliberately manipulate the genetic constitution of a species by introducing artificial mutations, which could then be artificially selected.

Today, it is often stated that genetic engineering is nothing inherently new since 'we have knowingly altered the genome of various species for

12,000 years through selective breeding'.[66] This is a misconception. Genetic engineering and selective breeding are very different technoscientific practices, although they are often complementary. Selective breeding – and, more generally, artificial selection – manipulates the environmental conditions by which random genetic variations are selected and propagated in a population. Breeders recognise desirable variations in a given population, and proceed to create the ideal conditions by which that variation may proliferate. For instance, a breeder may recognise a hen that lays more eggs than the average. The breeder may then attempt to conserve this useful genetic variation, making it spread throughout a population. Genetic engineering introduced a completely new way of manipulating heredity. For the first time in history, it became possible to manipulate both poles of the evolutionary process; combining techniques of artificial variation (genetic engineering) and techniques of artificial selection (selective breeding). Scientists could now induce novel variations through artificial mutagenesis, and select for desired traits. It was the birth of a technoscience of controlled evolution that would have profound effects.

A new conception of nature was starting to form in the confused debates surrounding Boyer and Cohen's 'molecular chimaeras'. Traditional separations – between nature and culture, artificial and congenital, industrial and organic – entered into crisis. Species were increasingly understood as contingent arrangements of genetic traits that may be infinitely disembodied, transferred and recombined. 'The Manipulation of Genes', a short article written by Stanley Cohen for the July 1975 edition of *Scientific American*, may be taken as paradigmatic of this rather sudden epistemic shift. The article starts from the premise that 'mythology is full of hybrid creatures such as the Sphinx, the Minotaur and the Chimera, but the real world is not. [...] This is because there are natural barriers that normally prevent the exchange of genetic information between unrelated organisms.' It continues: 'We called our composite molecules DNA chimeras because they are conceptually similar to the mythological Chimera (a creature with the head of a lion, the body of a goat and the tail of a serpent).' 'Genetic engineering', concludes Cohen, is a functional tool 'for creating a wide variety of novel genetic combinations'.[67] The promise of recombinant DNA was clear from the start: breaking down interspecies barriers, carving out new routes of genetic circulation, systematically producing difference and hybridity.

Most of the media were quick to embrace the new possibilities opened up by genetic engineering. *Fortune* predicted that 'transfers of DNA, those master molecules of life, [may be used] to improve production capabilities of microbes'. The *New York Times* envisioned recombinant DNA as a solution for 'some of the fundamental needs of both medicine and agriculture'. The *New Scientist* affirmed that 'bacteria could be turned into factories'.[68] And yet, enthusiasm was not the only emotion stirred by the dawn of recombinant bodies. Fears abounded too. In 1974, James Rose – then head of the molecular section of the Laboratory of Biology of Viruses at the National Institute of Allergy and Infectious Diseases – stressed the danger posed by 'novel hybrid viruses formed by linking SV40, a monkey virus known to be oncogenic in other animals, and human adenovirus, a virus responsible for respiratory disease' and, more generally, by the 'ever-increasing production in research laboratories of other mutant animal viruses'.[69]

The article prompted a number of prominent molecular biologists to call for a temporary moratorium on recombinant DNA research and the setting up of a convention tasked to discuss if the 'unfettered pursuit of [genetic engineering] might have unforeseen and damaging consequences for human health and Earth's ecosystems'.[70] Held in February 1975, the Asilomar Conference was organised by a scientific committee selected by Paul Berg, and approved by the National Academy of Sciences. The final report confirmed that genetically modified organisms posed new challenges, uncertainties and risks. Guidelines were proposed to regulate how genetically modified microorganisms were to be handled and disposed of. Yet, the potential benefits were considered too great to extend the moratorium. The conference concluded that 'the best way to respond to concerns [...] is for scientists from publicly funded institutions to find common cause with the wider public about the best way to regulate'. In retrospect, Berg suggested that many of the participants were convinced that 'once scientists from corporations begin to dominate the research enterprise, it will simply be too late [to introduce effective regulations]'.[71]

Today, the Asilomar conference is widely celebrated as a rare case of scientists taking upon themselves the responsibility of pointing out the risks implicit in their activities, and imposing restraints to the pursuit of technoscientific mastery. At the time, however, there were several critical voices. Erwin Chargaff, one of the most prominent biochemists of the time, noted that 'the edict published in due course, which lists the various forbidden items, reads like a combined curriculum vitae of the conveners of the conference' and painted the meeting as 'the first time in history that the incendiaries formed their own fire brigade'. He insisted that 'scientific research is falling more and more into the hands of entrepreneurs', who were likely to pursue 'an extension of the molecular nightmares that go under the name of genetic engineering'.[72] Robert Sinsheimer, then chairman of the biology division at Caltech, criticised the meeting for exclusively focusing on 'the potential biological and medical hazards', while eschewing controversial issues that went well beyond 'the vagrant lethal virus or the escaped mutant deadly bacteria'. Security concerns obscured fundamental political questions concerning the emergence of an unprecedented power on the biosphere: 'Do we want to assume the basic responsibility for life on this planet – to develop new living forms for our own purpose? Shall we take into our own hands our own future evolution?'[73]

On the other side of the Atlantic, the dawn of genetic engineering also sparked debates and controversies. Between 1974 and 1982, the governments of the UK, West Germany, Ireland, France, Belgium and the Netherlands commissioned reports on recombinant DNA and its potential biohazards. The European Commission and the Organisation of Economic Co-operation and Development established specialised committees to monitor the same issues. Labour Unions called for safer working conditions for laboratory workers. Religious organisations pushed for strict limitations on genetic engineering and the manipulation of life.[74] The distinction between somatic gene-therapy (which affects only the individual treated) and germline treatment (which could be inherited) emerged as central to these debates. While somatic engineering was presented as a possibility, germline modifications were strictly regulated in a number of countries. In 1982, for instance, *Council of Europe Recommendation 934* recommended an outright ban on all forms of human germline engineering, proclaiming 'the right to inherit a genetic pattern which has not been artificially changed'.[75]

Despite their rather limited focus, these debates soon invited broader reflections on the socio-political implications of genetic engineering. Michel Foucault's work on biopolitics, for instance, can and should be read as an attempt to interpret the new forms of knowledge and power enabled by molecular biology. Reviewing François Jacob's *The Logic of*

Life, he suggested that molecular biology represented 'the foundation, under our own eyes, of a theory as important and revolutionary of what may have been, in their own epoque, those of Newton or Maxwell'. He stressed, in particular, the rapid crystallisation of a new concept of life that transmuted 'living organisms' into 'beings determined by a program residing in the cellular nucleus'; 'bacteria' into 'chemical factories'; the 'cell' into 'a system of physico-chemical reactions', and the human body into 'a reproducing machine that reproduces its mechanism of reproduction'.[76]

This was not the last time that Foucault's writings touched on issues of molecular biology. The first volume of his *History of Sexuality* (1976) distinguished between two forms of 'power over life', which clearly mapped onto the emerging distinction between somatic and germline manipulations of the genetic code. Biopower may either effect an 'anatomopolitics' operating at the level of the individual body; or implement a 'biopolitics of the population' dealing with 'the species body, the body imbued with the mechanics of life and serving as the basis of the biological processes'.[77] On 17 March 1976, speaking at the Collège de France, Foucault noted that an 'excess of biopower appears when it becomes technologically and politically possible for man not only to manage life but to make it proliferate, to create living matter, to build the monster, and, ultimately, to build viruses that cannot be controlled and that are universally destructive'.[78] The birth of genetic engineering, in other words, constituted a critical threshold beyond which the production of life and the production of death – biopolitics and sovereignty – were starting to coincide and merge.

Foucault's reflections at once mirrored and framed the public fears of many protagonists of the molecular revolution. Three months after Foucault's seminal lecture on biopower, Erwin Chargaff published a short article in *Science* titled 'On the Dangers of Genetic Meddling'. The piece was striking, especially coming from an author that was, at the time, widely recognised as one of the most prominent biochemists of the twentieth century and a global authority in DNA research. It denounced early recombinant DNA experimentation for 'throwing a veil of uncertainties over the life of coming generations', interfering in biological processes that scientists did not fully understand. The production of new forms of life, according to Chargaff, while performed 'in a tremendous hurry to help suffering humanity' would likely spark a symmetrical production of death. 'You can stop splitting the atom; you can stop visiting the moon;

you can stop using aerosols; you may even decide not to kill entire populations by the use of a few bombs. But you cannot recall a new form of life.' Introducing genetically engineered varieties of plants and animals furthered 'a destructive colonial warfare against nature' and threatened to impact complex ecosystems in unpredictable ways.[79] Genetic accidents would be the flip side of the new genetic biopolitics.

If we offer this brief and partial sketch of the international debates surrounding the dawn of genetic engineering, it is certainly not to settle them once and for all. We would rather stress the extent to which the early 1970s were characterised by political struggles that threatened the possibility of turning genetic technologies into a profitable business and an instrument of power. Opposition was real and widespread, and the socio-political impasse was potentially paralysing. Only through an organised international campaign was genetic engineering depoliticised and kept out of democratic debates. The Committee on Genetic Experimentation (COGENE) – a corporate pressure group formed on the eve of the Asilomar Conference – played a key role in lobbying governmental agencies. They emphasised the value of genetic engineering, lobbying to restrict regulations to issues of safety while precluding wider debates on the social, political and economic effects of the new technologies. COGENE thereby erased the real conflicts taking place within the rather small international community of specialised molecular biologists; and projected the dubious image of a united scientific front, supporting genetic engineering against the irrational fears of an illiterate public. In 1992, Giorgio Bernardi – then chairman of COGENE – compiled a general report summarising the results of fifteen years of activities. 'It is clear', he writes, 'that the tasks originally assigned to COGENE were fulfilled'; and concluded that the group's 'actions played an important role in downgrading the excessively stringent conditions, which were initially imposed' on genetic engineering.[80]

The dismantling of the 1974 international moratorium opened the gate to the rise of the genetic industry. Two years later, Herbert Boyer founded Genentech in collaboration with Robert Swanson. Its name, a contraction of *gen*etic *en*gineering *tech*nology, spelled out its corporate agenda: turning recombinant DNA into an industrial system of production. Its first business plan sums up what would be the corporate strategy of many other biotech companies in the following decades: 'to engage in the development of unique microorganisms that are capable of producing products that will significantly better mankind. To manufacture and market those products.'[81] Like most business plans, it was a thoroughly promotional document, but it was sufficient to convince a number of financial investors to bet on turning genetically engineered organisms into living means of capital accumulation.

The first objective was to demonstrate that bacteria could be turned into molecular factories for the production of proteins, which might be sold as commodities on the global market. This was achieved in April 1978, when Boyer's corporate team managed to create genetically engineered bacteria capable of producing somatostatin – a small polypeptide produced in the human pituitary gland. Relying on the public research on genetic coding pioneered by Nirenberg, Genentech first designed a genetic sequence for the chain of fourteen aminoacids that compose somatostatin. The engineered DNA molecule was then spliced into a bacterial plasmid, and introduced in a selected population of *Escherichia coli*. The final result was a population of genetically modified bacteria that produced somatostatin as part of their modified metabolism. It was the first evidence that bacterial cells, and their microscopic cellular machinery, could be coaxed into manufacturing human proteins of commercial value. The day after, press coverage quoted a beaming Boyer proclaiming: 'We played a cruel trick on Mother Nature!'[82]

It was a trick that would be repeated countless times in the following years. After Genentech had proven the potential of its recombinant technique, it was able to raise another round of funding. In 1978, the company signed a partnership with the pharmaceutical giant Eli Lilly in order to develop a new, bioindustrial process for the production of insulin – a hormone which regulates the amount of glucose in the blood, and whose absence causes diabetes. Since the 1920s, Eli Lilly had profited from the commercialisation of bovine insulin. The hormone was extracted from the pancreas of slaughtered cattle and pigs, purified and sold to diabetics worldwide. This was a well-established market, with an annual worth of over $400 million, and a steep growth-rate. In fact, the number of type-2 diabetics was growing so rapidly that the market for insulin was expected to double every five years. The reason for this trend could be convincingly traced back to the growing production and consumption of industrial food full of sugars, fats and low-fibre carbohydrates.[83] Nevertheless, shifting this socio-economic tendency was not a feasible business strategy for Eli Lilly, or any other pharmaceutical corporation. What was desperately attractive was, rather, the prospect of developing a process of production that would shrink costs and increase profit margins on a pharmaceutical commodity.

Could Genenetech develop a new strain of genetically engineered bacteria, capable of synthesising human insulin as part of their metabolic cycles? As in the somatostatin experiment, Genentech started from the aminoacid sequence composing the insulin molecule. Then, guided by Nirenberg's genetic code, they reconstructed a DNA sequence that they believed underlied insulin. Finally, they spliced the synthetic DNA sequence into plasmids and transferred them into a population of *E. coli*. The cellular machinery of these genetically modified bacteria was thus coaxed to express the synthetic gene for insulin. On 21 August 1978, Genentech's scientists reported to have successfully produced synthetic insulin.[84] Four days later, the company had already licensed the worldwide rights to manufacture and market it to Eli Lilly, in exchange for 6 per cent royalties on product sales. In the following years, the pharmaceutical conglomerate turned this experimental technology into an industrial production process: manufacturing genetically modified bacteria, making them proliferate, extracting the product and commercialising it worldwide.

The partnership between Genentech and Eli Lilly pioneered a productive technique that would soon become ubiquitous. The *E. coli* bacterium, in particular, rapidly took a prominent role in the emerging biotech industry. Often referred to as the 'workhorse of molecular biology', this organism has been turned into a living factory for the production of a long list of recombinant proteins, experimental biofuels and other biochemicals.[85] We could say that, while horses helped power the early Industrial Revolution before the rise of steam engines, bacterial bodies fuelled the early stages of the ongoing Biotech Revolution. Biotech production often begins by extracting a gene of interest from a living organism – either bacteria, plant or animal – or by synthetically constructing an equivalent. The gene is then spliced into a selected strain of *E. coli*. By forcing artificial mutations in the bacterial genome, biotech companies can induce *E. coli* to dedicate its life – that is, the majority of its metabolic resources – to assembling a specific protein. In other words, the bacterium is turned into a single-minded microscopic industrial bioreactor: a living organism programmed to convert most of its feed into molecules for which it has no use.

Recombinant DNA furthered a new political and economic imaginary. *Science News* reported that 'human insulin has been produced at last by genetically engineered bacteria in a California laboratory – an achievement that catapults DNA technology into the major leagues of the drug industry'. The *Washington Post* noted that the somatostatin experiment held the 'practical promise' that engineered genes could soon manufacture biological commodities, while *Chemical and Engineering News* celebrated the event as 'a vindication of the utility of recombinant DNA research, which should further defuse a tiny group of scientific critics'.[86] During a Senate subcommittee hearing on recombinant DNA, Paul Berg and Philip Handler described Genentech's work as 'a scientific triumph [...] putting us at the threshold of new forms of medicine, industry and agriculture'.[87] It was a threshold moment also in another sense: the neoliberal counter-revolution of the 1980s was around the corner, and the tide was turning.[88] Biotech companies multiplied, and scientists flocked into private companies set up by venture funds keen to dump surplus capital into new frontiers of technoscientific infrastructure. Scores of biologists traded their lab coats for corporate suits, or at least placed one set on top of the other. Traditional boundaries between science and business, academia and finance, became increasingly porous.

The birth of recombinant DNA, and its rapid transformation into an industrial tool, represented at once the culmination of an historical trajectory and an opening toward a number of possible futures. As Richard Lewontin has convincingly argued, biology was profoundly shaped by the quintessentially modern quest to establish increasing control over hereditary processes. 'For nineteenth century biology', he writes, 'the program of providing a mechanical explanation of development was more than a question of producing a coherent explanatory scheme for living organisms. It held out the promise of success in the ultimate goal of biology, to produce living organisms artificially in the laboratory.'[89]

Capital investments in biological research have been often motivated by the search for technological means of steering evolutionary processes. Mendel's genetic studies were supported in the hope that they may generate a better understanding of hereditary processes, increasing the power and efficacy of the sheep breeding industry. A century later, the institutionalisation of molecular biology was indebted to the financial patronage of private interests that yearned the potential of '[breeding], in the future, superior men'.[90] In the 1920s, Muller pioneered the use of radiation as a technological means to greatly increase the frequency of genetic mutations in a population. Atomic energy accelerated the pace of natural history, forcing living organisms to mutate at an unprecedented pace. In the 1970s, Cohen and Boyer demonstrated the potential of recombinant DNA to induce more targeted genetic alterations, employing restrictase and ligase enzymes. It was now possible not only to induce random mutations, but also to design desired mutations by splicing DNA sequences isolated from a variety of living organisms. Genentech Corporation built an industrial and financial empire by using these early genetic engineering technologies to establish a profitable programme of bacterial breeding on an industrial scale.

For the first time, the production of a commodity was effectively molecularised. It would not be the last.

2

Manufacturing Lives

Corporate Genes, Genomic Rents and Living Assets

In early modern Europe, the first wave of enclosures transformed huge swaths of common land into privately held assets, directly controlled by a sparse landlord class. The Enclosure Acts that led to the rapid concentration of landed property in eighteenth century England, represent the paradigmatic form of accumulation by dispossession. Juridical reforms backed by police power forced entire communities of peasants off the land, transforming them into a landless proletariat with nothing to sell but their own labour power. The enclosure of forests, fields and common lands destroyed the legal and cultural regimes by which common resources had been socially managed and reproduced for centuries by multigenerational communities. It introduced, in their place, the absolute right of each landlord to dispose of their private property without external limitation or constraint.[1] In this way, the enclosures created a new juridical and social regime on the basis of which rents and profits could be systematically extracted from the labouring classes.[2]

The most recent wave of genomic enclosures has its own highly haphazard and thoroughly contingent historical genealogy, which is the subject of this chapter. And yet, there are similarities with the first wave of enclosures, which are worth emphasising from the start. The genomic enclosure movement was also formalised and enforced through a series of carefully crafted juridical acts, which were informed by scholarly texts of economic and legal doctrine, and vehemently policed. In this way, a complex proprietary grid has been imposed over the genomic commons, fragmenting it into innumerable, well-guarded, privately held pieces of property. The genetic contents of the cell have been excavated, worked up and turned into private financial assets from which profits and rents are extracted.

In this chapter, we follow the ongoing transformation of genomic science into an industry dedicated to the systematic production of rent-yielding bio-objects. First of all, we consider the creation of a neo-liberal legal and political framework for the accumulation of genetic property in the United States of the early 1980s. A series of controversial cases reclassified life-forms into objects of manufacture, extending patent protection to a new set of biological artefacts, including: bacterial, plant, animal and human genetic information; isolated biochemical compounds; and gene-edited organisms. Second, we analyse the political struggles that, throughout the 1990s, contested the rapid proliferation of patents on isolated human genes. We focus on the controversial history of Myriad Genetics' monopoly on breast cancer genes and how a political mobilisation from below was able to partially restrain and reshape the development of the genomic industry. Third, we consider the globalisation of the neoliberal regime through international agreements and the resulting enclosure of biological wealth and indigenous knowledges from a variety of global sites – from the thick of the Amazon to the depths of the ocean. This process of gene-grabbing has supported the creation of novel value-chains based on the industrial production of valorisable genomic knowledge. Through patenting, genomic biotechnologies were turned into a powerful machinery for the accumulation of private property: a transformation that dovetails with a general assetisation of the global economy and the emergence of a rentier capitalism underpinned by monopolistic, rent-yielding assets.

LIVING ASSETS

Proprietary Bacteria and Patented Mice

The privatisation of the planet's genetic commons has been a gradual, tentative and deeply contested process. A decisive step in this direction was taken in the early 1970s, when General Electric invited one of its microbiologists to apply to the US Patents and Trademark Office (PTO) for exclusive property over an unassuming bacterium with an unusual appetite for hydrocarbons. The patent application made three fundamental claims. It maintained that Ananda Chakrabarty had invented a new method of transferring genes across separate strains of bacteria. It affirmed that this innovative laboratory technique had allowed him to manufacture a new species of 'multi-plasmid hydrocarbon-degrading *Pseudomonas* bacteria', which now lived and replicated in the confined quarters of the General Electric Research & Development Centre in New York. Finally, the application claimed that this living invention was not only unprecedented and non-obvious, but also that it could be turned into a useful product. Thanks to its capacity to rapidly digest many of the hydrocarbons found in oil, claimed Chakrabarty, the genetically engineered bacterium may be employed as a valuable tool 'in oil spill clean up, or in the production of protein from petroleum'.[3]

The application was, at first, squarely rejected by the PTO. General Electric appealed, and the case reached the US Supreme Court in 1980. The debate focused on a series of unsettling questions: Was the *Pseudomonas* bacterium described by Chakrabarty really an 'artefact' designed by human labour? Can life be manufactured in a laboratory? Can an entire population of living beings be considered the property of a single company? And what would be the economic, social and political consequences of establishing such property? Sydney Diamond, the Commissioner of Patents and Trademarks, motivated his earlier rejection of the application on the basis of the 1930 *Plant Patent Act*, which suggested that 'manufacturing' was not a term applicable to living things. Living beings, according to this classic conception, cannot be 'manufactured', but only reproduced, tended to, and let live. This position was supported by a detailed *amicus brief* submitted by the People's Business Commission, a non-governmental organisation founded by Jeremy Rifkin. The document insisted on two fundamental points. First of all, it argued that, if patents were granted on simple microorganisms, 'there is no scientific or legally viable definition of "life" that will preclude extending patents to higher forms of life'. Furthermore, it stressed that the patent would not only grant the abstract possibility of patenting novel living organisms but would, in fact, provide a strong 'economic incentive to corporations in the field of genetic research and development' to do exactly so.[4]

General Electric, on the other hand, argued for a wider scope of the term 'manufacturing', which it construed as any process of 'production of articles for use from raw or prepared materials, by giving to these materials new forms, qualities, properties, or combinations, whether by hand-labour or by machinery'.[5] Chakrabarty's laboratory work, argued the corporation's lawyers, was not only a form of scientific experimentation; it was also a *manufacturing process* by which micro-organisms were given new forms and new qualities. This meant that the result of that scientific labour could be understood as a 'living artefact'. The ostensibly clear boundary between science and industry was directly challenged. The Regents of the University of California, together with seven other institutions with direct financial interest in the patentability of living organisms, submitted an *amicus brief* in support of General Electric. The document argued that there was no longer a scientific reason to maintain a juridical distinction between inanimate objects and living beings, since recent advances in molecular biology had erased once and for all 'the bright line between life and its absence'. This scientific fact, argued the Regents, 'effectively destroys the argument that life itself is not only the essential characteristic of any living being – even a microorganism – but the one which precludes patentability'.[6]

Ultimately, the Supreme Court ruled in favour of General Electric, backing many of the arguments presented by the University of California. It found that Chakrabarty had 'produced a new bacterium with markedly different characteristics from any found in nature' and that this life-form was therefore 'not nature's handiwork but his own'.[7] What had been for most of the twentieth century a rather obscure philosophical debate surrounding the nature of life, the function of genes, and the relation between the inanimate and the animate had suddenly become a political and economic issue, which was hastily resolved in a court of law. Embracing a radically mechanist ontology, a state institution affirmed that the US legal code would no longer consider living organisms as ontologically distinct from inert objects.

Life was declared to be nothing but a peculiar composition of matter. The conception of the genetic code as an industrial blueprint, and the living body as a relatively straightforward materialisation of that blueprint, was inscribed into law. William Tucker, manager for technology transfer at DNA Plant Technology of Oakland, succinctly summed up the new hegemonic vision: 'Just because it's biological and self-reproducing doesn't to me make it any different from a piece of machinery that you manufacture from nuts and bolts and screws.'[8] Genetic engineering was made to breed financial assets. This rather technical and obscure regulatory shift was the spark that fired the transformation of genomic science into a global industry.

This legal adjustment, later replicated in many other countries, instantly turned a relatively new technoscientific process into a machine for the accumulation of exclusive private property. Corporations around the world immediately understood the profound political and economic implications of the verdict. A spokesperson for Genentech – the company born to capitalise on Cohen and Boyer's research – commented: 'The Court's decision should accelerate the flow of investment capital into new, high technology ventures. By extending the reach of patents to encompass more than merely "traditional" fields of research, the Court has assured this country's technology future.'[9] The industry may have had a global outlook from the start, but the discourse had a distinctively nationalist flavour, aimed at securing a first-mover's competitive advantage for US industry in this newly minted field.

Now that genetically engineered organisms could be readily turned into patented assets, recombinant DNA was no longer simply a scientific technique. It was a means of capital accumulation. Biotech companies could now pursue two separate business strategies. On the one hand, recombinant DNA techniques facilitated the *commodification* of different life-forms. General Electric's oil-eating bacteria could be manufactured and sold around the world. Bacterial populations could be scaled up by employing industrial facilities, fermentation tanks and careful breeding labour. On the other hand, biotech companies could limit themselves to the systematic accumulation of patents on their living inventions. The patented genetic 'blueprints' could be offered to more established industrial groups, primarily in the pharmaceutical and agribusiness sector. A basic division of labour follows: industrial groups organise the production of living commodities and the extraction of profits from their commercialisation. Biotech companies capture a guaranteed stream of royalty-based income in the form of rents on their exclusive genetic patents. This corporate strategy of *accumulation by bioassetisation* effectively turned biotech companies into a new stratum of the global rentier class, leeching off a percentage of the profits realised by industrial corporations.[10]

So anxious was Wall Street to begin financing the biotechnological revolution that, when the first genetic-engineering firm offered its stock to public investors, it set off a buying stampede. On 14 October 1980, just months after the Supreme Court's decision, Genentech obtained its brand new Nasdaq ticker symbol GENE, and it immediately offered over one million shares of its stock to public investors. By the end of the day, the fledgling biotechnology firm had raised $36 million and was valued at

$532 million. It achieved this without having ever introduced a single product into the marketplace, and with a patent application on recombinant technologies that had been pending for years. Herbert Boyer, who held 925,000 shares in the company, became an instant multimillionaire, reaping nearly $70 million in one single day. Two months later, the company finally obtained monopolistic control over recombinant DNA technology.[11]

The financial craze sparked by early genetic engineering can be understood in the context of the crisis of capital over-accumulation that hit many advanced industrial economies in the 1970s and early 1980s. According to David Harvey, a series of alleviation strategies were promptly implemented in order to stave off 'a massive devaluation of both capital and labour'.[12] Financialisation represented one such alleviation strategy, by which surplus capital was channelled into increasingly speculative ventures.[13] The collapse of Bretton Woods Institutions, a general liberalisation of finance, financial disintermediation: all combined to create a speculative frenzy. The genomic industry was conjured into existence by regulatory fiat, during an economic conjuncture that urgently required novel frontiers of investment. It opened a new outlet for investments in technoscientific infrastructure, while offering new sorts of financial assets to the growing portfolios of investors.[14] Its growth, in other words, reflected and furthered more general processes of financialisation and neoliberal restructuring that played a key role in displacing the crisis. Rather predictably, this 'biotech bubble' met the same fate as the dot-com bubble: Nasdaq's Biotech Index soared by 700 per cent between its inception in 1993 and its peak in 2000, before bursting.[15]

In the following years, a series of rulings by the PTO significantly extended the number of patentable biological entities. In the context of lengthy discussions surrounding the patenting of genetically altered oysters, the PTO recognised that the logic first established by the Supreme Court could not be restricted to genetically engineered bacteria. In April 1987, the PTO issued a notice stating that it would now consider 'non-naturally occurring, non-human multicellular organisms, including animals, patentable subject matter'.[16] A single genetic modification in the billion base pairs composing the genome of multicellular organisms was now deemed sufficient for classifying them as manufactured objects.[17]

The claim that particular life-forms are *products of manufacture* became central to corporate strategies of appropriation, legitimising patents on countless genetically engineered organisms. It remains the key doctrine of genomic enclosure, even if it was always controversial from a scientific point of view. 'Recent advances in DNA techniques', writes a member of the National Academy of Sciences, 'are only modulations of biological processes. [...] The argument that genetically-modified organisms are the scientist's handiwork and not nature's wildly exaggerates human power, and displays the same hubris and ignorance of biology that had such devastating impact on the ecology of our planet.'[18] Patenting relies on an individualist conception of the creative process, which exaggerates the role of celebrated inventors and erases the role played by both biological processes and the collective intellect.

The rapidly multiplying menagerie of genetically engineered organisms spawned by corporate labs prompted many other legal and philosophical puzzles. Take, for instance, US Patent 4,736,866: a white and furry mammal with red beady eyes, owned by DuPont Corporation. OncoMouse™, an oncogenic rodent engineered by splicing cancer-promoting genes into fertilised mouse eggs, has become one of the most popular 'research models' for studying cancer: a living commodity sold to researchers at $50 apiece and prized by *Fortune Magazine* as '1988 Product of the Year'.[19]

The patents covered the gene-insertion methods, the resulting mice, as well as any other non-human mammal with the same cancer-promoting gene. They thus fulfilled two essential roles in sustaining the profitability of this rather puzzling product. Firstly, they prevented other biotech companies from producing competitive strains of oncogenic mice. As a result, OncoMice™ could be sold at higher prices, generating superprofits to DuPont. Second, the patents forbade researchers from propagating the OncoMouse™ in captivity, creating second and third generations free-of-charge. In 2001 the University of California, Davis received a letter of warning from DuPont for doing exactly that: letting their legally purchased OncoMice™ sexually propagate without the company's permission. In the topsy-turvy world of patent law, where genes are proprietary inventions, unapproved sexual intercourse between two patented animals has been turned into a form of theft. From the point of view of capital, in this case well represented by DuPont's corporate lawyers, the reproduction of life – insofar as it relies on DNA replication – appears as a criminal duplication of trademarked property.[20]

The Kafkian case of the pirate OncoMouse™ illustrates well some of the ways in which genetic engineering intensified the control exercised by corporate actors over living organisms. Most organisms cannot be patented and turned into financial assets. High-volume dog-breeding facilities churn out thousands of expensive Samoyed dogs to be sold for profit, but they could never claim exclusive ownership over the breed itself. Similarly, DuPont owned hundreds of thousands of laboratory mice before the 1990s, but it never exercised monopoly power over a whole strain. By contrast, genetically engineered organisms can be presented as novel manufactured artefacts and claimed as exclusive objects of monopoly ownership. The creation of genetically engineered bodies has granted unprecedented power to patenting corporations, extending their capacity to shape, control, and exploit animal bodies as living means of capital accumulation. This is one of the reasons why, starting from the early 1990s, hundreds of companies started investing heavily in genetic biotechnologies, with the explicit goal of accumulating patents as fast as possible.

Looking back at the cancerous body of the first OncoMouse™ from our present conjuncture, the array of hopes and dreams that this creature was able to generate appears perplexing. The genetically engineered mouse was hailed as a formidable weapon in the war against cancer, promising cures for the masses and profits for the few. In her influential writings, Donna Haraway suggested that its suffering body may be taken as a sign: announcing not only a revolution in medical techniques, but also the rapid emergence of a postmodernity inhabited by 'promising monsters, vampires, surrogates, living tools and aliens'.[21] This proliferation of hybrid technoscientific beings, Haraway argues, destabilises humanist assumptions about the sacredness of nature, challenges traditional binary separations between the normal and the monstrous, and troubles any essentialist notion of what is natural and what is not. 'Symbolically and materially', Haraway argues, 'OncoMouse™ is where the categories of nature and culture implode for members of technoscientific cultures'.[22]

There is much to be learned from Haraway's reflections. It is certainly true that our world is increasingly made up of hybrid nature/cultures, of which OncoMouse™ is a paradigmatic example. It may also be true, as Haraway claims, that the troubling hybrids generated by technoscience have eroded essentialism and cemented a conception of the world in which there is no longer any Nature to defend, but only contingent technobiological assemblages. Yet, we question her conclusion that the transgression of species-boundaries enabled by recombinant DNA should be perceived as a revolutionary moment signalling the end of modernity. Far from undermining the modern attitude to nature, Onco-Mouse™ remains a rather paradigmatic product of it. In fact, it represents an embodiment of Francis Bacon's project for a modern science, which would instrumentally and endlessly 'perfect nature'.

Generally hailed as one of the founders of the modern scientific method, Bacon claimed that 'the work and aim of human power' should not be directed at understanding nature, but rather should 'generate and superinduce a new nature'.[23] So, one might ask: How transgressive of modernity are the legions of gene-edited mice that live, suffer and perish in the laboratories described by Haraway? After all, is there anything more modern than the drive to instrumentally alter nature(s), and transform bodies into technoscientific artefacts? Certainly, OncoMouse™ would not have been out of place in that quintessential modernist Utopia that was Bacon's *New Atlantis* (1627): a dream-world where natural scientists could boast of managing 'parks and enclosures of all sorts of beasts and birds', where 'by art we make them greater or taller than their kind is; and contrariwise dwarf them, and stay their growth; we make them more fruitful and bearing than their kind is; and contrariwise barren and not generative. [...] We find means to make commixtures and copulations of different kinds; which have produced many new kinds.'[24]

When closely inspected, Bacon's vision of the modernity-to-come is not fundamentally different from the conceptions of postmodernity celebrated by 1990s cyberculture. From the very beginning, the locomotive of modernity was on a track thought to lead to ever-more powerful means of controlling heredity and shaping life in useful bodily forms. This is why, while we do not doubt Haraway's claim that we now live in a world of hybridity and endless technoscientific transgression, we are rather sceptical that this delivers us out of modernity. The OncoMice™ trapped in the gardens of this New Atlantis – far from being the expression of a postmodern politics – turns out to be the realisation of a thoroughly modern attitude to nature, which combines all too well with capital's tendency to transgress all boundaries that limit its self-valorisation. In fact, its genetic mutations are not random; they are thoroughly determined by corporate needs. Ultimately, the generations of OncoMice™ brought to market as value-bearing commodities betray the fact that the production of mutations and hybridity remains an activity guided by profoundly restrictive, and thoroughly modern, socioeconomic rationalities. If modernity culminates in the production of mutant lives, it is the logic of capital that shapes which mutants are made to be, and which are left dormant as inactualised virtualities.

THE GENETIC GOLD RUSH

The Molecular Politics of Genomic Rents

The 1980s and 1990s represented a time of juridical turbulence as the biotech industry proceeded to lay the legal foundations of its future profitability. In 1980, *Diamond v. Chakrabarty* transformed genetic engineering into a powerful means of capital accumulation. In the following years, another way through which biotech corporations may accumulate valuable patents was gradually established. 'An isolated and purified DNA molecule', writes an anonymous bureaucrat of the PTO in the winter of 1987, 'is eligible for a patent as a composition of matter or as an article of manufacture because that DNA molecule does not occur in that isolated form in nature.'[25] By then, the gene rush was already well under way. In 1981, the Upjohn Company was granted the first gene patent. The first patent on a human gene followed shortly. In 1982, a group of researchers based at the University of California were granted exclusive ownership over the *CSH1* gene – a DNA sequence instructing cells to produce a growth-hormone that is vitally important for foetal development.[26] It is an instructive case. The patent contributed to privatising the profits created by publicly funded research, and unleashed the first of many patent wars that have characterised the history of the genomic industry.

Four years before the patent was finally granted to the University of California, Peter Seeburg – one of the scientists that had isolated the *CSH1* gene – quit his academic position to join the ranks of Genentech. He smuggled with him a small quantity of the DNA material he had contributed to isolate. 'Just before midnight on 31 December 1979', writes a *Nature* reporter describing facts established in a trial almost twenty years later, 'Seeburg slipped into the university, removed a sample of human growth hormone DNA from a lab, and took it to Genentech.'[27] Seeburg would later testify that this stolen material was used to help Genentech develop a new strain of genetically engineered bacteria, which excrete human growth hormone as part of their modified metabolism. The bacterial excretion was subsequently harvested and commercialised as Protropin™. It would turn out to be the first blockbuster recombinant-drug: regularly prescribed to stimulate children's growth, and widely abused by athletes in an attempt to increase their muscle mass.[28] When Genentech agreed to settle its litigation with the University of California in 1999 for a compensation of $200 million, it was estimated that the drug had already generated almost $2 billion in revenue.[29]

In the four decades since the onset of the battle for *CSH1*, over 40 per cent of the human genome has been patented.[30] Another landmark case of this protracted 'gene rush' was sparked by a public announcement delivered during the 1990 Convention of the American Society of Human Genetics. Research by Mary-Claire King indicated that mutations on a single gene located on chromosome 17q21 appeared to be positively correlated with early-onset breast cancer. The study spurred an international competition to isolate, sequence and claim ownership of the *BRCA* gene. A research team at the University of Utah headed by Mark Skolnick rapidly gained the forefront of the race, securing the financial backing of Eli Lilly. In December 1997, the investment paid off. The PTO granted Skolnick a patent that covered 47 separate mutations in the *BRCA1* gene. Another eight patents followed in the next few months, granting Skolnick's newly formed Myriad Genetics full control over both the *BRCA1* and the *BRCA2* genes. Any diagnostic test based on those genes and aimed at assessing the risk of early-onset breast and ovarian cancer was also covered.[31]

Armed with these patents, Myriad began marketing its diagnostic tests. The most commonly used involved full sequence testing of the *BRCA1* and *BRCA2* genes for a whopping $2,400. No other US laboratory could offer an alternative test without risking a patent infringement lawsuit. As a result, Myriad saw its annual revenues grow steadily for over two decades, climbing from $59 million in 1998 to $771 million in 2013. Meanwhile, Myriad sought international recognition for its monopoly on cancer screening procedures, succeeding in Canada, Australia, Japan, New Zealand and the European Union.[32] Yet, a wave of opposition steadily mounted. In 2009, a motley coalition of patients, scientists, clinicians and activists – spearheaded by the American Civil Liberties Union – filed suit against Myriad in an attempt to shatter its monopoly over a potentially life-saving test. They asked the Supreme Court: 'Are human genes patentable?'[33]

The question was rapidly politicised. People gathered on the steps of the US Supreme Court, carrying signs that read: 'Your Corporate Greed is Killing My Friends!', 'Free Our Genes!' and 'End Patents on Life Forms'.[34] Leading genomic scientists wrote *amicus briefs* to the Court, expressing concern over the ongoing enclosure of genes into privately held assets.[35] Christopher Hansen – as counsel for the plaintiff – insisted that isolating a gene could not be equated to a process of manufacture but was rather a simple act of discovery. Isolating a gold mine, Hansen pointed out, never entitled anyone to patent gold. The Court's ruling – released on 13 June 2012 – largely mirrored Hansen's suggestion: it held that 'a naturally occurring DNA segment is a product of nature and not patent eligible merely because it has been isolated'.[36]

The decision is likely to cause an important shift in the future development of the genomic industry. It clarifies that 'separating a gene from its surrounding genetic material is not an act of invention'.[37] However, this shift should not be mistaken as fulfilling some protesters' desire to end patents on life-forms. To the contrary, the decision has been praised by many representatives of the genomic industry. Why? Three reasons seem prominent. First, the decision has reaffirmed the legitimacy of patents on recombinant DNA, synthetic proteins and gene-edited organisms. Second, patenting non-human genes remains possible. Third, the decision does not directly threaten the profitability of companies like Myriad Genetics. At most, it is likely to push them to shift their business models from the accumulation of patents (and the collection of royalties) to the accumulation of data (and the collection of access fees).[38]

Myriad Genetics has already announced that its business model will no longer rely on extracting rents from its patent portfolio, but rather on valorising the huge proprietary database on genetic information that it has built over the last decade as a monopolist in the sector. This new-fangled business strategy could be read as a paradigmatic instance of a more general transformation. New strategies of control over genetic information are already emerging. Gene banks enable centralised control over vast archives of genetic material from bacteria, plants and animals, and their preservation under controlled conditions. Virtual data banks enable the accumulation of abstract genomic information. Together, they constitute a global network of genetic Arks filled with cuttings and seeds, sperm and eggs, coral fragments and tissue samples as well as with sequenced information stored in terabytes of zeros and ones. This global network is a valuable source of knowledge and wealth. It also represents a site of profoundly uneven and unequal access to the means of genomic production, as well as a critical infrastructure for the further development of the global genomic industry.[39]

Overall, the struggles that surrounded gene patents throughout the 1990s shows that the development of the genomic industry has been shaped as much by scientific discovery and technological innovation as by legal shifts, juridical decisions and political mobilisations. (Bio)technologies are neither neutral tools, nor do they impose on society a single pattern of development. Certainly, the existence of a competitive economic structure dominated by a handful of multinational corporations has fundamentally influenced the form taken by genetic biotechnologies and their living products. And yet, this historical tendency has been troubled by an emergent form of molecular politics, whose contours we are only starting to decipher. Widespread resistance against genetic engineering has radically slowed down the adoption of genetically modified organisms in many countries around the world. Animal rights activists stole hundreds of mutant mice, stirring a political controversy over the use of patented laboratory animals in laboratory facilities. Peasant organisations uprooted genetically engineered crops. Indigenous groups have forced a partial redistribution of the value produced on the base of the genetic code contained in their very own cells. Feminist groups successfully contested Myriad's monopoly over cancer genes, disrupting the enclosure of the human genetic code. The genomes of all other species, nevertheless, remain very much up for grabs.

ON THE GENOMIC FRONTIER

Biopiracy and the Original Accumulation of Sequences

The contemporary genomic industry relies on the systematic transformation of genomic raw material into commodities and assets. Therefore, the extraction of genetic matter from microorganisms, plants, animals, and human beings is as essential to the biotechnology industry as the extraction of minerals, coal and oil were for the first industrial revolution. It is the first step to patenting isolated genetic sequences and to generating new recombinant bodies. It is not surprising, therefore, that genes have become an increasingly valuable resource that is bioprospected, mined and transformed into capital. As we have seen, the technological expertise and the capital investments needed to manipulate this new 'genetic resource' mostly reside in the scientific laboratories and the corporate boardrooms of the Global North. Yet, most of the genetic resources that fuel the industry lie in the ecosystems of the Global South: in tropical areas and 'biodiversity hotspots' that make up less than 2 per cent of the planet but are home to over 60 per cent of plant, bird, mammal, and reptile species.[40]

Corporations finance bioprospecting expeditions in biodiverse regions in search of rare genetic traits that may inform genomic research. In many important aspects, these collecting practices do not represent anything new. The history of modernity has been profoundly shaped by the mobilisation of biological wealth for the advantage of imperial markets. The so-called Age of Exploration was motivated by the desire of discovering, classifying and collecting new biological resources – such as crops, fibres, dyes and medicines – as much as by the prospect of accumulating inert minerals. In the imperialist quest for plunder, global ecosystems were as deeply mined for biological resources as the Cerro Rico de Potosí for silver, tin and zinc.

Colonial agents dedicated a great deal of time and resources to biological research with the hope of finding unknown living wonders, which could be transformed into lucrative global commodities. On occasion, the appropriation of a handful of seeds could shape imperial geographies – and global ecosystems – in profound and long-lasting ways. In 1876, for instance, Henry Wickham was contracted by the Royal Botanic Gardens in Kew to collect *Hevea brasiliensis* rubber tree seeds from Santarém in Brazil. Despite the country's laws against exports of the precious seeds, Wickham managed to smuggle 70,000 of them on a steam ship to England. They were then germinated and sent to British colonies in India and Southeast Asia. The resulting plantations broke the Amazon rubber monopoly, and dominated the rubber market until the invention of synthetic alternatives in the 1940s.[41]

Today, armies of genetic bioprospectors have taken the place of old-style colonial botanists. Since the early 1990s, we have been living through 'an historic revival of collecting'.[42] In search of rare genetic traits that might possess commercial value, corporate giants are tapping the huge reservoirs of biodiversity accumulated during the colonial era in botanical gardens and natural history museums, while financing novel expeditions across the Global South. The last decade of the twentieth century was characterised by the expansion of bioprospecting programmes in the last, green reserves of the world: from the Coastal Forests of Eastern Africa to the Brazilian Amazon, from the rainforests of Borneo to the landscapes of Madagascar. The first decade of the twenty-first century witnessed a dramatic increase in the number of patented genetic sequences resulting from corporate bioprospecting of the global oceans.

Genetic resources are systematically extracted in the Global South and shipped to high-tech laboratories in the Global North. They are then transformed into private assets protected by intellectual property rights.[43] The negotiations leading to the Uruguay Round of the General Agreement on Tariffs and Trade (GATT) have further entrenched this form of unequal exchange, leading to the erection of a uniform global framework for intellectual property protection. In particular, the Agreement on Trade Related Aspects of Intellectual Property Rights (TRIPS) assured that genetic resources would remain freely accessible for corporate bioprospecting, while providing worldwide protection to industrial patents on recombinant molecules and gene-edited organisms.[44]

This reflected the interests of the Advisory Committee on Trade Policy and Negotiation, a large coalition of international corporations that played a direct role during the GATT negotiations.[45] James Enyart, then a lobbyist for Monsanto, sums up well the key role played by major industries in shaping the global regulatory regime: 'Industry has identified a major problem for international trade. It crafted a solution, reduced it to a concrete proposal and sold it to our own and other governments'. In this context, he recalls, the 'industries and traders of world commerce have played simultaneously the role of patients, the diagnosticians and the prescribing physicians'.[46] Corporations forged alliances with state authorities to influence the negotiations and globalise the US regime of intellectual property rights over isolated genes and genetically engineered bio-objects.[47] In this way, the molecular frontier was effectively turned into the latest *terra nullius*: an unowned empty space that can be freely appropriated by effective occupation-through-patenting. Owing to a neocolonial legal framework, private companies have been able to extract naturally occurring biochemical and genetic material and, subsequently, enclose it through intellectual-property monopolies. According to

Vandana Shiva, 'at the heart of the GATT treaty and its patent laws is the treatment of biopiracy as a natural right of Western corporations, necessary for the development of Third World communities'.[48]

The 1992 Rio Earth Summit saw highly politicised protests against corporate biopiracy. The resulting UN Convention on Biological Diversity (CBD) represented an attempt at compromise. The CBD no longer presented genetic resources as 'a common heritage of mankind', which 'should be available to anyone without restriction'. Instead, states were given control over their so-called 'genetic resources' – that is, over the DNA molecules of each and every living organism dwelling in their sovereign territories. This unprecedented doctrine of genetic sovereignty affirmed an instrumentalist view of living organisms, while providing states with the juridical power necessary to secure at least a fraction of the economic returns generated by the rising genomic industry. The treaty promised to create new economic incentives for biodiversity conservation by turning biodiversity into a standing-reserve of molecular raw materials, and thus into a national resource worthy of protection.

The credibility of this neoliberal approach to conservation – promoting ecological governance through market mechanisms – was doubtful from the start. Despite its grandiose rhetoric, the CBD remains a non-legally binding, soft law instrument. Its terms and conditions are not subject to review by any independent regulatory body, and no sanction is prescribed for those who violate them. It is hardly surprising that the royalties offered by bioprospecting companies have largely remained an empty promise. Despite the publicity given to a handful of cases in which major corporations have gracefully granted very minor returns to collaborating states, most 'access and share agreements' have been absolutely futile bureaucratic exercises.[49]

Given the increasing ease by which sequenced genes can be synthetically reproduced, and the simple fact that most plants and animals do not respect national borders, companies can easily cover up the exact source of their biological raw materials. According to one of the most attentive studies on the subject, completed more than a decade after the signing of the CBD, 'all royalty payments still remain projected; nothing has yet been disbursed to supplying countries in recompense for the use of collected materials through this mechanism'.[50] In the few cases in which royalties have been disbursed, they were mostly channelled into the treasuries of collaborating sovereign states with little recognition for the essential contribution made by indigenous groups to the production of biological knowledge.

The granting of sovereign rights over national biological resources has done little to challenge genetic enclosure. It has, rather, facilitated the transformation of biodiversity into a private economic resource. The CBD, moreover, does not apply to spaces beyond national jurisdiction: an increasingly significant legislative gap since global oceans are rapidly becoming a frontier of genomic extraction. Recent projections suggest that the value of the emerging global marine biotechnology market could reach $6.4 billion by 2025.[51] In particular, bioprospecting ventures at sea have been targeting so-called extremophiles: deep-sea creatures that can survive under extreme environmental conditions and often metabolise unique biochemical compounds. As of 2020, 12,998 genetic sequences from marine species have been patented with 47 per cent of those patents belonging to a single multinational chemical giant: BASF, based in Ludwigshafen in Germany – a figure that indicates growing corporate control over marine genetic resources.[52]

Digitised forms of genetic collecting have proven even more difficult to regulate. Bioprospectors increasingly avoid collecting and storing physical genetic materials altogether. Rather, they rely on portable sequencing devices. Once translated into data, the information can be circulated instantaneously across borders. The shift towards new forms of digital genomic collecting makes the CBD, which regulates only the circulation of genetic materials, rather obsolete. This raises the question of whether there are effective ways of monitoring how this abstract bioinformation will be used and valorised. If the digital genomic sequences accessible on databases such as Incyte were to inform a technological breakthrough, would anyone ever know? Is there any reason to assume that the country from which that information was originally extracted would be contacted and rewarded? Or that the communities residing there would be compensated? There is now talk of including digital sequencing information in the Convention. It may be, once again, too little and too late. As pointed out by the chief operating officer of Shaman Pharmaceuticals, in the last twenty years corporations have accumulated immense banks of genetic materials 'and by the time they have gotten [regulations] all figured out it won't matter because these corporations will have gotten some huge quantities of genes stockpiled, and it will be too late to do any sort of retroactive regulation'.[53]

Indigenous communities around the world have often opposed bioprospecting projects and the conception of nature they both presuppose and reinforce. The Leticia Declaration (1996), Kimberley Declaration (2002) and Nibutani Declaration (2008) have denounced bioprospecting operations both as a form of plunder and as an imposition of Western conceptions of nature. They emphasise the extent to which international law obscures the role played by indigenous labour and knowledge(s). Indigenous communities not only play an essential role in the conservation of natural biodiversity, their understanding of local ecologies is also often mobilised directly by bioprospectors. In this sense, bioprospecting may be seen as a form of accumulation by dispossession. 'The patenting and licensing of genetic material', writes David Harvey, 'can now be used against whole populations, whose practices had played a crucial role in the development of those materials'.[54] This captures one of the rivendications made by indigenous struggles against bioprospecting. Yet, from this point of view, it may appear that biopiracy raises only questions of *redistribution*, which may be solved by a radical socialisation of the value generated by the commodification and assetisation of genomes and life-forms.

It is, however, essential to consider another aspect that is central to indigenous struggles. A member of the Nga Puni Whakapiri movement – a network of Māori groups resisting bioprospecting and genetic engineering – explains how this reconstruction of nature as a functional reserve of material resources and economic value constitutes a force of alienation: 'genetic engineering threatens all, everything cultural, everything Maori. It threatens mauri, which is the life essence of every single living thing on earth'.[55] Victoria Tauli Corpuz, a spokesperson of the Kankanaey people and the UN Special Rapporteur on the rights of indigenous peoples between 2014 and 2020, summarised her years of dialogues with indigenous organisations stating that 'the genetic determinism of biotechnology conflicts with the holistic worldview of many indigenous peoples'. When this is the case, she insisted no economic reparations can retribute the alienation caused by bioprospecting ventures since 'for indigenous peoples to accept the genetic determinist view, they have to radically alter their worldviews, their ways of knowing and thinking, and their ways of relating with nature and with each other'.[56] From the perspective of indigenous struggles against biopiracy, the colonisation of the genetic commons is not only a form of imperial plunder.

It is also a colonising force, which projects molecular biology's peculiar conception of life onto the earth, while erasing alternative cosmologies and philosophies of nature.

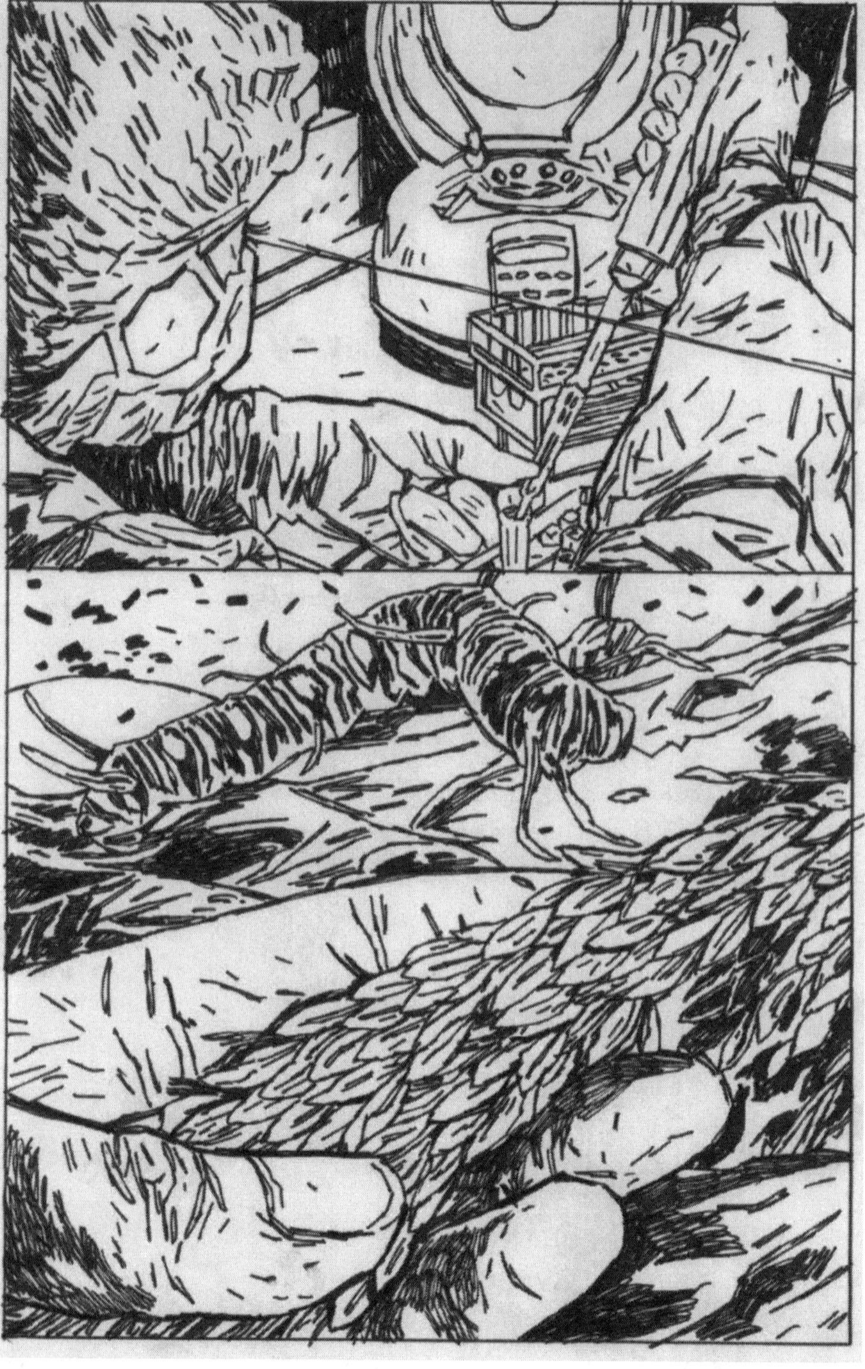

3

Genomic Infrastructures

Banking the Biosphere and Genomic Big Data

In 1953, Francis Crick and James Watson identified the molecular structure of DNA. Only a year later, the first functional silicon transistor was built. The simultaneous development of informatics and computer coding provided biologists with a new conceptual toolbox, a new set of mechanist metaphors, and a new way of thinking about the molecular processes that sustain life. The classic Cartesian conception of living organisms as mechanical machines swiftly mutated into cybernetic visions of living computers running a subtle genetic script. Already in 1961, evolutionary biologist Ernst Mayr expressed the new doxa in rather simple terms, writing that the 'purposive action of an individual, insofar as it is based on the properties of its genetic code, is no more or less purposive than the actions of a computer that has been programmed to respond appropriately to various inputs'.[1] In the 1970s and 1980s, François Jacob and Jacques Monod popularised the concept of 'genetic program'.[2] By the early 1990s, the description of living organisms as biological computer systems had become mainstream. According to Richard Dawkins, what 'lies at the heart of every living thing is not a fire, not warm breath, not a spark of life. It is information, words, instructions', while the genetic code is 'almost exactly like a computer tape'.[3] In the last decade, Watson and Crick's seminal conceptualisation of 'genetic information' has evolved into a bioinformatic theory of 'life as evolving software'; and scientists routinely talk of DNA as a code that may be copied, stored, transmitted, edited and hacked.[4]

The discipline of biology has passed through a watershed. No more than two centuries ago, the study of life was largely based on the direct observation of living organisms. In the twentieth century, biochemistry and molecular biology offered new methods for the study of living organisms by analysing the multitude of constantly interacting chemical molecules that compose their organic bodies. Today, digital approaches to the study of life are becoming dominant. Contemporary science's capacity to understand and manipulate living organisms is increasingly dependent on the mediation of complex digital technologies. It is not difficult to understand why. The printed version of the human genome, produced by the University of Leicester, consists of a single string of digits spanning 130 volumes. A single scientist would need years just to read it aloud. Only computers have the capacity to store, search through, and analyse the billions of DNA 'letters' that compose the genome of even the simplest of living creatures. Computing capacity has enabled the foundation of a new sort of science: genomics, which studies living organisms as complex information systems, and takes the genome as its privileged object of analysis.

This chapter is mainly concerned with this paradigm shift, and its economic, social and political implications. It is a history of how life-forms are translated into digital information, of how the extraction and storing of genomic data has become a global industry, and of how this data is being used to create new forms of knowledge and power over life. In short, we set out to investigate how global processes of genomic extraction are being carried out, and how genomic Big Data is being valorised in the global economy. We begin by discussing the first, foundational, global genomic enterprise. The Human Genome Project represents one of the largest scientific projects carried out in history. Formally launched in 1990, it aimed at identifying and mapping all 3.2 billion base pairs of the human genome, while developing a global technoscientific infrastructure for the future growth of the genomic industry. As a result of these massive public investments, the cost of developing new biotechnologies was socialised, and the cost of sequencing plummeted.

The second section recounts the rise of personal genomics as an industry and as a consumerist practice of the self. In 2007, the sequencing of James Watson's genome was completed for $1 million. In 2014, the sequencing giant Illumina offered its $1,000 genome, a milestone celebrated as the threshold of a brave biotech era in which DNA sequencing would become a routine operation and genomic data would flood the market. In early 2018, it was estimated that over 12 million people had their DNA analysed by direct-to-consumer genetic tests offered by companies such as Ancestry and 23andMe. How should we interpret this rapid accumulation of sensible genomic data? How is this data turned into capital? What novel politico-economic configurations are heralded by rapidly intensifying practices of genomic extractivism? Finally, we turn to how new sequencing technologies are being deployed beyond the human realm. Focusing on the recently announced Earth Biogenome Project – which intends to sequence all eukaryotic life-forms on Earth, from single-celled organisms to complex mammals – we consider how genomic data is collected, stored and valorised. We consider this scaling up of genomic efforts as signalling an epochal shift that can be described in terms of a drive towards hyper-genomics, and we are brought to ask: What logic drives this new genomic rush? And what will be its socio-ecological effects?

ON THE MOLECULAR RAILROAD

Mapping the Genomic Frontier

1 October 1990 marked the start of the biggest programme in the history of biology: determining the DNA sequence of the entire human genome within 15 years. With a price tag of over $3 billion, the scale and scope of the Human Genome Project (HGP) remains unparalleled in the history of biology. The idea of focusing research funds on DNA analysis originally emerged from two of the largest weapon research centres in the United States: Los Alamos and Lawrence Livermore laboratories.[5] Haunted by the fall of the Soviet Union, the laboratories were seeking novel ways to use the enormous technical and digital machinery accumulated throughout the Cold War period. They proposed to investigate whether the nuclear bombs exploded in the past half a century had caused any genetic mutations in the exposed populations. The first step would be to sequence all the three billion subunits that make up a 'normal' human genome, and then look for genetic mutations among the populations most exposed to nuclear radiation.[6]

The sequencing venture appeared as the culmination of a tradition in biological research that had long presented DNA as the 'master molecule' of life, and the human genome as 'nature's complete genetic blueprint for building a human being'.[7] In the new genomic dreamscapes, sequencing the human genome meant nothing less than reading 'the book of life' and finally revealing the secrets of what it means to be human.[8] Writing in 1990 as the newly appointed director of the HGP, James Watson argued that 'a more important set of instruction books will never be found by human beings', since 'the genetic messages encoded within our DNA molecules will provide the ultimate answers to the chemical underpinnings of human existence'.[9] Similarly Walter Gilbert, Nobel Prize-winning biology professor at Harvard University, biotech entrepreneur and chief proponent of the project, prophesied: 'we will finally identify the regions of DNA that differ between the primate and the human – and understand those genes that make us uniquely human'.[10]

In spite of the fervour of its proponents, the project soon garnered fierce critique from several fronts. Some concerns were disciplinary in nature. Many biologists worried that the US Department of Energy (DOE), dominated as it was by physicists, was not the appropriate agency for coordinating a research project that openly presented itself as a revolutionary turning point in biology and medicine. Others referred to the project as the 'DOE's program for unemployed bomb-makers'; little more than a power-grab by an institution populated by prominent phycisists and nuclear scientists.[11] Others were ontological in nature, and critiqued the cybernetic-mechanist conception of life that the project both presupposed and furthered. Evelyn Fox Keller and many others – including Hilary Rose, Steven Rose, Richard Lewontin, Ruth Hubbard and Leo Kamin – argued that the entire endeavour was based on a form of genetic determinism, which exaggerates the role of DNA in determining the development of living organisms.[12] Despite their differences, these authors converge on a vision of the HGP as 'the ultimate product of an extreme reductionist vision of biology'.[13] 'The probable social consequence' of this form of genetic determinism, wrote Fox Keller summarising one of the main insights of this line of critique, 'is the direction of technological fixes at the genome'.[14] The HGP, in other words, was criticised for reinforcing a culture of genetic reductionism, while raising expectations that the accumulation of genomic knowledge would offer new ways of tackling a variety of physical, medical, behavioural and socio-ecological problems by manipulating their supposed genetic underpinning.

Other critiques concerned the role of 'Big Science' in public health provision. In the context of rising inequality, the unprecedented price tag raised more than a few eyebrows, with critics questioning the rationale for spending public funds on a project so far removed from the immediate needs of most people. Many argued that the project had been sold on hype and glitter and that it would drain money and labour from smaller, worthier biomedical efforts. Wouldn't the money be better spent on universal healthcare? some asked.[15] Insofar as proponents did engage with such critical concerns, they did so by reframing social issues in genomic terms. To the question of why governments should spend billions on the genome project, rather than directing those resources towards ending poverty and homelessness, the editor of *Science* quipped that 'people don't realize that the homeless are impaired' and that 'no group would benefit more from the application of human genetics'.[16] The genome was presented as a promising site for the exercise of a new form of government, which would improve society by targeting the genetic roots thought to determine health and social status.

Genetic determinism reached a fever pitch towards the end of the 1990s. Prominent scientists such as Renato Dulbecco suggested that 'the sequence of the human DNA is the reality of our species', while Gilbert hailed the genome as the 'holy grail' of biology.[17] On its April 1998 cover, *Life* magazine quizzed the reader with a striking question, which captures well the zeitgeist of the moment: 'Were you born that way? Personality, temperament, even life choices: New studies show it's mostly in your genes.' It could be tempting to dismiss this as journalistic hyperbole, but many prominent scientists promoted similar messages. Gilbert predicted that mapping the human genome would allow us to 'understand deeply how we are assembled, dictated by our genetic information'.[18] James Watson – then head of the National Center for Human Genome Research – concluded that 'children come into the world with fixed personalities' and that 'effective remedies for socially inappropriate behaviors' must be targeted at the genetic level. Legal scholars debated the creation of a 'genetically oriented criminal justice system'.[19] In this budding technoscientific imaginary, the genome represented not only an important object of knowledge – worthy of systematic study – but also an instrument of power and a lever of government. Taking possession of our 'biological instruction book' would offer a new understanding of human nature, blaze the way to a revitalised society, and illuminate new possibilities for economic expansion.

With the sobriety of hindsight, it is clear that producing a representative genome sequence for the human species did not provide the promised panacea for health and social problems. A comprehensive review recently published in *Frontiers in Genetics* succinctly concludes: 'over 15 years after the completion of the HGP there is no noticeable global impact on drug development and health'.[20] Nor did it resolve long-standing ontological debates surrounding the nature of heredity, identity and difference, nature versus nurture, or determinism versus vitalism. What it did, however, was to conjure a global sequencing infrastructure, which opened new frontiers of capital accumulation. The genome project was not only a scientific venture built on shaky theoretical foundations; it was also a gargantuan infrastructural undertaking: a state-funded technological expansion. In the context of the late Cold War, and on the heels of a severe economic recession, the project appeared as a perfect 'technoscientific fix' in all three senses of the term elucidated by David Harvey: a 'fixing' in place of large sums of over-accumulated capital; a 'quick fix' to complex social issues; and a 'fix' to the growth-addiction of capital and the capitalist state.[21] Researchers involved in the project saw it as the herald of a new era in which industries would be built on the organised manipulation of the genetic code and the rational design of 'biological systems'.

Midwifing this new era, however, would require massive investments well beyond the capabilities of any single corporation. The HGP implied channelling considerable public capital into the creation of a vast technoscientific infrastructure: a 'molecular railroad' to open new frontiers for capital accumulation. Organising, interpreting, and storing large quantities of genomic data required the development of automated technologies, including computer servers, image processors, as well as 'computational requirements including database restructuring, the development of advanced algorithms and analytical capabilities, and national networking with a dedicated supercomputer at the central node'.[22] The project required the construction of an expanding array of super-sized laboratories, deep vaults of biological data, and increasingly complex computational machinery. It implied the creation of an international genomic infrastructure, which would inject new power into the engines of the biotechnology industry.

In turn, this fixed capital would allow for the exercise of power directly at the cellular level. The endeavour was framed as a vital public work of national interest, which would benefit national competitiveness across a number of key economic sectors. As Johnson & Johnson Corporation's Jack McConnell – who helped draft early US legislation on the HGP – testified to the US Senate: 'If we want the US to maintain its position as a dominant force in the pharmaceutical industry in the world, I cannot imagine our letting this opportunity pass us by [...] The group that first gains access to the information from mapping and sequencing the human genome will be in position to dominate the pharmaceutical and biotech industry for decades to come.'[23] Similarly, Wallace Steinberg warned the readers of the *New York Times* that 'there was an international race to lock up the human genome', and that if 'Americans do not participate, they will forfeit the race and lose the rights to valuable genes to Britain, Japan and other countries that are in the race to win'.[24] Just like public roads, canals and railways accelerate production across various sectors, the construction of a genomic infrastructure was presented as a question of national interest.

Genomics, in other words, was elevated to 'a general condition of production'.[25] As Marx suggests in Volume II of *Capital*, fixed capital investments in infrastructure not only provide a temporal and spatial displacement of overaccumulation crises; they also lay the groundwork for expanding the scale and scope of capitalist productive relations. Infrastructural projects of national relevance often require too much capital to be undertaken by any single company. In this context, the State often plays an important role in 'forc[ing] the society of capitalists to put a part of their revenue' into the realisation of 'generally useful works, which appear at the same time as general conditions of production'.[26] Marx argues that the involvement of the state indicates how generalised a particular condition of production is. Significant public investments in particular infrastructures can be taken as an indication that this is perceived as a significant technological basis for the economy. Writing in the nineteenth century, Marx had in mind public investments in streets, ports and railroads, but this analysis may be extended to the scientific infrastructure underpinning the contemporary genomic industry. The HGP involved massive capital investments, which were explicitly aimed at developing biotechnologies that would open new opportunities for capital accumulation across industrial sectors.

By the late 1980s, the idea of sequencing the genome had gained support in the US Congress. Critiques were glossed over, and the question of *whether* to invest in the project had been replaced by the questions of 'how big' and 'how fast'.[27] Right from the beginning, Watson stressed the necessity of embracing a Fordist approach to genomic sequencing, arguing that 'the cottage industry approach involving small groups of individuals, each working at a different site, seems unlikely to succeed'. He instead supported the establishment of 'DNA sequencing facilities that are far larger than any existing today and that more closely resemble industrial production lines than conventional university research laboratories.'[28] To make the human genome legible, laboratories had to be turned into 'industrial production lines', scientific research had to become a scalable labour process, and the production of biological knowledge had to be increasingly mediated by capital-intensive technologies. Much like the real subsumption of labour took place through the transformation of traditional artisanal workshops into mechanised factories, the subsumption of living processes has required the automation of sequencing labour and an increasingly industrial approach to scientific production.[29]

The HGP relied upon, and significantly furthered, an historical tendency towards ever faster and cheaper sequencing technologies. In 1965, it took several years of laboratory work to produce the first whole nucleic acid sequence: the 76-bases-long alanine tRNA from *Saccharomyces cerevisiae*.[30] At that rate, sequencing the three billion bases of human deoxyribonucleic acid would have been, in Watson's own words, an 'undreamable scientific objective'.[31] The endeavour would have required an army of scientists and lab workers, incessantly working to sample, identify, excise, and sequence. Starting from the 1980s, however, the combination of new sequencing technologies with industrial labour processes in ever-larger laboratories accelerated the production of genomic data. One of the technologies that facilitated this *genomic acceleration* was polymerase chain reaction (PCR) – often described as a sort of 'molecular photocopying', which allows molecular biologists to amplify the DNA in their samples.[32] The HGP was proposed in this context, with the explicit goal of scaling up existing technologies of molecular production through a well-funded and centrally coordinated international effort. Charles DeLisi, director of the US DOE, presented the official goal of the HGP as being 'not to sequence the human genome', but rather 'to develop technologies that would make sequencing a lot quicker than it currently is'.[33]

One of the major corporations set to capture the incoming flow of public money, thereby turning genomics into a business strategy, was Perkin-Elmer, Inc. and its two subsidiary arms: Applied Biosystems and Celera Genomics. Although seldom noted, the structure of this global corporation reveals quite a lot about the hidden political economy of the HGP.[34] On the one hand, Applied Biosystems focused on building, selling and servicing advanced sequencing machines, including many of the automated DNA sequencers used by the public consortium. On the other, those same machines were put to use by its sibling company Celera Genomics in order to sequence, database and annotate the human genome in direct competition with the HGP consortium.[35] In an interview with the *Chicago Tribune*, the chairman of Perkin-Elmer explained how the decision to start a race to map the genome was mostly aimed at expanding the market for its high-speed sequencing machines. After the announcement of Celera's aspiration to get to the fabled genome first, he explained, 'they [the HGP] got very competitive, and indignant, and were determined to win. That was great for us. We sold a lot more gene sequencers'.[36] By instituting Craig Venter's company and backing its genome sequencing efforts, Perkin-Elmer gave the Human Genome Project an identifiable enemy, and a reason to pressure states into escalating the investment.[37]

Given its position as a prominent supplier of the sequencing hardware used by *both* public *and* private genomic researchers, Perkin-Elmer was the only guaranteed winner in the competition. If Celera Genomics' business plan were to come to fruition, the company would enclose at least part of the data into a proprietary database, from which it could subsequently extract rents. Perkin-Elmer expected to make big profits by positioning Celera as the 'Bloomberg of biology': the preeminent source of data about genes and their function.[38] Celera would provide access to its genomic database for a subscription fee, costing thousands of dollars for a single academic user and millions of dollars a year for a pharmaceutical corporation. Moreover, Celera aimed to patent at least some of the resulting genomic data. By November 1999, Celera announced that it had already filed over 6,500 'provisional patent applications'.[39] The possibility that the human genome might end up locked away in private vaults posed a real threat to the vision embodied by the official genome project. A patented genome would impede the institution of the new genomic infrastructure as a general condition of production, which was intended to boost the profitability of several industrial sectors and shore up the US economy in the context of growing international competition.

If the public consortium decided to prevent Venter, it would have to do so by increasing its spending on sequencing machinery. This is exactly what happened. In response to Celera's competition, the public initiative released a new work plan aimed at accelerating the sequencing project from 90 Mb per year to 500 Mb per year.[40] The Wellcome Trust agreed to pump another £100 million into the sequencing venture, while the US government released a new line of funding for biological hardware and genomic sequencers. In March 2000, a joint statement by Bill Clinton and Tony Blair stating that the human genome data will be publicly and freely available sent the financial world into chaos.[41] Stocks across the biotech sector crashed, followed by the entire high-tech NASDAQ index. Celera's stock plummeted nearly 20 per cent overnight and continued to decline.[42] Celera's effort to become the 'Bloomberg of biology' eventually faltered, for the simple reason that the public project ended up offering much of the same information free of charge. In 2002, when only 25 companies and 200 academic institutions subscribed to Celera's 'Discovery System', the company ousted Venter, de-emphasised the information business, and began developing drugs instead. This failure, nevertheless, did not wipe out the almost $1 billion in capital that Celera had amassed in the bull genomics market of the 1990s, nor did it diminish the dominant position acquired by Perkin-Elmer in the global market for genomic machinery and laboratory infrastructure.[43]

In the 1990s, the biotech industry thus profited directly and indirectly from the so-called 'genome race'; just like the military industry profited immensely from the arms race between European nations in the 1930s; and the digital industry profited from the space race between the US and the Soviet Union. During the Cold War, the exploration of the cosmos was presented as a competition which would finally prove which economic system was more apt at organising social production, while stimulating scientific innovations that would further the commonwealth of mankind. Half a century later, the exploration of the human genome was presented in much the same way. It was turned into a theatrical confrontation between the public and the private sectors, but also casted as a scientific exploration that would advance a Universal knowledge of and for humanity. In both historical cases, the language was symbolic and ideological, the projected futures largely imaginary and overhyped, but the capital invested in infrastructure, hardware and technology was very much material.

In June 2000, the farcical race appeared to end with a tie. Clinton and Blair, flanked by director Francis Collins of the HGP and Craig Venter, announced the completion of the project. The publication of a draft the following year revealed that about 15 per cent of the genome remained 'unread'. While Celera abandoned the project, public consortia continued to work out of the spotlight. It would take another two decades before the publication of 'the first truly complete 3.055 billion base pair sequence of a human genome'.[44] Yet, Clinton's speech was a revelatory event. In comparing the HGP to the mapping of the fabled Northwest Passage in the early 1800s, Clinton tapped into a colonial imaginary of expanding imperial frontiers: 'Nearly two centuries ago, in this room, on this floor, Thomas Jefferson and a trusted aide spread out a magnificent map', the result of a 'courageous expedition across the American frontier, all the way to the Pacific'. Clinton continued, emboldened: 'Today the world is joining us here in the East Room to behold the map of even greater significance. We are here to celebrate the completion of the first survey of the entire human genome. Without a doubt, this is the most important, most wondrous map ever produced by humankind.'[45]

Clinton presented the completion of the HGP as an 'epoch-making triumph of science and reason', which opened up three 'majestic horizons': a more complete understanding of the molecular biology of living cells; the commercial development of new biotechnologies; as well as the promise of economic and social benefits trickling down to 'all citizens of the world'.[46] Clinton's neocolonial metaphors reveal much of the context in which the HGP emerged, as well as its epistemic baggage. Mapping and sequencing the genome represented both continuity and rupture: a continuation of the Baconian paradigm of technoscientific control over nature *and* a dramatic increase in scale, which engendered new molecular techniques for the control of living processes.[47]

The debates sparked by the completion of the Human Genome Project are far too wide ranging to be dealt with comprehensively here. What this cursory overview shows is that the HGP did not emerge from the internal workings of an independent republic of science, cordoned off from messy social worlds and sterilised from social struggles. Instead, it was from the very beginning a political and economic process, profoundly shaped by situated interests and unequal social relations. A controversial but widely cited study by the Battelle Memorial Institute estimates a 141-fold return on each dollar invested in the HGP. The study presents the race to the genome as a successful example of technoscientific Keynesianism that converted a $3.8 billion federal investment into $800 billion in private revenues between 1988 and 2010.[48] While the exact magnitude of the economic returns remains uncertain, there are few doubts that the dumping of vast public funds into the project expanded the genomic industry and the development of new biotechnologies that, in turn, reconfigured labour practices in laboratories around the world.

Over a decade of 'genome talk' had also profound cultural and political effects, supporting the progressive naturalisation of neoliberal society and the depoliticisation of a panoply of social issues: growing social inequality was increasingly presented as an epiphenomenal consequence of deeper natural differences hidden in our genes. In the writings of reductionist scientists such as Richard Dawkins and Steven Pinker, genetic fetishism reached a fever pitch.[49] In a neoliberal society dominated by corporations driven by capital's abstract tendency to self-replicate, molecular biology offered a rather congruous vision of nature: a world of fleshy automata driven by a selfish software code, whose single purpose is to secure its own conservation and endless self-replication.

As we learned to look for answers 'in our genes', everything – from inequality to homelessness to depression – was tentatively re-coded as the unfortunate result of genomic variations and genetic mutations. The biotech industry, which profited directly from the international genomic race and the billions of dollars thrown into it, was further boosted by this discursive 'geneticisation' of individual and social problems. If our destiny and identity is really crystallised somewhere deep 'in our genes', the only hope of getting things right is getting to know them, and hope for a technoscientific fix. This simple logic, as we will see in the next chapters, has not been displaced by the advent of epigenetics and post-genomics.[50] Wherever a new crisis is declared, its causes are sought in the genome and tweaking with genes is soon invoked as a possible answer.

'WELCOME TO YOU'

Watching Over Your Genomic Self

Data is life. This rather banal slogan of a famous marketing campaign may be taken as a sign of our bioinformatic present.[51] It is a symptom of the growing importance of data extraction in the global economy, and of how it ties together the digital and the biological dimensions of existence. Contemporary technologies – laptops, mobile phones, smart watches, satellites, CCTV cameras – relentlessly translate our biological and social lives into digital information. The resulting, endless series of zeros and ones is transmitted around the world through submarine cable wires, giant server hubs and corporate data banks. The constant production of data by means of surveillance, its accelerating circulation in ever-expanding networks and platforms and its valorisation through finance have engendered new pathways for capital accumulation. Contemporary capitalism endlessly renders life into data.

When pressed to describe the nature of the new beast, sociologists often talk of 'informational capitalism', 'platform capitalism', 'cognitive capitalism', 'communicative capitalism', 'surveillance capitalism', and so on.[52] This proliferation of concepts illuminates a fundamental aspect of the present conjuncture: the new economic order appears to be inexorably driven towards an ever-increasing extraction of biological and behavioural data. An increasing number of corporations are building their empires through the aggregation of data supplied voluntarily or involuntarily, wittingly or unwittingly, by all types of subjects. Everything and everyone can be subjected to this relentless process of 'dataification': the sick in their hospital bed, the healthy on their morning jog, the worker on their productive peak, the unemployed in their Netflix coma, the migrant in their crossings, the tourist in their journeys, the birds chirping, the plant sprouting, the bacteria infecting, the virus multiplying. Life itself appears as an endless process of production of data to be captured, stored and circulated. 'Our planet', shrieks an IBM commercial, 'is alive with data!'[53]

Maybe. And yet, what is this data flood really good for? Data is, first of all, a curious mirror of reality: an expansive record of people's lives, of animals' behaviours, of places, of processes, of objects, of relations. But this is not all there is to it. Data is also a predictive tool – a way of statistically forecasting subtle trends, ongoing tendencies and future developments. It is also a means of manipulation. Data is not simply passively collected; it is actively deployed as a way of predicting and manipulating the desires and needs of millions of consumers across the globe. The growing interest in the extraction, collection, and circulation of genomic data by governments and private corporations may be associated with the growing influence of behavioural genetics, a theory according to which 'most behaviours in animals and humans are under significant genetic influence', and it will soon be possible to 'identify individual genes that are related to behavioural outcomes'.[54] Genetic determinism stimulates genomic data collection. The growing value attached to genomic data production suggests that many economic actors do believe that there might be a way to turn this knowledge into an instrument of prediction and control over their consumers' future lifestyle choices. In the never-ending theoretical contest between nature and nurture as the primary determinant of behaviour, capital has been happy to spread its bets and support research on both fronts.

Over the last three decades, various biotech companies have specialised in the process of extraction, analysis and valorisation of large volumes of genomic data. In the late 1990s, a handful of small start-up firms such as University Diagnostics launched the first 'direct-to-consumer genetic testing' via mail orders and newspaper ads. By the early 2000s, the growth of digital commerce brought to the forefront new corporate players. In Europe, Sciona built a business providing tailored dietary recommendations on the basis of an individual's unique genetic profile. In the US, Myriad Genetics expanded its customer base for *BRCA* testing through both analogue and digital advertising campaigns. It was only the beginning of a decisive turn towards a business model based on relentless genomic data extraction. The 2010s were characterised by a dramatic increase in corporate genetic testing. A recent survey identified 246 firms operating on the direct-to-consumer genomics market.[55] In 2018, more than 26 million people purchased DNA tests.[56] Surging interest in discovering one's genomic self – propelled by heavy marketing – led to millions of customers spitting in a tube, in exchange for an expensive peek into their DNA.

This genome rush has fuelled the meteoric rise of two corporations: Ancestry and 23andMe. Ancestry was founded in the early 1990s and has now grown into a multinational corporation operating in over 30 countries, pulling over $1 billion in annual revenues. It markets DNA kits with the promise of giving each of their customers the sense of community and identity they long for. Genomic testing, one of their ads promises, 'can reveal your heritage and connect you to family past and present' – for $99, plus shipping costs and applicable taxes. The offer has convinced more than five million people to ship their spit, adding their saliva to the world's largest private collection of human drool – estimated in the hundreds of gallons. Most Ancestry customers consent to have their genomic data shared with a list of research partners. The data is then circulated to nonprofit entities and for-profit companies. In 2013, for instance, Ancestry announced it had cut a deal with Calico Life Sciences, a Google subsidiary focused on 'solving death' by 'harnessing advanced technologies to increase our understanding of the biology that controls lifespan'.[57] Bill Maris, Calico's founder and CEO of Google Ventures, dismissed concerns about genomic privacy, quipping that 'if we each keep our genetic information secret, then we're all going to die'.[58]

Google's shadow looms large over consumer genomics. Its collaboration with Ancestry is not an isolated case. In 2005, Anne Wojcicki – then spouse of Google's co-founder Sergey Brin – launched another major corporation in the genomic business: 23andMe. Unsurprisingly, Google was one of the major partners and investors in the firm. Its first DNA kit – a colourful box marked with the slogan 'Welcome to You' – promised to reveal the customer's 'genetic predisposition' to ninety characteristics from blindness to baldness. The company pushed genomics as a form of self-discovery, organising celebrities' 'spit parties', and commercials filled with solitary souls reconnecting with long-lost relatives. By 2017 its customer base grew from two to over twelve million people.

At this point, 23andMe started to mutate from a genetic testing start-up to what it was always meant to be: a major provider of genomic Big Data for global corporations around the world. 'The long game here', clarified one of 23andMe's board members, 'is not to make money selling kits, although the kits are essential to get the base level data. Once you have the data, 23andMe does actually become the Google of personalized health care.'[59] Indeed, 23andMe's business strategy appears to be modelled closely on Google's. While users can request to keep their personal information anonymous, the company reserves the freedom to share, sell and circulate the aggregated data. In fact, 23andMe has been selling customer data to Big Pharma companies such as Pfizer and GlaxoSmithKline for years. In January 2020, it licensed the rights to the first drug developed with the genomic data harvested from its users. Emily Conly, vice-president of 23andMe, described the development as one more way in which the company may profit from its genomic data.[60]

The concept of 'the genome' has considerably changed since the early days of the Human Genome Project. Some have speculated that we may have entered the age of post-genomics – a term that suggests a clear breakaway from a phase in which the genome was considered a linear set of instructions that determines the physiology and developmental life of an organism.[61] The post-genomic genome is no longer considered an all-determining script, but rather an organic and dynamic entity that responds to a variety of biological and environmental stimuli. Nevertheless, post-genomics is a rather misleading term. We are not witnessing a radical break away from genomics, nor a sudden abandonment of the genome as a significant object of knowledge. Rather, our supposedly 'post-'genomic age has seen an *intensification* of genomic research, the introduction of new methodologies, and the scaling-up of global genomic infrastructures. The genome retains an absolute centrality in both biological and popular discourse; increasing resources are expended in order to accumulate genomic data and develop new ways to interpret it. We are not living in a post-genomic, but rather in a hyper-genomic society.

Genetic testing kits have become an extremely effective technology for data extraction. They provide a shortcut into a future where corporations have access to large aggregates of genomic data from populations. Genomic marketing is already here. Spotify recently announced a new partnership with Ancestry in order to utilise listeners' genomic data in order to create playlists that 'match their genetic ancestry'. AirBnb partnered with 23andMe to offer holidays and experiences that are 'unique as your DNA'. Aéromexico offers discount rates according to the traveller's 'Mexican DNA percentage', while companies such as Gene Partner, Gene Future and Genemate 'match love seekers based on their genetic compatibility'.[62] Marketing firms begin targeting products not only on the base of customer's behavioural patterns – revealed by purchase patterns, product ratings, and Google searches – but also on the base of their genomic profile.

How can we estimate the market for genomic Big Data in light of these trends? In December 2020, Blackstone Group – the biggest private equity firm on the planet, boasting half a trillion US dollars in assets – bought a controlling stake in Ancestry for $4.7 billion. In this way, it acquired access to the largest genomic database ever assembled, storing information from over 18 million people. Blackstone's move is indicative of the growing corporate interest in controlling genomic data. Two months later, it emerged that 23andMe had also become a major prize in the global competition to acquire genomic knowledge. The company agreed to merge with Richard Branson's Virgin Acquisition in a $3.5 billion deal, and thereby entered into Virgin Group. A presentation by Virgin Acquisition announced that the 'vast proprietary dataset' offered by 23andMe will 'unlock revenue streams across digital health, therapeutics, and more'.[63]

Much of the critical debate surrounding consumer genomics has focused on the rather dubious efficacy of genomic counselling.[64] It has only scratched the surface of the new forms of power and exploitation that are increasingly central to Big Data capitalism. The business of direct-to-consumer genomics relies on the labour of millions of people, who provide free data in exchange for a promise: to discover their hidden genomic identities and thus their past ancestry, their present predispositions and their future health. Consumers of genetic testing kits are also producers of valuable data. They spit, swab, and post bodily excretions. They write on forums, fill in surveys and provide information on their state of health. They actively contribute to money-making research. They provide a form of free labour akin to the one performed by web-users feeding data into Google and Amazon. The digital labour of contributing data by participating in online networks, and the very material labour of providing genetic material, represent paradigmatic practices at the heart of contemporary capital accumulation. They are forms of production hidden within practices of consumption.[65]

In the last five years, the accumulation of sensitive genomic data has also conjured new forms of surveillance and police targeting. In the spring of 2018, the Californian police announced the capture of Joseph DeAngelo – suspected of murders, rapes and burglaries between 1973 and 1986. The police had long been in possession of DNA samples associated with the so-called Golden State Killer, but it had been unable to find any matching record in its databases. Thirty years after the last crime, investigators expanded the search by uploading the killer's DNA on GEDmatch, a public database where millions of people share and discuss test results obtained from companies such as Ancestry and 23andMe. The website identified a near-match for an undisclosed number of people sharing a common great-great-great grandparent. This led the police to focus their investigation on a number of potential suspects under the same family tree. On 18 April, a DNA sample was secretly collected from the door handle of one of the suspects. It was compared to the samples found on crime scenes and a match was established.[66] A report in *Science* concluded that around 60 per cent of US citizens of European descent may be tracked down in this way, since they have at least one relative whose genomic profile has been made available on a public or private database.[67] Given the dramatic expansion in genomic data extraction in the last two years, the percentage is likely to be considerably higher today.

While it is hard to determine to what extent police forces make use of existing genomic databases, the evidence points to increasing collaboration. Since 2018, the US police have confirmed that genomic companies have been instrumental in solving over thirty long-standing cases.[68] Given the complexity of human DNA, however, these genomic investigations often lead to the wrong door.[69] In addition to being error-prone, the use of genomic surveillance in criminal investigations raises other issues. In the US, the federal government has seized DNA from over 8 million people charged with a crime – one of the largest repositories in the world. Due to the rampant racial bias in US policing, Dorothy Roberts has argued that these databases 'effectively constitute a race-based biotechnology' which risks '[reinforcing] the racial order not only by incorporating a biological definition of race, but also by imposing genetic regulation on the basis of race'.[70] The addition of genomics to the ever-expanding surveillance toolbox reinforces and extends practices of policing and social control, whose effect on the health and safety of local communities is biased across lines of race, class and gender.[71]

The recent Covid-19 pandemic may considerably accelerate the trend towards increasing genomic surveillance. In the spring of 2020, the Trump administration ordered Immigration and Customs Enforcement to systematically collect DNA samples from all migrants taken in federal custody.[72] A year later, the Rockefeller Foundation published a report titled *Accelerating National Genomic Surveillance*. The document 'provides a blueprint for dramatically expanding genomic surveillance in the United States'. The main goal of the project is to stimulate investments in new genomic sequencing technologies that may help track the spread of Covid-19 in all its mutant variants and 'incubate a broad, data-driven platform so the world can better anticipate, visualize, and respond to future outbreaks'.[73] The creation of an extensive infrastructure for genomic surveillance is presented as an exceptional response to the current pandemic crisis, but it is unlikely to be scaled back in the future.

The United States' effort to harness genomic sequencing technologies for security and surveillance is not an isolated case. According to a recent report in *Nature*, China is following a similar course.[74] Malaysia has announced plans to create a national registration system that would store biometric and DNA data for anyone holding a national ID document.[75] Kuwait passed a law mandating DNA profiling for its entire population. The United Arab Emirates have recently unveiled their own Genome Program with the goal of eventually sequencing the entire population of Dubai.[76] These developments raise several urgent questions: Which futures are being constructed by the rise of genomic sequencing as an accumulation strategy, as a technique of self-exploration, and as a tool of surveillance? What new uses will be found for the mountains of genomic Big Data extracted from willing and unwilling populations across the globe? What forms of power will arise from the mastery of genomic data?

The last few years have also seen growing resistance against genomic mining. A sustained campaign led by critical collectives such as Gene Watch, Black Mental Health, Liberty and the Runnymede Trust has contributed to slowing down the dramatic expansion of the UK's National DNA Database. The campaign contests the systematic use of genomic profiling and the compulsory collection of DNA samples from those charged with a recordable offence. In 2014, a landmark ruling by the European Court of Human Rights condemned the practice as a direct violation of the European Convention on Human Rights. The verdict has forced the UK police to operate the largest destruction of genomic data in history.[77] May this event indicate the dawn of more widespread struggles over genomic data?

HYPER-GENOMIC DREAMS

Political Economies of the Earth BioGenome Project

Around the turn of the century, genomics was primarily a US-based science. The Human Genome Project (HGP) was conceived in a US context, it involved primarily US-based organisations, and it was concluded by a speech by Bill Clinton, then US president. The HGP, nonetheless, also represented an opportunity for the growth of large genomics corporations in other parts of the world. The Beijing Genomic Institute (BGI) was established in 1999 with the task of coordinating China's contribution to the mass-sequencing project. In the following years, the private company grew exponentially, investing heavily in its digital and genomic infrastructure. When BGI imported 156 of the fastest sequencing machines from the San Diego-based firm Illumina in 2010 for $1.58 billion, *Nature* quickly announced the rise of the next 'sequencing superpower'.[78]

Since then, the company has morphed into a sprawling international conglomerate, which employs over four thousand workers and manufactures about a quarter of the world's genomic data – more than any other institution on the planet. It recently acquired the American company Complete Genomics in a bid to develop its own brand of advanced sequencing machines.[79] It has also been entrusted with the management of China National Genebank: a public facility in Shenzhen that houses collected genetic material from people, animals, plants and microbes; hundreds of servers holding over 2221 terabytes of genomic data; and one of the largest experimental centres for cloning, genome editing and synthetic biology.[80] Most recently, the company has greatly profited from the Covid-19 pandemic and the technoscientific response to it. In the first six months of 2020, BGI sold millions of rapid Covid-19 testing kits to 180 countries and built 80 new laboratories worldwide. Meanwhile, the company's subsidiary listed on the Shenzhen stock exchange has doubled in price, reaching a market value of about $9 billion.[81] How can we understand the meteoric rise of BGI in the last decade? And what can it tell us about the fundamental role played by genomic sequencing in contemporary techno-capitalism?

In a speech delivered at the 6th International Conference on Genomics, Huanming Yang – founder and chairman of BGI since its foundation – defined the company's fundamental driving ambition. 'I have a dream', he asserted, mimicking Martin Luther King's most famous sermon, 'I dream that we are going to sequence every living thing on Earth; that we are going to sequence everybody in the world.'[82] In a follow-up paper, he clarified that the company's mission is 'to collect and preserve all living things on the planet before it is too late – before extinction'.[83] Two years later, a leading researcher for the company translated Yang's grand vision in lay terms, stressing that BGI's ultimate goal was to establish itself as a global 'bio-Google', which will help 'organize all the world's biological information and make it universally accessible'.[84] These may seem impossibly far-fetched visions, but they certainly help to make sense of BGI's activities in the last decade. The company has sequenced the genomes of thousands of living beings (from bacteria to plants, bats, rats, pigs and high-performing students) for all types of customers (including pharmaceutical corporations, governments, individual researchers, advocacy groups and non-governmental associations).

Delivery trucks roll in filled with DNA samples from labs around the world. Laboratory workers process the biological material, and the resulting genomic data is sent back to customers and collaborators. BGI has turned sequencing into a highly standardised industrial process, and

Shenzhen into the genomic workshop of the world.[85] Genomic sequencing is often portrayed as an abstract process conducted by fully automated digital machines. The reality, as Winnie Won Yin Wong recounts, is rather different: 'Video feeds from the surveillance cameras trained onto the peopleless rows of sequencers in Hong Kong – their efficiency apparently requiring the most minimal of human supervision, are displayed in the Shenzhen headquarters building. Meanwhile, on the Shenzhen factory floor, crowded rows of bioinformaticians work away on laptops between rows of potted plants placed to offset the noxious chemicals.'[86]

Leveraging its sequencing power, BGI has thrown itself at almost every single mega-sequencing project conducted around the world, including: the International HapMap Project, the 1000 Genomes Project, the 1000 Plant Genomes Project, the 10,000 Microbial Genome Project, the International Big Cats Genome Project, the Symbiont Genome Project as well as the Million Plant and Animal Genomes Project, the Million Human Genomes Project and the Million Micro-Ecosystem Project.[87] Most recently, it has embarked on its most ambitious venture, whose primary goal is 'to sequence, catalogue, and characterize the genomes of all of Earth's eukaryotic biodiversity over a period of 10 years'.[88] Announced triumphantly on the privileged stage of Davos during the 2018 World Economic Forum, the Earth Biogenome Project (EBP) has been presented as 'the most ambitious proposal in the history of biology'. First, because it involves a transnational network of public universities, private foundations and research institutes – including the Smithsonian Institution and the Department of Agriculture in the United States, the Wellcome Trust Sanger Institute and the Royal Botanical Gardens in the United Kingdom, the São Paulo Research Foundation in Brazil and many others. Second, because it has an estimated price tag of $4.7 billion.[89]

In order to justify this scale of investments, the EBP has been presented as much more than a purely scientific endeavour. According to the first article published by the EBP Working Group, investigating 'the dark matter of biology could hold the key to unlocking the potential for sustaining planetary ecosystems on which we depend and provide life support systems for a burgeoning world population'.[90] Similarly, an official statement released in August 2019 argues that the project 'aims to create a digital backbone of sequences from the tree of life that will serve as critical infrastructure for biology, conservation, agriculture, medicine, and the growing global bioeconomy'.[91] According to a recent paper published in the *Proceedings of the National Academies of Sciences* (*PNAS*) co-authored by Juan Carlos Castilla-Rubio, CEO of SpaceTime Ventures and collaborator of the EBP, genomics 'opens a new paradigm of seeing tropical regions not only as potential sources of natural resources and biodiversity but also as reserves of precious biological biomimetic knowledge that can fuel a new development model'.[92]

The venture furthers the logic of what Kathleen McAfee once called a neoliberal practice of 'selling nature to save it', which 'abstracts nature from its spatial and social contexts' and recodes ecosystems 'as warehouses of genetic resources for biotechnology industries'.[93] In the biopolitical economy of the Earth BioGenome Project, nature presents itself as an immense accumulation of biological assets – an 'endowment' that must be incessantly mined to yield stable streams of genomic data and economic value.[94] The fundamental objects of value to be captured are biochemical compounds and genomic matter: fragments extracted from frogs, ants and ferns, which pharmaceutical companies and biotech start-ups utilise in order to drive product development. It is this capacity of turning 'nature' into genomic data – and data into valorisable commercial products – that allows many observers to conceive of the genomic bioeconomy as a sustainable alternative to traditional extractive industries. 'Unlike prospecting for material commodities such as minerals and timber', Cori Hayden writes in *When Nature Goes Public*, 'biodiversity prospecting is not dependent on large-scale harvests of raw material': 'a few milligrams of extract might be all it takes to provide the lead to a useful compound'.[95]

An exclusive focus on the limited natural resources necessary for the production of genomic knowledge, nevertheless, obscures the fact that each genomic sequence is the end-result of a complex global chain of production. Genetic data is not a free gift of nature, but a socially produced bio-object. It is a technoscientific abstraction from biotic material, whose production mobilises waged scientists in research laboratories; freelance 'bioprospectors' in the Amazon; unpaid, volunteer 'citizen-scientists'; as well as biodiversity cataloguers and horticulturalists employed in botanical gardens. It relies on publicly funded research projects, whose knowledge-products are then enclosed and privatised. It exploits centuries of 'universal labour' crystallised in libraries of scientific textbooks and circulated through shared research methodologies.[96] We must leave behind the idyllic image of the bioeconomy as an effortless 'unlocking' of nature's latent value in order to analyse it as a gigantic, global labour process.

As soon as we shift the analytical gaze to the dark and earthly abodes of genomic production, the immaterial bioeconomy promoted by the Earth Biogenome Project reveals itself to be very much material: specimens must be gathered on the ground by armies of bioprospectors armed with 'portable genetic sequencers' and 'high-resolution camera drones'; the genetic material must be transferred through logistical networks and laboured in associated laboratories around the world; while the resulting genomic information must be stored in an enormous and ever-expanding infrastructure of energy-intensive server farms. Most importantly, no genomic sequence is a finished product, which can be valorised on the market as such. Rather, it functions as a 'source asset', an input mobilised in the production of biotech commodities such as synthetic and semi-synthetic flavourings and pharmaceuticals.[97] In the words of the EBP Working Group, 'organisms and their genomes will provide the raw materials for genome engineering and synthetic biology approaches to produce valuable bioproducts at industrial scale'.[98]

The EBP is developing ways to guarantee the rapid valorisation of the scientific knowledge produced by the mass-sequencing effort. The idea is to institute a single digital platform, which will host the genomic data produced by project members. This so-called Earth Bank of Codes (EBC) is described in official documents as a sort of genomic 'blockbuster' for pharmaceutical corporations and biotech firms: 'a one-stop shop for nature's assets from biomes around the world', which would regulate the systemic transformation of genomic data into financial assets expected to generate future streams of income.[99] The gene bank would allow verified bioprospectors to upload genomic information on its servers. By entering the digital ledger, a potential user – such as a pharmaceutical company – would automatically enter into a 'smart contract', which would then track which pieces of information are downloaded. If any commercial use of the genomic data is detected, a payment would be automatically transferred to the sourcing bioprospector. These payments would represent a paradigmatic case of *genomic rent*: a payment resulting from the transformation of genomic information into a financial asset.

Facilitating the institutionalisation of property over genetic information, the project prepares the ground for the inclusion of a new invisible frontier – at once vast and microscopic – into the expansive geography of the world market. In the words of the *Economist*, the Earth Biogenome Project and the Earth Bank of Codes have the potential to rewrite 'the rules of international trade by bringing the raw material of biotechnology into an orderly pattern of ownership'.[100] Genomics creates the technical conditions for a novel production paradigm: a 'bioeconomy' based on the systematic extraction of genomic data, and its subsequent utilisation as a means of production on an industrial scale. However, despite all its apparent newness, the EBP is likely to perpetuate and exacerbate historical inequalities. Genomics is a technology-intensive labour process in which the mastery of fixed capital represents the key for what is often represented as an abstract 'unlocking' of value. A high organic composition of capital is necessary: not only to produce genomic data, but to utilise that data as a factor of production. For those without capital, the door to the realisation of value in the bioeconomy will remain firmly locked.

The Earth Bank promises to redistribute a portion of the surplus value realised through the selling of bioproducts back to the 'custodians of nature'. In this way, the World Economic Forum hopes to diffuse social conflict and enrol 'indigenous and traditional communities' in the bioeconomy. The EBC, they say, 'could provide an immutable community asset' which 'indigenous and traditional communities can leverage to create wealth and improve their own wellbeing'. This new round of genomic enclosures therefore seems to present some rather original characteristics: as opposed to generating a body of labourers dependent on the wage relation for the reproduction of their lives, it explicitly aims to create an army of custodians of natural capital, dependent on the rent-payments allocated by the international corporations exploiting their so far unvalued natural assets. Through this mechanism, the sponsors of the bioeconomy promise to solve the ecological crisis: effectively transforming each living being into a receptacle of genomic capital worthy of preservation. 'By providing an alternative source of income to local communities' a document by the World Economy Forum explains, 'local incentives will shift to maintain these complex ecosystems. Indeed, why destroy an asset that could reveal further value as the Fourth Industrial Revolution unfolds?'[101]

Sequencing the genome of each eukaryotic life-form not only opens the possibility of transforming biodiversity into a financial asset; it is also supposed to provide a new set of incentives to preserve and protect life. This objective is already implicit in the programmatic title of the first official report of the EBP Working Group, which presents its mission as 'sequencing life for the future of life'.[102] In the words of its architects, the EBC ultimately aims at 'preserving biodiversity' by 'adjusting incentives'.[103] The neoliberal background assumption is that biodiversity is threatened by the combined activities of a multiplicity of individualised market subjects, whose actions are primarily guided by defined utility-functions. The bank presents itself as a market solution to this market problem, offering a mechanism that 'helps shift local incentives away from short-term clear-cutting towards longer-term preservation'. Attracted by the multiple flows of income to be obtained by the bioeconomy, local communities would abandon their traditional livelihoods and rapidly reinvent themselves into an army of 'custodians of nature', entrepreneurial bio-prospectors, and genomic miners.[104] In short, the EBP constitutes a paradigmatic instance of rent-based neoliberal governmentality by which society is reduced to an aggregation of rent-seeking social monads, managing the natural environment as a vast genomic pool waiting to be mined.

Ultimately, the project should not be understood as a purely scientific endeavour, but as part of a wider tendency to transform genomics into a capital accumulation strategy. In the 1990s, the HGP laid the foundation for the contemporary genomic industry by channelling public money into the construction of a technoscientific infrastructure. Increasingly efficient sequencing technologies stimulated an unprecedented accumulation of genetic information on millions of people around the world. Corporations continue to invest heavily to access this new form of data on the bodily existence of their customers and labourers. Genomic data is valued insofar as it may inform new marketing strategies: what customers are offered to purchase – medicines, music, trips, even potential lovers – can be tailored to their specific biological profile. But the value of a genomic sequence is not limited to the knowledge it provides. As we will see in the next chapter, it is also the starting point for emerging techniques of genome editing. Tweaking the sequenced genome of bacteria, plants and animals, corporations are constructing new generations of gene-edited bodies, purposefully designed to assist the accumulation of capital in a range of economic sectors going from agriculture to aquaculture, from cosmetics to pharmaceuticals.

KR 210 R2700

4

CRISPR Assembly Lines

Speeding up the Molecular Factory

New techniques of manufacturing life are transforming the molecular factory. Twentieth-century *genetic engineering* required sophisticated scientific expertise, substantial financial resources and considerable time investments. Today, a *genome editing* 'revolution' is well under way. New genomic tools, elaborated in the last decade, promise to make the industrial production of genome-edited organisms cheaper and faster. They represent the point of convergence of multiple historical trajectories, which were analysed in previous chapters. They are biotech assemblages that bring together developments in genetic engineering and synthetic biology (Chapter 1), in global bioprospecting (Chapter 2) and in bioinformatic genomic analysis (Chapter 3).

New genome editing technologies are rapidly transforming the way in which life is imagined, manipulated and exploited for profit. But their impact reaches well beyond the laboratories from which they emerged and the corporate boardrooms that bankroll the research. Genome editing is increasingly presented as a solution to many of the most pressing economic, social and ecological problems facing the world, from hunger to infectious disease. It has revitalised and reconfigured debates about the relation between nature and culture, ontology and technology, society and the self. It is ensnared in a biotechnological vision of the future that negates the necessity for radical political change.

In the following pages, we investigate and critique some of the structural political-economical tendencies that are shaping, and will continue to shape for the foreseeable future, the development of these genome engineering techniques. The first section offers a short history of CRISPR that emphasises the multiple social forces at play in its development, as well as the economic interests that are currently struggling to establish proprietary control over this emergent technology. The second section of this chapter focuses on how CRISPR is supposedly ushering in a new era of 'genetic command and control'. The third section situates CRISPR in a broader context by considering recent advances in synthetic biology and industrial bioproduction, while asking: What cascades of social, economic and ecological effects may be engendered by the widespread adoption of genome engineering techniques in the so-called Anthropocene?

'AN INCREDIBLE MOLECULAR MACHINE'

Industrialising Genome Engineering

Genetic engineering emerged from a rather obscure corner of science, concerned with bacterial immunity and viral infections. Fifty years ago, microbiologists determined that bacteria possess antiviral immune systems endowed with restriction enzymes that disrupt viral replication by slashing through the genome of any invading phage. They also established that bacteria possess ligase enzymes dedicated to continuously maintaining the DNA structure, repairing any break caused by viral attacks. The key intuition that led to early genetic engineering technologies was that these enzymes could be appropriated and mobilised to manipulate the genetic code of various organisms. Laboratory workers created the first generation of recombinant bacteria by splicing together fragments of DNA from distinct living beings, using restriction enzymes as 'scissors' and ligase enzymes as 'glue'. This approach turned immunity against itself: biochemical entities evolved to protect bacterial genetic integrity were transformed into biotechnical tools that induce genetic mutations and direct evolutionary change. Since then, restriction enzymes have proved to be powerful molecular tools for the mapping and modification of DNA molecules.

The recent development of genome editing techniques was also sparked by the study of bacterial immunitary systems. The discovery of CRISPR – 'clustered regularly interspaced short palindromic repeats' – can be traced back to the late 1980s, when Yoshizumi Ishino reported the existence of five identical segments of DNA, each separated by unique 'spacer regions', in the genome of a bacterium studied in the laboratories of Osaka Universities. In the following years, further research showed that these curious genetic repeats are far from exceptional. By analysing the rapidly expanding pool of genomic data available on bacterial strains,

Ruud Jansen showed that they appear in almost 50 per cent of sequenced bacterial genomes.[1] This suggests that CRISPR is not simply a genetic oddity, but rather plays an important role in microbial biology. The question became: What role? What do these genetic repeats 'do'? What is their biological 'function'?

The answer to this fundamental question emerged from the routine application of automated bioinformatic analysis onto the avalanche of genomic data produced by global sequencing efforts. In August 2003, Francisco Mojica inserted the variable parts of a bacterial CRISPR sequence into BLAST, a bioinformatic software – often dubbed 'the Google of biological research' – that allows researchers to compare a given nucleotide sequence with various genomic databases. He had tried this exercise numerous times before without success, but the databases are continuously expanded by genomic laboratories around the world and, for the first time, the software found a match. BLAST showed that the spacer sequence entered by Mojica corresponded to the DNA of a viral phage named P1. In other words, it suggested that the bacterial genome had somehow integrated bits of phage's DNA. It was the first indication that CRISPR operated as a sort of genetic record of past invading phages.[2]

A couple of years later, Eugene Koonin and Kira Makarova elucidated the functioning of the CRISPR system. They suggested that bacteria use a special class of enzymes called Cas to cut fragments of DNA from invading viruses.[3] They then insert these DNA fragments within the bacteria's own genome in the area containing CRISPR sequences. The bacteria can thus recognise and neutralise viruses characterised by that particular DNA sequence. Just as our immune system can remember past infections and respond to known viral agents, so bacteria possess a system for fighting off viral infections based on the combination of CRISPR genetic sequences and Cas-restriction enzymes.

Through the early 2000s, microbiologists explored CRISPR as a fascinating piece of natural history: an unexpected way for bacteria to defend themselves from pathogens. Yet, the discovery of the role played by CRISPR-Cas systems in bacterial immunity also had a number of immediate practical applications. In 2007, scientists at Danisco – a leading global dairy corporation and a subsidiary of DuPont – recognised that CRISPR might provide new ways of protecting the precious bacterial populations used to convert milk into yoghurt, cheese and other dairy products. The annual market value of commercialised cultures of

Streptococcus thermophilus – the bacterium most widely used for industrial milk fermentation – has been estimated as exceeding $40 billion. Dairy companies spend millions of dollars every year in order to shore up the immunitary system of their proprietary bacterial populations and help them proliferate. Yet, frequent viral outbreaks constantly threaten the profitability of the industry.[4]

CRISPR provides new ways of securing the productive power of bacteria. It is now possible to sequence the genome of a bacterial population to verify exactly which snippets of viral DNA have been captured and stored in the CRISPR region. This makes it possible to produce bacterial populations with certified immunity to a precise list of viral agents. This is exactly what DuPont did in 2012, when it marketed the first CRISPRised bacterial population. Called CHOOZIT SWIFT™, the first industrial application deriving directly from early research on CRISPR is expressly aimed at speeding up industrial production, shortening the labour process and increasing the extraction of relative surplus value. As noticed by Annie Mornet, a senior business director at DuPont, CRISPRised bacteria are 'designed to help factories under productivity pressure' since 'every gram of fat or protein and every minute lost during processing have an impact on performance'.[5] As stressed by Rodolphe Barrangou, a prominent microbiologist employed by Danisco, CRISPR-based immunisation has become so popular in the industry that 'if you've eaten yoghurt or cheese, chances are you've eaten CRISPRized cells'.[6]

No matter how widespread in our fridges, CRISPRised yoghurt was only a taste of much bigger things to come. The next major breakthrough in CRISPR science not only provided a better understanding of how bacterial immunity functions, it also suggested how the CRISPR-Cas system may be turned into a powerful gene-editing tool. In the summer of 2012, Jennifer Doudna, Emmanuelle Charpentier, Krzystof Chylinsky and Martin Jinek published a landmark article demonstrating that CRISPR works by producing an RNA-copy of viral DNA, which then directs the Cas enzyme. Armed with this RNA 'mugshot', the Cas9 enzyme can recognise any virus with a matching sequence. It then cuts it in pieces, disabling it. This research started to capture the molecular mechanism at work in bacterial immunity. In parallel, Rodolphe Barrangou, Philippe Horvath and Viginijus Siksnys demonstrated that CRISPR-Cas9 is both transportable and easily programmable.[7] Most importantly, both research papers clearly pointed to the possibility of turning CRISPR-Cas9 into a biotechnology. The first one concluded by duly noting 'the potential to exploit the system for RNA-programmable genome editing'.[8] The second boldly declared that its 'findings pave the way for engineering of universal programmable RNA-guided DNA endonucleases'.[9]

Traditional genetic engineering methods such as recombinant DNA enabled researchers to insert foreign genes into a host genome. CRISPR has been hailed to open a new age of 'genome editing'.[10] The introduction of this new concept – actively and aggressively promoted by the global biotech industry in the last decade – represents a directed discursive shift. As an operative metaphor, genome editing constitutes an attempt to mark a radical discontinuity in the history of genetic modification. The transition from 'genetic engineering' to 'genome editing' implies a two-sided transformation. Firstly, it indicates a change of scale: the target of genetic redesign is no longer single 'genes', but the entire 'genome'. Second, it suggests that the labour of genetic modification is now better described as a form of 'editing'. The 'engineering' metaphor, with its evocative linkages to large-scale infrastructure projects and the ecological disasters associated with them, is no longer as popular as it used to be in the 1970s. Targeted mutagenesis is now most often presented as editing the 'book of life', a simple labour process performed with a bioinformatic 'word processor'.[11]

The history of the biotech industry, at least when told by those who control and handsomely profit from that industry, goes roughly as follows: there were once genetic engineering technologies with great potential but, unfortunately, they raised all sorts of controversies, health concerns and public resistance. Luckily, this is all behind us because genetic engineering has been displaced by something radically different. While genetic engineering was complex, imprecise and politically contested, 'genome editing' will be simple, precise and uncontroversial. The websites of countless biotech companies – from Corteva to Synthego, Biocytogen and SynbioTech – present some version of this soothing history. This rhetoric obscures the fact that there have been no 'radical discountinuites' in the history of genomic biotechnologies, no eureka-moments, but rather a gradual and methodical search for technoscientific mastery over evolutionary patterns.

There is one key characteristic of the CRISPR-Cas assemblage that makes it particularly attractive for the biotech industry. As opposed to the recombinant technologies central to genetic engineering 1.0, it can be industrially produced, commercialised as a standardised tool, and put to work on virtually any DNA segment. The only thing necessary to get started is a copy of the basic CRISPR-containing artificial plasmid, which can be conveniently ordered on Addgene, a highly successful and ever-expanding global repository for biotools. Once in possession of one of these plasmids, a researcher simply needs a bespoke guide-RNA guiding the 'molecular scissors' to the chosen DNA sequence. CRISPR is also *cheaper* than genetic engineering 1.0 – a feature that follows directly from the fact that its production can be scaled up industrially. James Haber recently estimated that the combined cost of Cas9 and a guide-RNA comes to about $30, making it over 150 times cheaper than previously popular systems of genome modification, such as zinc finger nucleases (ZFNs) or transcription activator-like effector nucleases (TALENs). As Emmanuelle Charpentier pointed out, the selling point of CRISPR is rather straightforward: 'With CAS9, anyone can use the tool. It's cheap, it's fast, it's efficient, and it works in any size organism.'[12]

It is because of its potential to trigger a Fordist revolution in the biotech industry that CRISPR has also been dubbed 'the Model T of genetics'.[13] Just like the moving assembly line enabled the mass production of cheap automobiles, CRISPR promises to radically reduce the labour time embodied in genetically modified organisms. This is why many proponents have celebrated this means of biotech-production as a form of democratisation. Jennifer Doudna claimed the technology 'became a democratising tool that allowed labs to do experiments that in the past had been prohibitive'. Bayer calls CRISPR the 'most democratic' gene-editing tool, which is so 'cheap and simple' that it can be used by virtually anyone.[14] In reality, CRISPR biotechnologies are already firmly under the control of a small group of powerful economic actors fighting over the rapidly growing market. This competition is fundamentally shaping how this tool will be put to use.

In this sense, the story of CRISPR is rather indicative of how biotechnologies are developed in late capitalism. Elements of natural history – such as bacterial immune systems – are remade into proprietary tools and turned into means of capital accumulation. Publicly funded basic research is appropriated to generate private profit. In the decade since 2012, CRISPR-Cas9 has ceased to be a puzzling biological phenomenon (captured by a cryptic acronym) and become a bio-technology endowed with an endless string of endearing epithets: 'a molecular scalpel for genomes' (Doudna), 'an incredible molecular machine' (Doudna and Sternberg), 'a search-and-replace function for the genome' (Zhang), 'a nano-sized sewing kit' (Kaul).[15] But it was the Royal Swedish Academy of Sciences who popularised what has become the most successful metaphor, awarding the 2020 Nobel Prize in Chemistry to Doudna and Charpentier for developing 'one of gene technology's sharpest tools: the CRISPR/Cas9 genetic scissors'.[16]

As shown by one of the images chosen by the Nobel Committee to illustrate the award, the trope is undoubtedly suggestive. On her knees, floating over an abstract background, a genomic tailor is busy trimming, customising and reshaping a long string of DNA. Through this molecular machinery, biotech laboratories are suddenly transfigured into garment factories tailoring dazzling genomic fabrics with threads extracted from eukaryotes and prokaryotes from around the world. Meanwhile, specialised research teams make space for a deskilled labour force working with tools as simple as scissors and needles. A recent article titled 'Quilting Plant Chromosomes with CRISPR/Cas9' greatly expands the metaphor encouraging the reader to 'imagine a one-size-fits-all sewing machine that can do all kinds of stitching, from small embroideries to patching large pieces of fabric into a quilt. Such a machine exists, except the fabric is DNA, and the needle is CRISPR/Cas9.'[17]

When the first sewing machines appeared in nineteenth-century factories they rapidly became, to borrow Karl Marx's description, 'the decisively revolutionary machine' that enabled the 'transition to the factory system proper'.[18] Similarly, CRISPR is hailed as the central apparatus of a coming biotech revolution. Just like the first sewing machines radically accelerated, automated and simplified the production of textiles; the genomic sewing machine promises to accelerate, automate and simplify the production of tailored genomes and genome-edited lives. Just as the sewing machine enabled the passage from artisanal labour to industrial labour, CRISPR abolishes traditional laboratory practices in favour of an industrial strategy of production aimed at tailoring life to the needs of capital accumulation. Ultimately, a number of questions are suggested by these metaphors: Where does the genomic textile come from? Who are the selected customers that get to choose the new genomic designs? What kind of labour practices dominate the genomic tailoring industry? And what kind of commodities are produced and commercialised by it? The next chapters will explore these and other questions.

'A NEW ERA OF GENETIC COMMAND AND CONTROL'

Conjuring CRISPR Futures

In the first years of the new millennium, molecular biology elucidated how the CRISPR mechanism may be turned into a genome-editing tool. Since then, research has focused on two fronts. On the one hand, researchers worked to modify the molecular structure of the CRISPR-Cas9 system to improve its efficiency, precision, and versatility.[19] On the other hand, an avalanche of papers has experimented with the potentials and limits of the new tool by manipulating the genomes of a growing list of living organisms. In early 2013, the number of research articles presenting some form of genetic experimentation with CRISPR began to skyrocket – a trend that continues to this day. By 2014, plant biologists had produced gene-edited rice, sorghum and wheat, soybeans, tomatoes, oranges and corn. By 2016, the gene-edited frontier advanced to include everything from cucumbers to potatoes, cabbage and mushrooms. Meanwhile, researchers have reported successful genome-editing in a variety of organisms across the animal kingdom – including microbes and yeasts, nematode and silkworms, fruit flies and mosquitoes, beetles and butterflies, zebrafish and salmons, mice and rats, cows and goats, pigs and horses, cats and dogs, monkeys and humans. Today, the list of genome-edited living beings could constitute a book in itself. Even viruses – those entities straddling the ambiguous border between life and non-life – have had their genomes rewritten using CRISPR.[20]

Most of these early experiments have been based on three genome editing strategies. 'Gene knock-out' represents the simplest, most common use of CRISPR-Cas9: 'cut' the genome at a specific sequence and allow the cell to repair the damage. Since the 'cut' (known as a double-stranded break) poses a serious threat to cellular life – potentially causing a terminal loss of genomic stability – cells will prioritise rejoining the DNA strands even at the cost of causing an incorrect repair (also known as 'nonhomologous end joining'). Most of the time, the result will be a frameshift mutation that disrupts the functioning of the targeted gene. In this case, CRISPR functions as a *gene disruptor*, an approach primarily used to understand the role of specific genes. Disabling a gene may give a direct clue of its functioning, especially if it causes physiological dysfunctions in the targeted organism.[21]

A more elaborate form of CRISPR genome editing is 'gene knock-in'. Once the Cas enzyme has caused its characteristic double-stranded break in the targeted DNA strand, an alternative sequence is inserted in the cell. In this second case, CRISPR is used as a way of inducing targeted mutations.[22] Finally, 'gene-expression control' is used as a way of affecting the epigenome without manipulating the genome itself.[23] This method exploits a deactivated version of the Cas9 enzyme, influencing how targeted genes are expressed without cutting them off altogether. A gene may be turned off completely, or just slightly dimmed, while conserving its place in the genome. By systematically producing thousands of 'knock-out', 'knock-in' and 'gene controlled' organisms, technoscience is furthering the exploration of the genome. Genomic knowledge is currently produced by observing how cutting off, silencing or substituting a single gene affects the physiology, behaviour, health and life-span of a multitude of model organisms.

Looking back at the first decade of CRISPR trials, Jennifer Doudna wrote: 'I could already see a new era of genetic command and control on the horizon – an era in which CRISPR would transform biologists' shared toolkit by endowing them with the power to rewrite the genome virtually any way they desired. Instead of remaining an unwieldy, uninterpretable document, the genome would become as malleable as a piece of literary prose at the mercy of an editor's red pen.' 'Our new technology', she specifies, 'offered scientists the remarkable ability to rewrite the code of life with surgical precision and astonishing simplicity'.[24] Doudna's euphoria for the technoscientific futures opened by her own research is perfectly in touch with the biotech utopianism characterising media discourse.

In 2013, *National Geographic* published an entire issue titled 'The DNA Revolution', while the prestigious cover of *Time* announced the coming of 'The Gene Machine'. In 2014, the editors of *Nature* doubled the hype, putting CRISPR repeatedly on their cover accompanied with titles such as: 'CRISPR in Action', 'Seek and Destroy!' and 'Dawn of the Gene Editing Age: CRISPR Everywhere'. In 2015, Doudna and Charpentier pocketed the $3 million 'Breakthrough Prize', a science award set up by tech-billionaires such as Mark Zuckerberg (Facebook), Jack Ma (Alibaba), Sergey Brin (Google) and Anne Wojcicki (23andMe). In 2016, *Wired* produced headlines such as: 'The Genesis Engine: No hunger. No pollution. No disease. And the end of life as we know it' and 'Easy DNA Editing Will Remake the World'.[25] These media accounts paint an alluring picture of the new era of genetic command and control, welcoming CRISPR as a technology that will finally enable the creation of functional ecologies adapted to a global industrial society.

Financial interest in genomic biotechnologies started to heat up in the early 2010s, when a new generation of scientists suddenly turned into biotech entrepreneurs and billionaires started bankrolling their attempts to turn gene-editing into an accumulation strategy. In 2011, Doudna founded Caribou Biosciences with the slogan 'Engineering any genome, at any site, in any way'. Soon, it established ties with major agrochemical corporations such as DuPont and Corteva with the expressed goal of developing CRISPR for agribusiness. Other companies – including Bayer-Monsanto, Cibus, Catalyxt, Syngenta, Arcadia and BH Biosystems – entered the rapidly expanding market for gene-edited crops. In parallel, another group of companies emerged with the aim of developing new CRISPR tools for the pharmaceutical and clinical industry. In 2013, Doudna teamed up with George Church, Feng Zhang and David Liu to launch the private company Editas Medicine with $43 million in financing from three prominent venture capital firms; while Emmanuelle Charpentier founded CRISPR Therapeutics, with a similar business plan and $25 million in venture capital. In 2014, Doudna – together with a roster of CRISPR pioneers such as Barrangou, Maraffini and Rossi – broke away from Editas to join Intellia Therapeutics. By the end of 2015, these three companies alone had raised over half a billion dollars in investments.[26]

Yet, amid the corporate excitement, there is one question over which roars of battle continue to rumble: Who owns CRISPR? The ongoing clash for CRISPR patents can be traced back to 2012, when two separate conglomerates claimed exclusive ownership over the newly minted molecular machine. In May 2012, Jennifer Doudna and Emmanuelle Charpentier – backed by the University of California and the University of Vienna – were the first to file a provisional patent application on the use of CRISPR-Cas9 as a genome-editing tool. Yet, it was the research group led by Feng Zhang – backed by the Massachusetts Institute of Technology (MIT), Harvard University and the Broad Institute – who first obtained a number of patents on CRISPR (thanks to an application filed in December 2012, which was pushed faster through the patenting process). By early 2016, the US PTO issued 23 CRISPR-related patents, mostly controlled by the Broad Institute and the MIT. In October 2021, a search in the published documents of the PTO revealed over 6,200 CRISPR-related patents under review and, according to a recent study, an average of 200 are added every month.[27]

The ongoing 'Battle for CRISPR' is partly juridical, partly scientific, partly cultural, partly political, partly economic and partly historiographic. Since the answer to the question 'Who owns CRISPR?' very much depends on the answer to the question 'Who invented CRISPR?', the conflict is rapidly exceeding the walls of specialised courthouses. It has spilled onto a multitude of unexpected terrains from academic journals to science magazines, documentaries, websites and Nobel Committees. For instance, the publication of Lander's historiography of CRISPR rapidly turned into a clamorous political case. The piece was widely seen as an attempt to influence the patenting battle in favour of the MIT-Harvard-Broad faction by recounting twenty years of CRISPR research as a succession of small, incremental steps, finally culminating in a decisive experiment performed by Zhang, which successfully deployed the technology to edit a eukaryotic cell *in vivo*. In response, the circle of scientific, institutional and financial supporters of the Doudna–Charpentier faction produced a deluge of alternative accounts in which Zhang's experiments appeared as an obvious, inevitable corollary to their all-important, foundational research.[28]

Despite the obvious differences, both narratives represent the history of science as punctuated by sudden inventions, born from the minds of special individuals worthy of monopolising the profits created by their single-handed efforts. We are confronted with two, rather unconvincing, technoscientific versions of Thomas Carlyle's infamous dictum according to which 'the history of the world is but the biography of great men'.[29] A radically different account of the history of CRISPR would emerge from a 'history of technoscience from below'. Such a history would emphasise the fact that technological innovations do not result from individual 'stroke of geniuses', but through complex processes of co-production of knowledge that always involve a multitude of nameless people across time and space. The global rush for CRISPR patents would then appear as part of the long history by which powerful groups have enclosed and privatised the wealth produced by diffused processes of research and knowledge creation. In this sense, CRISPR patents can be read as one more example of neoliberal privatisation by which a handful of companies are privatising the benefits – while socialising the costs – of the new technologies emerging from social cooperation on a global scale.[30]

What is ultimately at stake in these debates is not only who will cash in years of IP revenues, but also who will control future developments in genome editing. Establishing ownership over CRISPR is the key to control who will get to shape the genomic composition of countless biological organisms and thereby transform global ecosystems. Precisely because it has been so loudly celebrated as the central machine of an oncoming bioindustrial revolution, the battle over CRISPR has opened a number of political questions. Who will control the new power to edit genetic information and manipulate evolutionary history? What kind of regulatory guidelines and economic interests will influence its applications?

The advent of CRISPR has not only created new struggles over property; but also new forms of uncertainty and growing calls for global regulations. The possibility of experiments in germline editing – changing heritable DNA in human sperm, eggs or embryos – has proven particularly controversial. In 2015, a group of scientists – including Jennifer Doudna, George Church and Paul Berg – called for a 'prudent path forward for genomic engineering and germline gene modification'.[31] Four years later, in a gesture that mimicked the Asilomar conference of the 1970s, many of the same experts published a 'call for a global moratorium on all clinical uses of human germline editing [...] to make genetically modified children'.[32]

But the very real possibility of unapproved experiments in germline editing is not the only source of anxiety among the scientific community and the public. Already in 2016, the US President's Council of Advisors on Science and Technology published a warning concerning the multiple security implications of new genome editing technologies. According to the document, CRISPR opens up a historical phase in which both intentional acts of biological terror and accidental biohazards become a future possibility that demands pre-emptive action in the present.[33] The prospect of an accidental, synthetic pandemic has been increasingly discussed since the recent resurrection of the 1918 influenza strain by means of synthetic biology.[34] The experiment demonstrated that it is now possible to synthesise viral agents such as smallpox in controlled laboratory conditions. A first step in this direction was taken in 2017, when a small team led by David Evans of the University Alberta managed to synthesise the horsepox virus for $100,000. 'The world', Evans declared to the press a few days after the experiment, 'just needs to accept the fact that you can do this, and now we have to figure out what is the best strategy for dealing with that'.[35]

The new uncertainties created by new gene editing technologies are not limited to apocalyptic scenarios triggered by intentional or accidental misuse. Growing evidence indicates that unintended alterations may be a normal and systemic feature of genome editing with CRISPR-Cas9. Off-target effects may be impossible to avoid, given how the molecular scissors work. The Cas enzyme necessitates a precise indication of where to cut the targeted genome. Scientists provide such a guide by synthesising an RNA strand that mirrors the DNA sequence to be edited. The trouble is that more than one site in a genome can present the same string of nucleotides. One study, published by *Nature Biotechnology*, has shown that when CRISPR is directed to edit the *VEGFA* gene in chromosome-6 of the human genome (which stimulates the production of blood vessels) the scissors may hit corresponding genes on each of the other twenty-two chromosomes, causing unpredictable effects. Even if the enzyme cuts its intended target, it may still go on scanning the genome for corresponding sites, causing off-target cuts later on.[36]

In the context of these debates, George Church, one of the putative fathers of the new genomic biotechnology, recently surprised the international scientific audience calling CRISPR/Cas9 a 'blunt axe, whose use should not be labelled "genome editing" but rather a form of "genomic vandalism"'.[37] Off-target effects are particularly dangerous when genome editing is used in clinical settings, where unplanned cuts might disable a tumour-suppressing gene. Scientists may try to predict the possible off-target effects on the basis of detailed analyses of available genomic maps – such as the one provided by the Human Genome Project – but these simulations can be misleading, since no two people's genomes are identical. Because of random mutations and standard genetic variations, CRISPR may cut the genome in areas where the individual genome differs from the standard map.

'Every technology', Paul Virilio once pointed out, 'carries its own negativity, which is invented at the same time as technological progress. [...] When you invent the ship, you also invent the shipwreck; when you invent the plane you also invent the plane crash; when you invent electricity, you invent electrocution etc.'[38] In much the same way, genome editing brings with it 'off-target effects' and 'unexpected mutations' as its inescapable doppelgängers. CRISPR opens up an unstable frontier in which technoscience's ability to manipulate genomes systematically surpasses its capacity to understand and predict the cascading effects of those very manipulations. Certainly, the genetic integral accident can be perpetually warded off, contained, predicted through simulations, and limited in its destabilising effects. And yet, it continuously threatens the illusion of total and complete control over complex biological processes, which technoscience is only starting to understand in highly limited and precarious ways.[39]

'THE ORGANISM IS THE PRODUCT'

Synthetic Biology and the Making of Genomic Tailorism

In the 1990s, the Human Genome Project promised to deliver humanity's very own instruction book: a document that would clarify once and for all how the human organism develops and operates.[40] Yet, once the sequencing race was over, the human genome appeared messy, incomprehensible and noisy. Over 98 per cent of the sequenced DNA appeared to have no recognisable function: an ocean of nonsense, quickly dismissed by molecular biologists as 'junk DNA'.[41] At the time, this was largely understood as representing the accumulated debris of evolutionary history, an immense archive containing obsolete fragments of forgotten ancestors, scars of past viral invasions, weird and ancient mutations gradually accumulated over millennia. This conception – which portrayed the genome as an excessive, confusing and noisy piece of software – would give birth to contemporary aspirations of civilising the genomic frontier: cleaning up its junk, reducing it to an orderly and legible totality that would be easier to interpret, control and manipulate.

If molecular biologists once dreamed of 'reading the book of life', contemporary projects such as Genome Project Write aspire to move 'from passively reading genomes to actively writing them'.[42] In this context, the discipline of synthetic biology has taken upon itself the goal of rationalising biology by designing technoscientific life-forms that may be easier to measure, assess, predict and direct.[43] According to Drew Endy – a member of the bioengineering faculty at Stanford University and the author of *Foundations for Engineering Biology* – 'evolution isn't selecting for things that are easy for scientists to understand and interact with, that always do what we wish. So, we want to make modellable biological systems that we understand and can use as engineers, to rebuild the living world to make it modellable.'[44] Synthetic biology, he affirms, offers a technoscientific 'liberation' from 'the tyranny of evolution' that imposes 'random mutation without representation'.[45] Evolutionary history could then be directed according to rational principles, erasing the unpredictable effects of random genetic mutations filtered by natural selection.

Similarly, Thomas Knight – a senior research scientist at MIT and founder of the synthetic biology company Ginkgo Bioworks – points out that to rationalise genomes, it is necessary to 'eliminate unnecessary genes; remove overlaps; standardize promoters' and 'recode proteins to use a reduced portion of the coding space'. The final goal is to engineer 'simple organisms' that are 'modular; understood; malleable' and characterised by a 'rationalized infrastructure'.[46] Synthetic biologists, in short, promise to impose the instrumental rationality that characterises the Enlightenment project onto biology. According to Max Horkheimer, modern technoscience has long aimed at reconstructing the world according to principles of efficiency, calculability, predictability and control, thus degrading nature into 'mere material, mere stuff to be dominated, without any other purpose than that of this very domination'.[47] This modern tendency culminates in contemporary synthetic biology, a research venture aimed at rationalising the physiology of living beings, transforming them into functional components in bio-production systems.[48]

By reconstructing the genome of a particular bacterium, synthetic biologists pursue the goal of making its metabolic processes 'better' from a functionalist perspective. For a group of researchers, the 'rationalisation' of a particular bacterium could entail making its genome more legible, more understandable, more predictable. For a pharmaceutical corporation, it might mean making its metabolic processes more efficient, maximising the production of a desired molecule and minimising the excretion of futile by-products. Genome editing, in other words, enables the embodiment of instrumental reason and its inscription in the flesh of the living. Long treated as means of production, living beings are now designed as such. If random mutations appear to many synthetic biologists as the irrational whims of tyrannical evolution, the targeted mutations introduced by synthetic biologists appear as the enlightened enfleshment of economic reason.

A paradigmatic example of the role played by instrumental rationality in shaping the forms of life engendered by synthetic biology is *Syn3.0*.

In 2010, synthetic biologists at the J. Craig Venter Institute announced the creation of 'the first self-replicating, synthetic bacterial cell'. In a publication in *Science*, the researchers outlined the steps to synthesise the genome of the bacterium *Mycoplasma mycoides* from basic chemicals and transplant it into a recipient cell. They nicknamed the resulting bacterium *Synthia*, and celebrated it as representing 'the proof of principle that genomes can be designed in the computer, chemically made in the laboratory and transplanted [...] to produce a new self-replicating cell controlled by the synthetic genome'.[49] Venter hopes that the experiment may promote genome synthesis as an alternative to genome editing. 'If you want to make a few changes, CRISPR is a great tool', he said in a recent interview in *Nature*, 'but if you are really making something new and you are trying to design life from scratch, CRISPRs are not going to get you there'.[50]

The next step pursued by Venter's team was to synthesise a 'minimal cell', containing only the genes necessary to sustain life in its simplest form. In 2016, the institute declared the task accomplished and presented to the public *JCVI-Syn3.0*, a new version of *Mycoplasma mycoides* endowed with a synthetic genome stripped of all 'unnecessary' DNA sequences. According to Venter, this 'minimal bacterial cell' will soon provide 'a system where we know and understand all the components, so that when we add specific ones to it, we can do it in a logical design way'.[51] Ultimately, the goal is to use *Syn3.0* as a standardised platform to construct highly specialised life-forms, whose metabolism focuses on manufacturing a desired product. For instance, Synthetic Genomics Inc. – which Craig Venter founded in 2005 – has received $600 million from ExxonMobil, and further investments from British Petroleum, to synthesise bacteria 'to function as cell factories producing energy-dense oils that can be easily refined into renewable diesel and jet fuels'.[52] Fossil capital, in other words, is fostering the creation of bacteria whose organism is 'rationalised' to serve the peculiar irrationality of a system running on petrol.[53]

Early works in synthetic biology have been celebrated by many prominent molecular biologists as opening an age in which, as George Church writes, 'we too can turn the inorganic into the organic. We too can read and interpret genomes – as well as modify them. And we too can create genetic diversity, adding to the considerable sum of it that nature has already produced'.[54] At the centre of the new discipline is, undoubtedly, the MIT Synthetic Biology Working Group, an academic centre whose self-described mission is to 'make life better, one part at a time'. In this context, 'better' means easier to predict, to manipulate and to control. 'Synthetic biology' writes Drew Endy, one of the early pioneers of the Working Group, '[teaches] me how to: design and build living organisms that behave as expected' and 'write new genetic programs to do my bidding'.[55]

Synthetic biology promotes a drastic approach to redesigning life, combining genome-editing techniques – including CRISPR – with methods of DNA synthesis. It is a discipline inspired by the hypermodernist aspiration to impose rationality and legibility on the genomic frontier, and finally produce a biological world that is at once predictable and amenable to scientific manipulation. Paraphrasing James Scott's reflections on scientific forestry, we could say that synthetic biology represents an experiment in the functional simplification of organic life, which aims to refashion cells into 'commodity machines' by 'dismembering an exceptionally complex and poorly understood set of relations and processes in order to isolate a single element of instrumental value'.[56]

Perhaps the most influential and ambitious part of this programme involves efforts to develop a so-called 'Registry of Standard Biological Parts': an open library of functional genetic components – also known as 'BioBricks' – that can be easily combined to build synthetic devices and systems. The imposition of the 'BioBrick Assembly Standard' ensures compatibility between these functional units and defines how they can be combined into more complex assemblages. The stated goal is 'to simplify and standardize the process of creation', making genome engineering a simple and straightforward process of mechanical assembly. Playing with BioBricks, synthetic biologists aspire to create new synthetic life-forms. This is presented as an initial step 'towards an imagined future: synthetic biology without DNA manipulation'; a future in which biotech labs will no longer need to genetically modify natural organisms, 'because they could synthesize their designs entirely'.[57]

Two companies are at the forefront of the new frontiers opened up by synthetic biology: Amyris and Ginkgo Bioworks. The first has as its stated goal to 'make infinite what is finite in the world'.[58] Launched in 2003 with a grant from the Gates Foundation, it initially focused on creating a synthetic metabolic-pathway for artemisinin, an antimalarial drug. Amyris' scientists re-engineered yeast cells by inserting synthetic DNA fragments for amorphadiene synthase, an enzyme that catalyses the production of artemisinic acid. Then, they further modified this engineered yeast to focus its metabolic activity on the production of the desired molecule, increasing production almost 500-fold. The new technology offers a biotech alternative to the traditional method to produce artemisinin, which has long been extracted from sweet wormwood.[59] It has thus increased global production of the drug, reconfiguring peasant labour processes into bioindustrial ones. Until the early 2010s, artemisinin was mostly sourced from small farmers and wild harvesters in Vietnam, China and Eastern Africa. As synthetic and semi-synthetic artemisinin have started to become available, the Netherlands Royal Tropical Institute projects that 'pharmaceutical companies will accumulate control and power over the production process; *Artemisia* producers will lose a source of income; and local production, extraction and (possibly) manufacturing of ACT in regions where malaria is prevalent will shift to the main production sites of Western pharmaceutical companies'.[60]

In the following years, the company created similar synthetic pathways for many other lucrative molecules, which are currently used in commodities made by more than 3,000 global brands. Its products include RealSweet™ (a sweetener produced by engineered yeast-strains); Neossance Squalane™ (a fermentation-derived molecule bioidentical to shark-derived squalane used in cosmetics) and Santalols™ (a synthetic substitute for Indian sandalwood extract).[61] In developing its industrial units, Amyris explicitly took as its model the Joint Genome Institute, one of the main sequencing facilities of the Human Genome Project. The objective is to automate as much as possible the production of new bacterial strains. 'Creating a new life-form', according to Amyris' chief scientific officer, 'has become as simple as this: a scientist types out a DNA sequence on his laptop, and clicks "send". And nearby in the laboratory, robotic arms start to mix together some compounds to produce the desired cells. [...] You can now build a cell the same way you might build an app for your iPhone.'[62]

The automation of laboratory processes hides from view the messy materiality of genetic synthesis and modification. While biological materials are handled in the deserted robotic assembly, genetic designers can experience 'creating new life-forms' as a purely intellectual process conducted by editing (genetic) code on a computer screen. Automation also enables a dramatic acceleration of bio-production. Amyris boasts of creating new organisms, mostly forms of genetically modified bacteria and yeast, at the dizzying rate of more than 1,500 a day. These are then deposited in the company's repository: one of the world's largest collections of techno-natural organisms, counting over 3 million living items.

Amyris is not the only company that has profited from recent advances in synthetic biology. Ginkgo Bioworks – whose telling motto is 'The organism is the product' – was started in 2008 by Thomas Knight, one of the faculties at MIT synthetic biology programme, along with four graduate students. By 2017 the company was already valued at over $1 billion. In a recent interview, Jason Kelly – currently CEO of the company – compares Ginkgo to 'an online app store, except the apps are programmed cells'.[63] The company develops synthetic bacteria to be employed as living means of bioproduction by other corporations. The promise is: place an order with Ginkgo Bioworks, and its labs will spit out an organism that does whatever you want it to do. Do you need a cheap equivalent of an expensive natural product such as vanilla, rose essence or THC? Ginkgo will design a microbe to excrete it. 'We're sort of like cell programmers for hire', Kelly says in another interview, 'our job is to make the cell do what our customers want'.[64] In exchange, Ginkgo usually charges a flat-fee at delivery, plus a royalty on any profit deriving from the use of its proprietary molecular machines. Over two dozen industrial partners currently collaborate with Ginkgo – including Moderna, Roche and Bayer/Monsanto – to develop microbes used in the production of fragrances, drugs and food. Through synthetic biology, these corporations hope to transform their processes of production by employing armies of genetically engineered bacteria as programmable living means of production.

The heart of the company is a 100,000 square foot facility in Boston, which a visiting reporter from *Science News* described as 'a biological assembly line' in which 'synthetic biologists aim to reinvent biology in the same way Henry Ford revolutionised automobile manufacturing. Instead of installing standardised spark plugs or carburetors as a car moves down the line, the scientists tuck brand-new biological parts into the body of a bacterium'.[65] Since it was founded, Ginkgo has spent nearly half a billion dollars on gene sequencers, mass spectrometers, computers and robots in an attempt to automate its production of synthetic bacteria. 'Inside Bioworks 3', reports Amy Feldman after a visit to Ginkgo's facilities in 2019, 'a robot does pipetting, moving fragments of DNA suspended in liquid into a tray with eight rows and 12 columns at a speed beyond human capability. After the cells grow in plastic containers, another robot photographs them and uses that image to accurately pluck the irregularly shaped colonies from the surrounding jelly. The facility is quiet, with relatively few people. Largely, the machines do the work.'[66] 'The foundries', boasts a manager, 'bring economies of scale to biological engineering for the first time'.[67] Thanks to this capital-intensive programme of automation, Ginkgo claims that it has the potential to create 50,000 different genetically modified cells in a single day.

Biotech companies like Ginkgo automate the artisanal laboratory practice of genetic engineering, subjecting it to organisational principles traditionally associated with Fordism and Taylorism. In the early twentieth century, the introduction of the Fordist assembly line revolutionised the industrial production of mechanical goods, dramatically increasing labour productivity through automation. Meanwhile, the introduction of Taylorism – a management system that aimed to increase labour productivity by breaking down production into specialised repetitive tasks – led to a growing separation between a small stratum of 'skilled' workers (dedicated to product design, management and distribution) and a mass of 'unskilled' workers employed in the assembly line. Manufacturing costs plummeted, contributing to turning cars into products of mass consumption. This restructuring of production profoundly transformed urban geographies, national economies and global ecologies, leading to a period of mass production that has also been dubbed the 'Great Acceleration' in industrial impact on Earth. This history inspires the vision of twenty-first-century molecular biologists, who dream of turning genetic engineering into an automated labour process, and genetically modified organisms into products of mass consumption.

The bio-Fordist dream is currently pursued on two scales. On the one hand, synthetic biologists explicitly aim to refashion living organisms into efficient and rationalised molecular factories designed to transform cheap feed into valuable products. On the other hand, biotech companies such as Amyris and Ginkgo Bioworks focus much of their capital investments on mechanising the human labour of genetic engineering, transforming scientific labs into largely automated, high-tech biofoundries.

The industrialisation of genetic engineering, nevertheless, cannot be reduced to a simple repetition of what happened to the mechanical industry in the twentieth century. Today's social and economic context is dramatically different. Finance plays a pivotal role in the economic strategies of most biotech companies, including Amyris and Ginkgo. In 2020, Amyris reported $173 million of revenues, but a net loss of $151 million; yet its stock went up 10-fold, reaching a market capitalisation of $5.9 billion.[68] In September 2021, Ginkgo announced a SPAC merger with Soaring Eagle. The deal, worth $17.5 billion and including a $15 billion valuation for Ginkgo, was celebrated as 'one of the largest SPAC mergers to date'.[69] Forbes asked its readers: 'Could this Be the Next Apple or Microsoft?'[70]

Undermining the hype, Scorpion Capital – a short-seller company focusing on over-valued publicly traded companies – released a damaging 175-page report accusing the corporation of artificially inflating its valuation.[71] The document maintains that Ginkgo's business model represents a 'Frankenstein mash-up of the worst frauds of the last twenty years'; and that the majority of the company's current income is based on 'a dubious shell game' through which Ginkgo – or some of its largest investors such as Bill Gates's Cascade Investments – systematically invest in its corporate customers, sending its money on a round-trip journey back into Ginkgo's accounts.[72] A couple of hours after the report's publication, Ginkgo's stock price dropped almost 20 per cent. Eight days later, after reassurances by the company management and a few hyped predictions for the future, it soared up 56 per cent.

The volatility of Ginkgo's stock – which is characteristic of many biotech companies – is indicative of the extent to which its market valuation depends on the investors' confidence in its potential to pioneer the industrialisation of biology. The financialised context in which contemporary biotech companies operate helps to explain their careful PR strategies, designed to continuously manufacture hyped-up expectations for the impending bioindustrial revolution. Ginkgo Bioworks, for instance, publishes its very own monthly magazine, which 'seeks to tell creative stories that reflect our wonder for synthetic biology' and 'imagine collective biological futures'.[73] Much of the enthusiasm surrounding synthetic biology is, undoubtedly, the product of carefully designed discursive strategies.

Nevertheless, it would be short-sighted to dismiss the whole sector as a financialised scam, or a sophisticated act of conjuring value out of thin air.[74] According to a recent market review, over 600 companies – based mainly in the US, Japan and Europe – operate in the synthetic biology market, growing at an average rate of 5 to 10 per cent each year.[75]

These firms, no matter how over-valued and over-hyped, are pioneering new ways of manipulating biological matter. By doing so, they design living organisms that do not fit neatly in the classifications of traditional taxonomy, and problematise the very distinction between life and non-life. They are recombinatory sites in which genes are continuously sliced up, redesigned and sewn together in novel genomic combinations. They are places of labour and centres of capital accumulation, automating traditional scientific activity and turning it into an industrial strategy. Ultimately, they are sites in which the principles of Fordism and Taylorism are embedded in social relations and embodied in countless living organisms.

As Sophia Roosth remarks, the synthetic organisms proliferating in corporate laboratories around the world 'do not fit neatly into trees of life based on descent, ancestry, or lineage'.[76] Their artificial genomes trouble the classical division of life-forms into bacteria, archaea and eukarya. Some even debate whether a fourth branch of the tree of life should now be added. The cover of a recent special issue of *Nucleic Acids Research* features an image drawn by Alexandra Ginsberg titled 'The Synthetic Kingdom'. The work intends to provoke discussions on the ways in which genome engineering and synthetic biology are transforming the world of the living. It proposes 'to insert an extra branch into the Tree of Life' to 'make sense of this novel menagerie of engineered life'.[77] In this vision, genetically engineered organisms (whose genome has been purposefully tweaked) and synthetic beings (whose genome has been designed from scratch) compose a Fourth Kingdom, whose emergence has nothing to do with classic Darwinian evolution. Industrial design and industrial breeding, in other words, have taken the place of random mutation and natural selection.

In this emerging scenario, who decides which synthetic lives are made to live, which designed organisms are made to thrive, which are artificially revived, which are abandoned and let to die? Who exercises sovereignty over the synthetic kingdom? While many synthetic biologists promise to build a better world through their biological designs, we must be wary of the fact that there is no Universal 'better design'. Design has always been a value-oriented, political activity; and biological design is no different.[78] Marxist critiques of ideology have shown how social relations are continuously reified in the books authors write, in the movies directors shoot, in the music singers create, in the buildings architects design.[79] From this perspective, it is not surprising that today – when synthetic biologists and genetic engineers imagine how to redesign bacteria and engineer new forms of life – they do so in ways that reflect the social context in which they work, think and live. Ideology is increasingly embodied in the living flesh of millions of molecular factories, each designed to operate as a metabolic pathway for capital accumulation. A critical appraisal of synthetic biology and genome engineering, exposes the way in which social relations of power and market forces shape the 'synthetic kingdom'.

Recognising the political dimension of genetic engineering and biological design means recognising that these technoscientific labour processes operate within a socio-historical context defined by the structures of competition and exploitation associated with a capitalist world-market. It means elucidating and critiquing the structural political-economical tendencies that have shaped (and continue to shape) the development of genetic biotechnologies, and the way in which these biotechnologies are deployed in industrial production. It also means asking some uncomfortable questions. Who are synthetic biologists and genetic engineers designing their living-products for? What criteria guide biological engineering projects? Which values are pursued through genome editing? What does it mean to 'improve' the genome of a living organism? Does it mean to make it more beautiful, more entertaining, more predictable, more profitable, more exploitable, more fit to its own environment? Who has the power and the resources to decide? Who will establish the criteria that will guide the design of single organisms, entire species and whole ecosystems? How can people marginalised from the laboratory and the corporate boardroom participate in the decisions made within them and negotiate the mutant ecologies they will have to coexist with? How are the living beings subjected to genetic modification affected by these operations of redesign? To what extent are their inhuman values, interests and desires considered in the decisions made? The next chapters will engage with these questions by investigating how genome editing is currently being applied around the world by a multiplicity of public and corporate actors.

Threshold

The study of the genome has shaped scientific understandings of life and living processes, while spawning techniques aimed at engineering these processes. The first half of this book has shown how the conditions of possibility for a capitalist genomics were historically fabricated. It has offered a history of our present moment, of the messy developments that produced the unprecedented power over life currently yielded by biotechnology companies. Like all histories, it is partial and situated. Although it has emphasised the continuous consolidation of (bio-)power in the hands of a diminishingly small number of corporate enterprises and entrepreneurial scientists, it has also highlighted how unstable and crisis-prone the constellation of genomic capital truly is. Everywhere there are cracks in its epistemic and material foundations.

In the second half of the book, we offer a preliminary cartography of the ways in which this wealth of genomic information is currently being employed in different sectors: from agricultural and pharmaceutical production to climate change adaptation. We focus on the manufacturing of mutant life-forms as commodities and as means of production. We conceptualise 'manufacturing life' as a two-faced process. On the one hand, the contemporary genomic industry relies on the extraction of genomic information from living organisms. Through genomic sequencing and bioinformatics, concrete, living bodies are worked into 'abstract life' – mountains of genomic big data stored in public and private code banks. On the other hand, this genomic information is the precondition for the manufacturing of proprietary life-forms. Through genetic engineering, abstract life is worked into concrete, living bodies.

5

Molecular Factory Farms

Engineering Living Means of Production

Herbicide-resistant crops, non-browning mushrooms, heat-resistant cattle, disease-resilient chickens and jacked-up pigs: contemporary biotechnologies are transforming capitalist agriculture by engineering life-forms designed to accelerate, expand and secure the accumulation of capital on a global scale. What kind of logic shapes the life-forms produced by biotech laboratories and employed by global agribusiness? Why are CRISPR crops and genome-edited animals represented as being at once nothing radically new *and* as something revolutionary – a genomic fix that will finally make the economy more equitable, ecologies greener, and the population healthier? And why are these 'innovations' forcefully opposed by growing social movements throughout the world?

To answer these questions, this chapter sketches a critique of the biopolitical economy of genetically modified organisms in industrial agriculture.[1] It interrogates the dominant assumption that there is no alternative to large-scale industrial agriculture, and the related conclusion that GMOs are necessary to save this system from impending collapse. It spins a web of historical threads to situate genome editing technologies in centuries-long trajectories of dispossession, exploitation and struggle. It maps the different ways in which new genome editing techniques have enabled agribusiness to manufacture a new ecology, increasingly adapted to the requirements of capital accumulation.

The first section of the chapter places GM crops in the context of a protracted history of technological fixes in industrial agriculture. Since its inception in the nineteenth century, industrial agriculture has been beset by periodic crises, spurring the search for ever-new fixes. Yet, far from resolving once and for all the ecological and social problems associated with industrial monocultures, these technological fixes have only exacerbated the underlying contradictions. This dialectical history of crises and fixes has fed a toxic treadmill, which steadily led to the present ecological crisis. Moreover, modern technological revolutions – from the fertiliser revolution in the mid-1800s through the Green Revolution in the 1960s and to the Gene Revolution in the 1990s – have concentrated the means of ecological (re)production in the hands of ever-fewer corporations and institutions. These very same corporations – together with philanthropic foundations, venture capitals, and state bureaucracies – are driving a genomic transformation in agriculture, introducing new generations of life-forms engineered to withstand the deadly conditions of the factory fields and farms of our times.

The remainder of the chapter charts how genome editing is being transformed into a means of capital accumulation for agribusiness corporations around the world. The second section focuses on the creation of a new generation of 'CRISPR crops', designed to withstand climate change, support industrial harvesting, and entrench monocultures. It contextualises the emergence of CRISPR genome editing in an historical conjuncture shaped by ecological crisis and climate change: a time of emergency in which many have started to demand new technological fixes. The third section descends into the overcrowded quarters inhabited by the genome-edited animals employed by molecular factory farms. Targeted mutagenesis is increasingly mobilised to shape a new generation of living means of production, designed to conduct their existence in concentrated animal feeding operations. Overall, we wish to emphasise the extent to which the genetic mutations imposed by technoscience are guided by structural market forces, which increasingly shape the mutant ecologies of corporate farms and barnyards.

FROM THE GREEN TO THE GENE REVOLUTION

GM Crops in Capital's World Ecology

The history of industrial agriculture has been a history of crises and fixes, followed by ever-bigger crises and fixes. As capitalist productive relations spread across the surface of the Earth, these crises only increased in magnitude and severity. In the first decades of the nineteenth-century, Britain pioneered an economic model based on a growing separation between urban economies and a rapidly depopulating countryside. By the middle of the century, Marx already suggested that this new spatial order 'disturbs the metabolic interaction' between society and the environment, distorting the biogeochemical cycles that characterise life on Earth.[2] Channelled into rivers and seas, important nutrients – such as nitrogen and phosphorus – were removed from the soil without returning to it. This fundamental contradiction led to two complementary ecological crises, emerging together towards the end of the nineteenth century. On the one hand, the water cycle was increasingly polluted, leading to local and global processes of eutrophication and deoxygenation. In the summer of 1858, the stench of human excrement rising from the River Thames and seeping through the Houses of Parliament became so unbearable that Britain's political class fled to the countryside. 'Through the heart of the town a deadly sewer ebbed and flowed, in the place of a fine fresh river', Charles Dickens wrote that year.[3] On the other hand, the soil was gradually impoverished, causing a structural decline in agricultural production. Soil fertility became an acute concern among European societies, 'comparable only to concerns over the growing pollution of the cities, deforestation of whole continents, and the Malthusian fears of overpopulation'.[4]

Imperial trade temporarily suspended the crisis. The ports of Britain regularly received cargo ships loaded with precious sources of nitrogen and phosphorus: bird dung from Peru, organic remains from Egyptian pyramids, human bones from the battlefields of Leipzig, Waterloo and Crimea. 'Great Britain', wrote Justus von Liebig, the agricultural chemist who first identified the critical role of phosphorus in agriculture, was then 'like a ghoul, searching the continents for bones to feed its agriculture'.[5] By the end of the century, however, the demand for nutrients to supplement industrial agriculture had already grown unsustainable. Guano deposits were too small to sustain years of aggressive extraction. Crushed bones could cover only a fraction of the rising demand for phosphorus. The colonial displacement of the ever-widening metabolic rift engendered by capitalist agriculture and industrial monoculture was no longer sufficient. In the 1840s, Leibig could still assert that 'in the remains of an extinct animal world, England is to find the means of increasing her wealth in agricultural produce'.[6] Half a century later, no amounts of bones and coprolites were sufficient to keep at bay the spectre of soil erosion and declining harvests.

Malthusian fears were revived, sparking a frantic search for technoscientific fixes.[7] The early twentieth century saw heavy investments in chemistry and an international competition to develop nitrogen-rich chemicals by artificial means. The approach that 'solved', or at least temporally displaced, the nitrogen crisis was developed by Fritz Haber, a German scientist driven by the promise of a lucrative contract with the chemical conglomerate BASF. In 1909, he discovered that it was possible to synthesise nitrogen fertilisers – in the form of ammonia – by forcing atmospheric nitrogen to combine with hydrogen under conditions of high heat and pressure. In the following years, Carl Bosch scaled up the procedure for industrial production.[8] This has been good news, since agribusiness has plenty of fertilisers to replenish the soil. It has also been a catastrophe, insofar as industrial nitrogen production has driven the deoxygenation of rivers, lakes and seas and is currently responsible for roughly 1.2 per cent of global CO_2 emissions.[9] The technological fix to the nitrogen crisis caused by intensive industrial agriculture was *also* one of the main drivers of the impending climate crisis. By enabling industrial agriculture to continue on its course, it protracted the metabolic rift and further warped global biogeochemical cycles.

Synthetic chemistry received a further impulse in the early twentieth century. Chemical herbicides and toxic pesticides were developed in the midst of the First World War, to be then turned into ubiquitous means of agricultural production. Chlorine gas was developed by Fritz Haber as a weapon of chemical warfare in the 1910s. A few years later, as head of the chemistry department of the National Office for Economic Demobilisation, he directed the transformation of chemical weapons into pesticides to exterminate the many life-forms that threaten crop production and munch into landlords' profits.[10] The first commercial herbicide, 2,4-dichlorophenoxyacetic acid (2,4-D), was developed in the 1940s by Allied Forces to destroy agricultural fields and impose famine on the enemy.[11] Industrial agriculture has been entangled with military production from its early days. In a morbid paradox, the production of the means of life has been conducted through means of death. The quest for a technoscientific fix to the fertility crisis engendered by industrial monoculture has fueled a permanent war against weeds and pests. Yet, once again, the drive to increase land productivity by manipulating biogeochemical cycles engendered a number of unintended effects: only an estimated 0.1 per cent of applied pesticides reach the target pests, the rest linger in the soil, endanger biodiversity, and contaminate rivers, lakes and the sea.[12]

With the end of the Second World War, US hegemony provided the international framework for an imperial extension of industrial monocultures throughout the world. This transformation was partly motivated by the pressing need to offer an alternative model of development to the one advocated by growing socialist movements in the Global South. 'Developments in the field of agriculture contain the makings of a new revolution', William Gaud, asserted in a 1968 speech as head of the US Agency for International Development, 'not a violent Red Revolution like that of the Soviets, nor a White Revolution like that of the Shah of Iran. I call it the Green Revolution.'[13] Yet, it also provided an opportunity to open new markets for the chemical products offered by US industry. Cheap financial credit and a variety of developmental programmes enabled farmers to acquire new capital-intensive means of agricultural production, produced primarily by chemical corporations in the Global North. Through the Green Revolution, farms around the world were turned into receptive markets for synthetic fertilisers, herbicides and insecticides; but also for a new type of biological commodity: hybrid seeds.[14]

Plant breeders had long learned to select for particular genetic traits by inbreeding over several generations. The resulting inbred varieties were mostly low-yielding. In the 1960s, agronomists such as Norman Borlaug developed high-yielding hybrid seeds by crossing two pure inbred lines – a phenomenon known as 'hybrid vigour'. Hybrid seeds were presented as a spectacular achievement of scientific research, which promised to dramatically increase agricultural output.[15] And yet, the new varieties had some distinct disadvantages. The new hybrid crops were highly dependent on a constant supply of water and nutrients in order to support their exuberant vitality. Perhaps most importantly, they were functionally sterile seeds: while the first harvest may provide 20 to 30 per cent more yield than traditional cultivars, the second harvest is often erratic and prone to early failure. Hybrid seeds had to be bought again and again, thus transforming a self-reproductive life-form into an artificially scarce commodity.[16]

The long-term results of the Green Revolution's techno-chemical fixes have been, at best, paradoxical. Chemical fertilisers temporarily increased soil fertility, but they also polluted soil and aquifers. Pesticides and herbicides emptied the fields of undesired forms of life, but they also produced extensive side-effects for human and other animal lives.[17] High-yield varieties increased land productivity while reducing biodiversity, as thousands of traditional cultivars were abandoned. Hybrid seeds demanded intense watering, spurring the construction of environmentally catastrophic infrastructural projects.[18] Finally, high yields drove down the price of many commodity crops, creating masses of indebted farmers and furthering corporate control over agricultural production.[19]

While the overall effects of the Green Revolution on small farmers and urban populations remain hotly contested, two groups undoubtedly profited from the global transition to more capital intensive agricultural practices: input suppliers – mainly chemical factories, seed companies and agribusiness corporations; and buyers of commodity crops – mainly transnational grain processors such as Cargill and the growing 'intensive meat complex'.[20] After pressure from the seed industry, the Plant Variety Protection Act (1970) granted 'patent-like' protection to developers of new plant varieties. This assetisation of life-forms was a formidable force in restructuring the seed industry. Emboldened by the possibility of patenting new varieties of plants, large chemical corporations – increasingly owned by private asset management firms – began purchasing seed companies.[21] This trend intensified in the 1980s and 1990s with corporate giants snapping up seed companies and biotechnology companies to capture the newly created market. In this period, for instance, Monsanto acquired over thirty seed companies before being itself acquired by Bayer.[22]

The global tendencies towards further concentration, mechanisation and chemicalisation of agriculture opened up by the Green Revolution, were further intensified when the 'Gene Revolution' started making headways in the early 1990s. In 1994, the Agreement on Trade-Related Aspects of Intellectual Property Rights (TRIPS) established a global minimum standard for intellectual property protection, and thereby internationalised the US model. This sparked another wave of mergers and acquisitions (M&A) as corporate giants now had a truly global market as their playing field and as these same corporations 'sought to enhance complementarity between seeds and agrochemicals'.[23] The result of the M&A frenzy of the 1990s and early 2000s was a truly oligopolistic market in which only six gigantic firms controlled over 75 per cent of the market in agricultural inputs.[24]

In the early 1990s, the corporate labs of these newly formed agribusiness giants were rushing to develop proprietary varieties of important food crops. When the first GM crops hit the market in the mid-1990s, it would have been difficult to predict the wide-ranging changes they were to have on the mass production of commodity crops. In 1994, the Flavr-Savr™ tomato made world-ecological history when the US Food and Drug Administration (FDA) approved it as the first genetically modified food to be commercialised for human consumption.[25] Conventionally, industrial tomatoes are picked green off the branch in order to transport them to far-away markets without rotting. They are then chemically treated with ethylene to ripen, resulting in a watery, flavour-less tomato. Flavr-Savr™ was genetically modified to slow down the ripening process, which would allow it to be picked ripe and still withstand transport. The intention was to create a tastier tomato, which could nonetheless conform to the exigencies of industrial logistics. Yet, the costs were too high to compete with the increasingly cheap products crowding supermarket shelves. After failing to make a profit during its first three years, Flavr-Savr™ was discontinued in 1997, and Calgene, the producing company, was acquired by Monsanto.[26]

The targeted mutations in Calgene's tomatoes were aimed at creating a recognisable product, whose innovative profile would entice consumers and increase demand. By contrast, the first generation of GM crops to take root in the global market was aimed at maximising land productivity by exterminating every vegetal life-form that might compete with cultivated crops. In 1996, the commercialisation of Monsanto's RoundUp Ready® soy – genetically engineered to resist large-scale applications of Monsanto's brand of herbicide – marked the beginning of a new form of war on weeds. Glyphosate – the active ingredient in RoundUp™ and many other herbicide brands – targets an essential enzyme found in most vegetal life-forms. It is a chemical agent that readily penetrates leaves and roots, blocking an enzyme involved in the synthesis of several essential aminoacids and vitamins. While plant growth stops within hours of application, targeted weeds can survive several days before shrivelling away for lack of nutrition and metabolic failure.[27] Glyphosate was first commercialised by Monsanto in the 1970s, but its use was long limited by its very efficacy. As it destroys most vegetal life-forms on which it is applied, it was mostly utilised to control weeds in urban environments or applied to fields before planting any valuable crop.[28]

Glyphosate-resistant soy opened a whole new market for Monsanto's most popular herbicide: making it possible, for the first time, to blanket-spray increasing quantities of herbicide directly on agricultural land. Since the genetically modified crops are not affected by glyphosate, fields can be sprayed during the entire growing season, exterminating any competing vegetable life. In the following years, Monsanto launched RoundUp Ready® varieties of all major commodity crops. By 2010, an estimated 122 million hectares of agricultural land were turned into monocultures dominated by herbicide-resistant crops.[29] A decade later, 90 per cent of all corn, cotton, soybeans and sugar beets planted in the United States are genetically modified to tolerate one or more herbicides.[30]

This has created the precondition for a boom in the herbicide market. The Gene Revolution, just like the Green Revolution, has led to a net increase in the global market for agricultural chemicals. Glyphosate, in particular, has gone from being a marginal commodity to the world's most widely used herbicide. Global purchases rose 1,500 per cent from 1995 to 2014, and currently over 750 million kg of glyphosate are sprayed on agricultural land every year.[31] Throughout this period, farmers growing newly introduced, genetically engineered, herbicide-resistant crops were responsible for over 50 per cent of the global demand for this herbicide.[32] When we consider the impact of GMOs on the global production and use of herbicides the numbers are rather unambiguous. Agribusiness corporations have supported the development of genetically modified organisms that, far from offering an alternative to the herbicides produced by those same companies, enabled a further expansion of the market for deadly chemicals.

Today, transgenic herbicide-resistant crops have colonised global ecologies, taking root in the US and Canada, in Brazil and Argentina, in India and China, in South Africa, Australia, and many other countries.[33] They are, at least from a commercial point of view, the most spectacular success story in the history of targeted mutagenesis: a living product of genetic engineering technologies that has radically transformed rural ecologies across the world. Herbicide-resistant crops have created ecological rifts and shifts whose long-term effects remain unknown. According to a recent review, the adoption of herbicide-resistant crops has been found to 'contribute to biodiversity loss in several ways': glyphosate-based herbicides are toxic to a range of aquatic organisms and negatively affect the microbial communities that maintain soil fertility.[34]

Most importantly, glyphosate has become so widespread that it has become a major transformative force in evolutionary history. Just as the intensive application of antibiotics led to the development of antibiotic-resistant bacteria, the growing use of glyphosates has engendered new strains of 'superweeds'. The first glyphosate-resistant weeds were identified in soybean farms over twenty years ago. Since then, mutant superweeds have infested millions of acres planted with RoundUp Ready® crops. According to the US Department of Agriculture, there are now 383 known weed varieties with genetic resistance to one or more herbicides.[35] 'What we're talking about here', a weed scientist at Iowa State University has argued, 'is Darwinian evolution in fast-forward'.[36] The selection pressure imposed by herbicides is dramatically reshaping plant ecologies. Herbicide-resistant variants are taking over entire territories at unprecedented speed.

Increasingly resilient weeds are forcing farmers to either return to more labour-intensive techniques of mechanical weeding and ploughing, or resort to spraying fields with ever-different cocktails of herbicides. The value of the first generation of glyphosate-resistant crops is thus rapidly declining. They are about to become outdated living machines. According to many experts, 'the RoundUp Ready® revolution is over'.[37] What agricultural practices will emerge from its demise? As we write, agribusiness is already starting a fresh round of accumulation by introducing a new generation of gene-edited crops, engineered to survive the application of multiple herbicides in new deadly cocktails. This is not a theoretical possibility; it is a well-established trend. The use of dicamba, a herbicide considered to be 400 times more toxic to broadleaf plants than glyphosate, has grown considerably since the introduction of new dicamba-tolerant crops in 2017.[38]

Genetically modified crops have played a key role in the ongoing war against weeds, and they are set to continue doing so in the near future. They have also enabled an unprecedented escalation in the chemical mobilisation against agricultural pests. The widespread use of synthetic pesticides in the 1980s led to the creation of growing populations of pesticide-resistant insects, and a growing awareness of the catastrophic effects of chemical overuse. Commercial interests in alternative pest management technologies grew. New microbial pesticides, based on the insecticidal proteins synthesised by many soil bacteria, began to be developed around this time.

Commercial production of *Bacillus thuringiensis* (*Bt*), a soil-dwelling bacterium that synthesises several insecticidal proteins, took off. The first *Bt* sprays had already appeared on the market in the late 1930s, but these early microbial pesticides were only effective against caterpillars.[39] In the early 1980s, however, microbiologists started isolating previously unknown *Bt* strains from around the world. Many of them were found to possess genes encoding for unique toxic crystals. Some were found to synthesise proteins toxic to flies, others to beetles, others to ticks, nematodes and other groups of pests. This led to the production of hundreds of *Bt* insecticides targeting a growing number of insects, including bollworms, stem borers, beetles and leaf worms in cereal and grain fields; cabbage loopers and diamondback moths in vegetable fields; gypsy moths and spruce budworms in forests.[40]

Supported by an influx of funding from agribusiness corporations, molecular biologists focused on advancing research on *Bt* genetics and its metabolic synthesis of insecticidal proteins. In 1981, Ernest Schnepf and Helen Whiteley determined and patented a *Bt* DNA fragment associated with a protein that is toxic to lepidopteran insects.[41] With this method, scientists could construct a synthetic version of the gene for the insecticidal protein and, eventually, new recombinant *Bt* strains.[42] Genetic recombination has since been used to create new strains of *Bt* with a higher rate of toxic proteins and/or the capacity to synthesise several types of insecticidal toxins.[43] In the early 1990s, several biotechnology firms developed insecticides containing strains of this genetically modified bacteria; and genome editing technologies are now stimulating a new wave of investments intended to genetically 'improve the insecticidal activity of *Bacillus thuringiensis*' Cry Toxins'.[44]

Research on microbial pesticides, nevertheless, has been largely overshadowed by another product deriving from research on the molecular biology of *Bt*. The original gene patent filed by Schnepf and Whiteley in 1985 already emphasised that it would be possible to splice *Bt* genes directly into a plant's genome, obtaining transgenic crops whose cells would express insecticidal proteins as part of their engineered metabolism.[45] Several companies soon began investing in turning this abstract idea into a material transgenic organism.[46]

Genetically modified crops expressing *Bacillus thuringiensis*'s characteristic insecticidal proteins have been adopted worldwide and are presently grown on more than 100 million hectares.[47] The most commercially successful *Bt* crops have been corn and cotton. As of 2021, the International Service for the Acquisition of Agri-biotech Applications (ISAAA) database reports over 250 GM varieties of pest-resistant corn and 50 varieties of pest-resistant cotton. Additionally, Monsanto has obtained authorisation to commercialise strains of soy, tomato and potato spliced with *Bt* genes. *Bt* rice varieties have been developed by Huazhong Agricultural University and the Agricultural Biotech Institute. The Centro de Tecnologia Canavieira has obtained the green light for growing and selling *Bt* sugarcane in the United States and Brazil. The Maharashtra Hybrid Seed Company is authorised to grow *Bt* aubergines in the Philippines and Bangladesh.[48]

This transgenic strategy has several advantages for agribusiness corporations. Once plants produce *Bt* pesticides within their cells, it is no longer necessary to produce them in factories, transport them to agricultural fields and spray them repeatedly over crops. This substantially reduces labour costs, bio-automating the war against pests. The labour of thousands of workers employed in pesticide factories, logistical firms and agribusiness corporations is substituted by a metabolic process of protein synthesis engineered directly within the plant's cells. Industrial production is delocalised within the cells of countless living organisms. In the nineteenth century, Marx described automation as a process of real subsumption by which labour processes were re-organised and accelerated by the introduction of machines. *Bt*-crops indicate one of the ways in which this process of automation is now pursued by molecularising entire labour processes. This genomic strategy of capital accumulation, which is performed by engineering metabolic processes within living cells, represents a paradigmatic form of real subsumption of life under capital.

The ecological impact of *Bt* crops has been ambiguous. On the one hand, they have enabled a significant reduction in the application of damaging industrial pesticides.[49] However, as a recent review of the available scientific literature highlights, 'it must be recognized that alongside these benefits, there are concerns over the wide-scale growing of such crops, including potential effects on human health, and the environment at large.'[50] Twenty-first century fields are no less imbued with chemicals designed to bring death to undesirable lives. When researchers have taken into account the gallons of chemicals produced by the engineered metabolism of *Bt* crops themselves, the application of pesticides per acre appears to have dramatically increased almost nineteen-fold.[51]

The global diffusion of *Bt* crops, sparked by the rise of mutant insects resistant to synthetic pesticides, has driven the evolution of *Bt*-resistant organisms.[52] Faced with the emergence of specialised superpests, agribusiness and technoscience are developing 'pyramided *Bt* crops'; plants that produce not one, but multiple toxins.[53] Yet, evidence indicates that many pests are now evolving resistance to 'pyramid' traits as well.[54] Summing up this endless spiral of fixes and crises in the pesticide industry, a recent article in *Science* has pointed out that, while agribusiness generally treats resistance as 'a temporary issue that will be solved by commercialization of new products with novel modes of action', 'current evidence suggests that insect and weed evolution may outstrip our ability to replace outmoded chemicals and other control mechanisms'.[55] The war on pests appears to be locked in a feedback loop by which an ever-expanding arsenal of pesticides is driving the accumulation of pesticide-resistant traits in insects that is, in turn, driving the release of even more pesticides.

Pesticides are short-term technological fixes to two underlying problems. First, arthropod pests 'have proven incredibly successful in exploiting the artificial and highly simplified ecosystems that modern agriculture creates'.[56] Second, climate change has created an ideal environment for the exponential growth of these agricultural predators.[57] As long as these fundamental problems – the spread of industrial monoculture and climate change – remain unaddressed, contemporary genomic fixes will likely only spur the toxic treadmill fueled by the endless modern war against pests.

GM crops have generated ripple effects throughout world ecologies. They deflected the evolutionary history of arthropods and other insect species, while introducing targeted genetic mutations in plants that now drift around the globe. Ever since the first generation of GM crops, scientists have warned that they would inevitably contaminate other crops. 'Gene flow', explains a plant geneticist at the University of California, 'is a regular occurrence among plants. So if you put a gene out there it's going to escape. It's going to go to other varieties of the same crop, or its wild relatives [...]. It's clear that zero contamination is impossible.'[58] Genetic drift affects ecosystems in unpredictable ways, while sparking novel socio-ecological struggles and geo-political tensions.[59]

In 2001, a widely publicised study reported that farmer's maize fields in Oaxaca, Mexico contained genes thought to originate from *Bt* corn grown in the US.[60] The findings caused immediate controversy, not only because the planting of GM crops is not permitted in Mexico, but also because of the cultural importance of maize landraces. In March 2004, hundreds of farmers entered a meeting organised by the Commission for Environmental Cooperation (CEC) protesting the global trade in GM maize and asking for regulatory measures to prevent further contamination.[61] Zapatista and indigenous communities based in neighbouring Chiapas set up a seedbank, calling the project 'Mother Seeds in Resistance', to reclaim sovereignty over their traditional food.[62] Although the question of whether genetically modified material had entered local landraces was never conclusively answered, the ambiguity itself continues to stir resistance.[63] Genetic drifts challenge territorial borders: countries may decide not to cultivate GM crops on their land, yet they remain unable to stop their spread. According to a global review, almost 400 cases of GMO contamination occurred between 1997 and 2013 in 63 countries.[64]

In 2005, for instance, herbicide-tolerant canola weeds were found to grow spontaneously all around major Japanese ports. The finding caused widespread protests against 'genetic pollution' and the emergence of activist groups of weeders on the lookout for feral transgenic weeds.[65] In 2013, members of the Technical Office of the Indian Supreme Court denounced *Bt* crops in Bangladesh as posing a direct threat to Indian agriculture.[66] Genetic drift transgresses both public and private boundaries, creating new forms of conflict. GM crops can appear into fields not meant for them, exposing the unwitting farmers to fines.[67] Newly coined crimes of 'gene piracy' and 'seed piracy' have already been the subject of legal contentions.[68] To ensure that no farmer grows its patented seed without permission, for instance, Monsanto hired the notorious Pinkerton Detective Agency and 'pays its agents to comb the countryside looking for cheaters, and if necessary, it seeks out informants. The company set up a toll-free number where anyone can denounce his neighbor. It spends a lot of money to enforce its rule in the fields.'[69] With the help of its 'gene police', between 1998 and 2005 Monsanto exacted a total of $15 million from farmers accused of witting or unwitting gene piracy. 'There's a gene in there that's the property of Monsanto, and it's illegal for a farmer to take that gene and create it in a second crop', explained the CEO of Monsanto: 'It's necessary from the point of view of return on investment, and it's against the law.'[70]

Much like patent law has turned unapproved sexual encounters between OncoMice™ into theft, it turned farmers who replant seeds – a practice as old as agriculture itself – or whose fields have been contaminated with GM material into 'pirates'. If genes are a coded text and the genome is the book of life, then Monsanto can refashion itself as a prolific author threatened by a deluge of unauthorised pirate-copies of its oeuvre.[71] The category of 'piracy', which Monsanto has explicitly and repeatedly mobilised, indicates the extent in which the reductionist conception of the 'genome' has assisted the enclosure of the genomic commons and the policing of newly established – and highly contested – forms of private property.[72] Bayer, which has acquired Monsanto, has taken up the same strategy and it now threatens legal action against seed piracy: 'Remember: Planting patented seed saved from a prior harvest without the patent owner's permission is illegal.'[73]

This sketch of the socio-ecological relations that contributed to the rise of GM agriculture reveals a number of persistent patterns. The past two hundred years of agrarian change have been constituted by dizzying, ever-escalating cycles of crises and fixes. Writing only decades after the first 'fertiliser fix', Marx remarked that 'All progress in capitalist agriculture is a progress in the art, not only of robbing the worker, but of robbing the soil; all progress in increasing the fertility of the soil for a given time is a progress towards ruining the more long lasting sources of that fertility'.[74] Technological 'progress' not only increased the rate of exploitation of workers by intensifying the pace of work, but also led to ever-more destructive practices of extraction from the Earth. Crises and fixes endlessly escalate. Biotechnology rushes to patch over longstanding metabolic rifts, only to create new kinds of ecological shifts and genetic drifts. Today, CRISPR crops are presented as an innovative solution to the many social and ecological problems generated by industrial agriculture. Will this latest fix resolve once and for all the centuries-long rollercoaster of crises, and slow down the Great Acceleration in industrial impact?

CRISPR CROPS

'An Arms Race between Plants and Pathogens'

Agribusiness corporations, entrepreneurial scientists, and policy-makers are eyeing genome editing as a prime tool to secure large-scale industrial monocropping in the context of rapidly shifting environmental conditions. Advocates claim that these novel tools will finally resolve what Raj Patel has called the 'big, fat contradiction' at the core of the global food system, which has produced a jarring juxtaposition of over-abundance and scarcity, mountains of waste and empty bellies.[75] 'Agriculture is at a crossroads. Plants are under attack from changing weather, drought, floods, heat waves, diseases and pests', warns Corteva Agriscience, the agricultural branch of corporate giant DowDuPont. 'At the same time, our population is growing and consumers are increasingly demanding food that is healthier for their families and the planet. This means we need tools – like CRISPR – to grow more food that is better for people and the environment using fewer resources'.[76] The Malthusian alarm rings loud and clear; the population is growing, so there is no alternative but to funnel money to new technologies that would lubricate the circuits of accumulation. Corteva presents us with an impossible equation: growing population + increasingly precarious food ecologies = ?. Such a sinister algebra forces upon the public an extremely limited choice: CRISPR crops or socio-ecological disaster.

It is thus not surprising that Jennifer Doudna has repeatedly asserted that the most immediate and revolutionary applications of genomic biotechnologies will take place in agriculture. In a 2019 interview with *Business Insider*, she stated that 'in the next five years the most profound thing we'll see in terms of CRISPR's effects on people's everyday lives will be in the agricultural sector'.[77] In *A Crack in Creation*, Doudna similarly argues that CRISPR will allow those with financial stakes in the industry to 'not only make profit' but also create a 'more humane' and 'environmentally friendly agriculture'.[78] In spite of widespread popular resistance to GM food, Doudna confidently conjures up a future of molecular farming by 2024. The Innovative Genomics Institute, of which Doudna is founder, is working to create this biotech future by pushing a multitude of research projects aimed at turning genomic biotechnologies into means of agricultural production.

If CRISPR crops are coming soon to both fields and supermarket shelves near you, what will be on the menu? Genome-engineered plants are often divided into those that are 'input enhanced' and those that are 'output enhanced'. Input enhanced crops have their main impact on processes of production by increasing yield, accelerating growth, improving complementarity with chemicals, and making crops more amenable to mechanical harvesting. These include all the familiar examples – resistance to herbicides, pesticides, and viruses – as well as novel experimental traits designed to stabilise yields in the face of drought. By contrast, output-enhancement is aimed at altering the physical characteristics of the final commodity. Slow-ripening fruits and biofortified crops are the most paradigmatic examples in this category.

The possibility of using new genome editing technologies for crop improvement was first demonstrated in 2013, when researchers at the Institute of Genetics and Developmental Biology in Beijing successfully engineered the genomes of rice and wheat.[79] Two years later, researchers used CRISPR to 'knock-out' genes in rice, lettuce and tobacco, creating disease-resistant plants *without* inserting foreign DNA. The authors highlighted that because the resulting product is not transgenic, it could potentially circumvent existing regulations.[80] Industry rapidly realised the potential of this technological tweak. That same month, the vice president for agricultural biotechnology at Pioneer Hi-Bred – part of DuPont's $11 billion seed business – expressed optimism that CRISPR crops could provide the basis for a new round of accumulation: 'We have no doubt that genome editing is going to have a material impact on the value proposition. We think another whole cycle could come from genome editing.'[81] Shortly after, DuPont signed an agreement with Doudna's Caribou Biosciences to develop CRISPR-Cas9 as a crop production tool. Meanwhile, researchers associated with the International Rice Research Institute and Syngenta utilised molecular breeding tools to develop a variety of rice resistant to drought.[82]

CRISPR crops are presented as a potential technological solution to the many ecological challenges facing agriculture. The spectre of climate change haunts the future of agribusiness not only by increasing the frequency of severe droughts, but by creating favourable environmental conditions for the proliferation of plant pathogens, whose capacity to grow into devastating pandemics is further aggravated by the monocultural paradigm. Evolutionary theory suggests that populations with low genetic variation are more vulnerable to changing environmental conditions. Take bananas, that emblematic fruit of early Caribbean plantations, which are presently under existential threat from a soil fungus. Bananas on the global market are almost exclusively of a single variety – the Cavendish – and this ecological simplification has rendered the fruit incredibly vulnerable to specialised pests. Already in the 1950s, the previously dominant export variety Gros Michel was driven to extinction by the rapid spread of the very same soil-inhabiting fungus.[83] Today, agribusiness – rather than moving away from monoculture – is hoping to strengthen the Cavendish by knocking out the receptor through which the fungus invades the plant.[84]

The ongoing attempt to create disease-resistant bananas exemplifies how CRISPR is being used to mitigate the pathologies of industrial monocultures, which are otherwise undermining their own ecological conditions of existence. It is not the only example. Millions of dollars are currently invested to create genome-edited, blight-resistant varieties of potatoes.[85] Once again, these genome editing experiments appear to perpetuate historical tendencies that may be traced back at least to the nineteenth century. In the 1800s, the British Empire turned Ireland into an export-oriented economy, largely centred around the cultivation of a single variety of potato. This monocultural model brought comparative advantages to Imperial economies for decades, but it also set the stage for the 1840s Irish famine that brought the country to its knees. By then, the island's agriculture was so specialised – and the genetic variation of the potato cultivated so low – that when an outbreak of *Phytophthora infestans* found its way through the Irish countryside, it wiped out millions of acres of cultivation. At least a million people died in what remains one of the deadliest famines in recorded history.[86]

For years, historians and plant biologists have taken this history as indicating the dangers of industrial monoculture and the need to protect rural biodiversity.[87] Contemporary agribusiness has, nevertheless, taken a different developmental route. Following a toxic treadmill, potato farmers now treat their crops dozens of times a season with an ever-changing cocktail of poison. And yet, despite billions being spent worldwide each year to control the pathogen by chemical means, *P. infestans* keeps coming back. In the 1990s, farmers were dismayed to discover that new types of potato blight had found their way to Europe and America. They were more virulent – and more resistant to pesticides.[88] In 2009, *P. infestans* wiped out most of the tomatoes and potatoes on the East Coast of the US.[89]

While this may be taken (once again) as indicative of the dangers of industrial monoculture and the necessity to protect rural biodiversity, technoscience is taking (once again) a different route. In 2014, a team of researchers from the Swedish University of Agricultural Sciences identified a number of genes that make potatoes susceptible to blight. Using CRISPR, they knocked out the corresponding sequences. The resulting varieties appear to be more resistant against fungal pathogens, and thus offer a new technoscientific fix to entrench and sustain industrial monoculture.[90] 'It's essentially an arms race between plants and pathogens', suggests Sophien Kamoun of the Sainsbury Laboratory, 'we want to turn it into an arms race between biotechnologists and pathogens by generating new defences in the lab'.[91] The global war against *P. infestans* led to the aggressive use of chemicals designed to turn agricultural land into a toxic environment inhospitable to blight. Today, it continues with the use of CRISPR to re-design living organisms to withstand the shifting ecologies created by industrial monoculture. CRISPR potatoes are but one example of growing investments aimed at controlling the pathogens that riddle monoculture plantations. In less than a decade, gene editing technologies have spawned a growing portfolio of patented 'resilient' plants, each with resistance to a range of fungi, viruses and other pathogens.[92]

In the 'arms race' between plants and pathogens, the US Defense Advanced Research Projects Agency (Darpa) is investing in a new tactic. Darpa is the research branch of the US military and it aims 'to make pivotal investments in breakthrough technologies' with 'a constant focus on the Nation's military Services'.[93] In 2016, it launched an initiative called Insect Allies as part of an effort to produce 'scalable, readily deployable, and generalizable countermeasures against potential natural and engineered threats'.[94] Plant pathogens represent one such threat to national security. Darpa hopes to enlist three bugs – aphids, leafhoppers and whiteflies – as allies in this battle. Insects routinely spread viruses to plants. Darpa intends to use them as living vectors to deliver gene-edited viruses into crops. The GM viruses would infest the fields, injecting their modified genetic cargo into the crops's cells. Such engineered horizontal gene transfer would represent a radical turning point, allowing scientists to bypass sexual reproduction and introduce modified traits directly into adult organisms.[95]

Shortly after the programme was announced, a group of scientists published an alarmed critique in *Science*, arguing that this entomological biotechnology could easily be turned into a biological weapon. The collective – mostly based at the Max Planck Institute of Evolutionary Genetics – stressed two essential points. They argued that the dispersion of 'infectious genetically modified viruses that have been engineered to edit crop chromosomes directly in fields' represents an inherently dangerous form of genetic experimentation, whatever the final goals may be. Because of its inherent unpredictability, the technology 'appears very limited in its capacity to enhance US agriculture or respond to national emergencies (in either the short or long term)'. Yet, the study argues, the 'biological, economic, and societal implications of dispersing such horizontal environmental genetic alteration agents into ecosystems are profound'.[96]

According to the authors, 'the program may be widely perceived as an effort to develop biological agents for hostile purposes and their means of delivery, which – if true – would constitute a breach of the Biological Weapons Convention (BWC)'.[97] Shortly after the publication, the programme manager of Insect Allies, entomologist Blake Bextine, published a response arguing that the alarm was unfounded, confirming that the research is exclusively directed towards developing new lines of defence. Yet, in an interview with the *Washington Post*, Bextine recognised that the techniques developed for Insect Allies could be deployed for both defensive and offensive purposes: 'I think anytime you're developing a new and revolutionary technology, there is that potential for dual-use capability', Bextine said. 'But that is not what we are doing. [...] We want to make sure we ensure food security, because food security is national security in our eyes.'[98] Whatever the intentions of the military establishment, it is widely recognised that the development of this new generation of genome editing technologies has the potential to be deployed both as a defensive and an offensive bioweapon.

Insect Allies – and more conventional programmes of input enhancement outlined above – are designed to stabilise production in the face of increasing socio-ecological precarity. By contrast, output enhanced crops have traits designed to play a role after harvest and outside the farm. Output enhancement is meant to facilitate global logistical transport, support longer shelf life, and confer attractive traits to the final commodity itself. For instance, a non-browning white button mushroom developed at Penn State University was the first CRISPR crop to be cleared from US Department of Agriculture's (USDA) oversight in 2016. In the mushroom, a gene for polyphenol oxidase, an enzyme that causes browning, was knocked out. The result is a sturdier product designed to withstand mechanical harvesting and the long delays associated with global logistics. In a statement confirming that the non-browning mushroom would be exempt from regulation, the USDA explained that 'the anti-browning trait reduces the formation of brown pigment (melanin), improving the appearance and shelf life of mushroom, and facilitating automated mechanical harvesting'.[99] Similarly, CRISPR has enabled the cultivation of harder tomatoes, more adapted to the necessities of global logistics, through the inactivation of DNA demethylase.[100] Genome editing has also been used to create 'value-added' products by increasing size and weight of several economically significant crops, including rice, wheat, maize, tomato, potato, soybean and brassicas.[101]

Genome-edited crops are also increasingly presented as a fix to another contradiction of industrial agriculture: between what Raj Patel has called the 'stuffed and starved'.[102] In spite of ever-increasing agricultural output throughout the twentieth century, hunger remains a dramatic reality. Could corporations design 'charitable' crops to deliver nutrients to people who have been violently cut off from the land and the means of life? Already in the 1990s, amid mounting criticism that GMOs were narrowly serving corporate interests at the expense of both farmers and consumers, the Rockefeller Foundation started developing what one of their lead scientists described as a 'purely altruistic use of genetic engineering technology'.[103] Early research efforts in this area strived to design transgenic rice cultivars enriched with vitamin A. This new GM crop, according to representatives of the Rockefeller Foundation, has the potential to integrate the diet of the global poor. After years of research, 'Golden Rice' – a variety of genetically engineered rice enriched with B-carotene – was finally flaunted as a ready-made solution to vitamin A deficiency. Subsequently, other genetically modified crops were presented as a way to pack all the necessary nutrients into an otherwise deficient diet.[104] Today, the Innovative Genomics Institute has developed a CRISPR-based Golden Rice variety, while other groups have developed varieties of barley, pulses, wheat, and other commercially important crops with enhanced nutritional value.[105]

Yet, Golden Rice triggered a series of global controversies and a growing opposition. In 2013, early field trials in the Philippines prompted hundreds of farmers and local environmental activists to tear down the fences surrounding the experimental station and uproot the genetically modified rice plants. MASIPAG, one of the largest farmers' organisations in the Philippines, supported the protests and accused promoters of Golden Rice of 'using the victims of poverty and malnourished children as excuse to promote a product that will pave the way towards more profit and control of biotech corporations to food and agriculture'.[106] In 2014, the Asian Conference Against Golden Rice led to the creation of the Stop Golden Rice Network, which currently counts over thirty farmers' organisations across Asia. These civil society organisations have been arguing for over a decade that Golden Rice 'will only strengthen the grip of corporations over rice and agriculture and will endanger agrobiodiversity'.[107] They accuse the Golden Rice Federation of depoliticising poverty and malnutrition, peddling easy technoscientific solutions to complex social problems, while diverting resources away from organisations that directly address those underlying social and political problems. 'Golden Rice', writes a member organisation, 'sends the wrong message on how to address malnutrition and hunger. Instead of increasing diversification of food sources, which also provides much needed nutrition, the people are being offered a single crop that addresses only Vitamin A deficiency.' They warn against the use of Golden Rice as a 'Trojan Horse' for the expansion of genetically modified crops in global agriculture and as an advertising tool to rebrand GMOs as a humanitarian technology in the interest of the Global South.[108]

Contemporary critiques of Golden Rice extend and deepen a long-standing global opposition to the Gene and Green Revolutions. Over the last thirty years, non-governmental organisations, civil society coalitions, and farmers groups – among others, ETC Group, Vía Campesina, Movimento dos Trabalhadores Rurais Sem Terra and Navdanya International – have challenged the very premises of GMOs and the ways in which they entrench a corporate-dominated agri-food system.[109] These critical voices highlight that the Golden Rice Project risks cementing the radical dietary inequality created through centuries of capitalist and colonial expansion. Rather than challenging inequality head on, it furthers a divergence between foods designed for the poor and foods designed for the rich.

From this point of view, the project – first promoted by the Rockefeller Foundation and now by the Gates Foundation – is part of a larger tendency: creating highly specialised genome-edited crops aimed at isolated segments of the global population. In September 2021, for instance, Japan approved its first CRISPR crop to be sold directly to consumers: a genome-edited tomato developed by Sanatech Corporation and Pioneer Ecoscience Ltd.[110] The crop has been genome-edited to express high levels of gamma-aminobutyric acid (GABA) – an amino acid believed to help relaxation and lower blood pressure – and has been presented as a high-tech, 'healthy' product for conscious urbanites. It joins the ranks of the many high-GABA superfoods that have reached the Japanese high-end market in the last decade. According to Sanatech's president: 'This tomato represents an easy and realistic way in which consumers can improve their daily diet.'[111] While the first generation of GM crops responded to Malthusian fears by singularly aiming to increase crop yields and multiplying the food commodities brought to market, genome editing is being developed in ways that segment the world population through a global dietary segregationism. Specialised genome-edited crops are produced to provide different classes with different gene-edited foods catering to profoundly unequal lifestyles.

Genomic biotechnologies do not impact everybody in the same way. They can also create new forms of difference, distance and social divergence.[112] What type of world ecologies of food do these technologies rely on and further? Although it is often claimed that CRISPR is more 'democratic' than previous generations of gene-editing tools, it is also widely known that it requires tremendous bioinformatic resources to develop a successful cultivar.[113] It is not surprising, therefore, that most CRISPR-related patents have been obtained by a handful of actors that largely control the molecular means of production: agribusiness and pharmaceutical corporations, prestigious universities in the Global North, and the multitude of biotech-startups that link those two worlds.[114]

These actors are using their significant influence to lobby for exempting genome editing from regulations, already making great strides in Japan, in the US and in the UK.[115] In the context of the European Union, which has traditionally imposed stringent regulations on the cultivation and commercialisation of GM crops, powerful groups are pushing to liberate CRISPR crops from existing regulatory regimes. In 2018, the European Court of Justice (ECJ) ruled that genome-edited organisms such as CRISPR crops *are* GM organisms and therefore subject to risk-assessment, monitoring and labelling regulations.[116] The decision appears to reflect the wishes of the wide majority of Europeans; recent polls indicate that a mere 3 per cent of voters oppose the decision and support deregulation.[117] It also reflects a tradition established over twenty years ago by the European GMO Directive 2001/18/EC. It reiterated that any organism whose genome has been modified by artificial mutagenesis is *ipso facto* a genetically modified organism. From this rather straightforward EU law perspective, it does not matter if the genetic mutation has been induced by recombinant techniques, or by employing a bacterial enzyme such as Cas9. It does not matter if the induced mutation is large or small.[118]

A wide front of industry representatives and technical experts immediately came out in opposition, adopting a revisionist position. While this group cannot deny that the genome of CRISPR crops is modified by means of targeted mutagenesis, they argue that they are also fundamentally different from previous generations of GM crops. Unlike recombinant organisms, the new generation of CRISPR crops does not necessarily contain foreign genetic material. For example, in 2014 a powdery mildew-resistant wheat was created by 'knocking out' the *mlo* gene using CRISPR. In this case, the Cas9 enzyme has induced a single double-stranded break in the plant genome, which is then repaired by the natural plant mechanisms. This induces a random mutation that inactivates the gene. No additional DNA is added, so the resulting mutant organism is not transgenic.[119]

This fact has no legal relevance in the context of EU law, where the relevant distinction is between standard crops and 'GM crops'. All GM crops are subject to the same regulations, independently from their classification as transgenic or non-transgenic. A few months after the decision by the ECJ, *Nature Biotechnology* published 'A call for science-based review of the European court's decision on gene-edited crops'. The article argued against the ECJ's decision to classify 'genome-edited plants as genetically modified organisms (GMOs)' because 'genome editing produces genomic alterations that are similar to those that occur through spontaneous and induced mutation'.[120] This position, echoed by many representatives of agribusiness and the biotech industry, is singularly byzantine since it suggests that 'genetically modified organisms' are not defined by the fact that they have been produced by laboratory techniques of mutagenesis, which is the reasoning followed by the ECJ. Rather, 'genetically modified organisms' would be only those organisms whose engineered mutations could *never* spontaneously occur.[121]

The recently published 'CRISPR-files', assembled by the Corporate Europe Observatory, has uncovered a range of covert tactics used by private interests to push these revisionist perspectives and force a change in policy. The files show that the European Plant Science Organisation held a series of lobbying meetings with hand-picked EU officials focused on two main issues: working out the best legal strategy to deregulate gene-editing and promoting a number of climate-ready 'flagship products' designed to win the 'hearts and minds of the public'. The files also detail a $1.5 million grant by the Gates Foundation to the think-tank

Re-Imagine Europa with the aim of setting up an 'expert committee' – composed of representatives of agribusiness, researchers working directly on gene-edited crops and pro-biotech lawyers – driven by a single goal: 'engaging with a broad set of European stakeholders on genome editing in the 21st century' in an attempt to manufacture public support for deregulation.[122]

Fundamental democratic debates over what type of social ecologies should be collectively constructed through science and technology are systematically side-tracked by well-funded marketing propaganda that pushes a restricted and depoliticised agenda. Political questions concerning what collective futures should be built, what form of social regulations should be imposed on the modifications of delicate socio-ecological metabolic interactions, what forms of life should be constructed through targeted mutagenesis, which political ecologies should be fostered, are erased to focus mostly on two technical questions: Should all forms of genetic manipulation be considered processes by which organisms become 'genetically modified', or do only transgenic organisms merit that label? Are gene-edited crops equally safe for human consumption as conventionally bred crops? Technocratic management is presented as an alternative to democratic deliberation.[123]

Debates over CRISPR crops, however, far exceed technical assessments over definitions and 'health safety levels'. What defines an epoch is not so much what is produced, but how and by what means. What are the 'hows' and 'whats' of genomic breeding? How do GMOs shape the world-ecology of food? What are the opportunity costs of directing funding towards technoscientific fixes to complex socio-ecological problems? What futures does this approach open, and which ones does it foreclose? Do farmers and the general public have a say in whether or not CRISPR crops will be introduced? What kind of agricultural models are to be pursued? Whatever position one has on the potential and risks of CRISPR Crops, no amount of technical knowledge can solve these fundamental political questions.

CAPITALOGENIC CATTLE

Living Machines in the Animal Prison System

'In the earliest period of human history', writes Marx in *Capital*, 'domesticated animals, i.e. animals that have undergone modification by means of labour, that have been bred specially, play the chief part as instruments of labour'.[124] Farm animals, in other words, have always been both living raw material and living means of production. Their individual bodies and species-bodies have been worked, transformed, modified by different forms of human labour, from domestication to slaughter. They have also been employed as living means of production: animal-power has been until fairly recently the prime mover of industry. As the term 'horse-power' suggests, animals still played a central role in early modern industry. Animal agency, however, continuously threatens the disciplined rhythm of production: 'Of all the great motive powers handed down from the period of manufacture', writes Marx, 'horse-power is the worst, partly because a horse has a head of his own.'[125]

By the nineteenth century, new mechanical techniques of production rapidly transformed human relations with other animals. 'The starting point of modern industry', explains Marx, 'is the revolution in the instruments of labour, and this revolution attains its most highly developed form in the organised system of machinery in a factory'. The introduction of Watt's steam engine represented a revolutionary moment. Machines started to rely on a 'prime mover' that 'drew its own motive power from the consumption of coal and water' and 'was entirely under man's control'. Industrial establishments could finally void themselves of animal agency and resistance – with the notable exception of human agency and resistance – and enter the age of mechanical automation. Animals were replaced as instruments of labour by an 'organised system of machines': 'a mechanical monster whose body fills whole factories, and whose demon power, at first veiled under the slow and measured motions of his giant limbs, at length breaks out into the fast and furious whirl of his countless working organs'.[126]

Taking factories as the central space of modern production, Marx's subsequent analysis of the 'real subsumption of labour under capital' stresses how capital applies scientific knowledge to the labour process to increase labour productivity and progressively escalate the extraction of relative surplus-value. Admittedly, animals do not feature as prominently as machines and workers in Marx's critique of the capitalist mode of production. But there are pages in which Marx steps out of the factory in order to look into barnyards and onto fields. In those pages, the shifting role of animals in capitalist societies returns to centre-stage. What characterises areas of production such as agriculture, forestry and livestock farming – separating them from mechanical factories – is that they employ living organisms as means of production. This does not mean that machines do not play an increasingly important role in these sectors: tractors considerably speed up the tilling of land, rotary tillers reduce the time necessary to prepare seedbeds, and mechanical harvesters accelerate the reaping of crops. Yet, in these sectors, the production process must adapt to the metabolic rhythms of the living organisms it uses as instruments of production: the plants growing in the fields, the animals fattening in the stables and the bacteria multiplying in the fermentation tanks.

Living organisms are characterised by metabolic processes of organic growth and decay that most often cannot be manipulated nor mechanised. In livestock production, machines can hardly accelerate natural processes such as the gestation period of a calf and the ageing period of cattle. 'Naturally', Marx points out, 'it is impossible to deliver a five-year-old animal before the lapse of five years.'[127] Similarly, in agriculture, 'grain needs about nine months to mature' and, as a result, between 'the time of sowing and harvesting the labour-process is almost entirely suspended'. Finally, in industrial forestry: 'after the sowing and the incidental preliminary work are completed, the seed requires about 100 years to be transformed into a finished product'. During this long maturation time, the capital invested in the growing seeds remains congealed, and 'it stands in comparatively very little need of the action of labour'. The use of microbial metabolic processes in industrial fermentation present similar challenges to the circulation of capital as 'grape after being pressed must ferment awhile and then rest for some time in order to reach a certain degree of perfection'.[128]

In these sectors, Marx stresses the ways in which the material characteristics of non-human participants in the labour process impact the temporality of production. Automation and a more efficient division of labour may reduce the socially necessary labour time required for production, but it cannot reduce turnover time which remains tied to the metabolic process of the living means of production. Capital can push labourers to work faster – imposing new ways of organising labour cooperation, introducing new machines, etc – but it cannot make chickens grow faster. It must attend to these natural processes and adapt to the characteristic temporality of their organic unfolding. [129]

In production processes that utilise living organisms as instruments of production, the lack of control over their metabolic processes forces a disjuncture between production time and labour time. Production time can be considerably longer than labour time, especially when the production process is characterised by significant 'interruptions independent of the length of the labour-process, brought about by the very nature of the product and its fabrication, during which the object of labour [...] must undergo physical, chemical and physiological changes'. Production time must be divided in two parts: one period when labour is engaged in production, and a second period – which we will call 'maturation time' – when the unfinished commodity is 'abandoned to the sway of natural processes'.[130] Capital has long been able to reduce the first portion of production time, applying Taylorist principles and automation technologies. But the existence of a significant maturation time has represented an obstacle to capital's valorisation. As Marx writes, 'there is no expansion of the value of productive capital so long as it stays in that part of its production time which exceeds the labour-time [...]'.[131] The longer the maturation time, the longer capital is confined to the sphere of production and its value delayed from materialising into commodities. The longer the maturation time, the slower the turnover, and the lower the valorisation of capital.

What capital can do, if within certain limits, is to get the 'animals ready for their destination in less time by changing the way of treating them'.[132] In Marx's time, cattle breeding turned this artisanal practice into a formalised science.[133] In the century and a half since, the landscapes of farming have changed dramatically. Yet, one tendency remains constant. Capital systematically strives to accelerate the biological processes mobilised in any given production process: in agriculture and forestry attention will be given to scientific and technological innovations that open new ways to accelerate the growth-rate of crop and trees; in oenology and pharmacology, the most prized research will focus on accelerating fermentation and other microbial processes of metabolic synthesis; in the livestock industry research will dream of reducing gestation time and ageing time.

Genome editing technologies offer new ways in which this type of industry-driven research is pursued and realised. In July 2021, American shoppers got to buy and taste AquAdvantage® salmon, the first genetically engineered animal ever approved for human consumption.[134] The inaugural harvest of genetically modified salmon was sold to restaurants and diners in the Midwest, where labelling of the fish as genetically engineered was not required.[135] Some celebrated it as an historical moment: the culmination of a commercial project that has been cooking for decades and that might spark a revolution in animal farming. While the commodification of fish and the damming of rivers has led to a rapid decline in global populations of wild salmon, industry was ready to launch a biotech fix that would flood the market once again with cheap fish.[136]

The development of AquAdvantage® salmon can be traced back to 1989, when researchers at the Memorial University of Newfoundland produced an experimental gene construct with the assistance of recombinant DNA technologies. They isolated the DNA sequence that induces the synthesis of growth-hormone in Pacific Chinook salmon. They spliced it together with a promoter and terminator regions of the antifreeze protein gene from ocean pout. Then, they injected this chimeric assemblage into fertilised eggs of a wild Atlantic salmon. Finally, they waited for the eggs to develop into a new breed of transgenic salmon, which they hoped would be better fitted to the environmental conditions and economic needs of industrial aquaculture.[137]

The reason for making this particular genetic modification is simple: wild salmon produces growth hormone only under specific environmental circumstances, with water temperature and light playing particularly important roles. The promoter region of ocean pout, instead, always activates the genes they regulate, so by associating it to the cDNA for growth hormone, the salmon keeps growing independently from its environmental circumstances.[138] As a result, the new transgenic salmon will grow even in cold waters, where wild salmon would not grow; and it will grow year-round, enabling it to reach market size twice as fast – with a 10 per cent improvement in gross feed conversion efficiency – than its wild ancestor.[139] Given that feed is one of the major cost factors in aquaculture, and given the rapidly growing global demand for salmon, these two engineered characteristics represent a substantial improvement in the 'efficiency' of the living means of production mobilised by the food industry.

Since 1996, the genetic code of this fish has been the exclusive domain of the producing company AquaBounty Technologies: a financial asset protected by US Patent No. 5,545,808 col.5 ls.12-15. Today, thousands of these gene-edited fish spend their genetically abridged lives in concrete tanks located in the AquaBounty farms in Indiana, USA and Prince

Edward Island, Canada to be then 'harvested' and sold on Canadian and US markets. AquAdvantage® is hailed as a 'better salmon', whose engineered body will trigger a 'Blue Revolution'.[140] This begs the question of what makes a life superior in the peculiar environmental context of high-tech capitalism. What political and economic principles guide this search for genetic improvements and 'better' life-forms? Ellion Entis, founder of AquaBounty, points out that 'increased growth rates, enhanced resistance to disease, better food conversion rates, alteration of breeding cycles, more efficient use of indoor water recycling plants are all aspects of this revolution'.[141] Genetically improving the salmon, in an industrial context, mostly consists in engineering its body into a better living means of converting cheap feed into valuable flesh.

Genetic engineering, in other words, modifies animal bodies to reduce the time that the capital invested in them remains dormant in concrete pools, 'abandoned to the sway of natural processes'. Not only does this genetically engineered metabolic acceleration lower costs in terms of feed, but it shortens turnover time and thereby increases the valuation of capital. Industry has engineered a functional genetic mutation, which drastically reduces the time that the capital embodied in the fish must stay swimming in concrete tanks before its value can be realised first through sale and finally in consumption. It has steered a species' evolutionary history on the basis of financial calculations. In this way, AquAdvantage® salmon *doubly* embodies capital. The fish embodies capital firstly in that its genetic code has been turned into a rent-yielding financial asset through patenting. More corporeally, the salmon embodies capital in that its flesh has been inscribed with capital's need for speed, its body has been engineered to accelerate turnover time. Much like a viral phage hijacks the cellular machinery turning the latter into a living means of viral proliferation, corporations transform animal bodies into increasingly efficient living means of capital accumulation.

Genetically engineered fast-growing salmon is not an isolated historical curiosity; it is the result of structural socio-economic trends, which drive a growing demand for technoscientific animal bodies. In February 2022, a fast-growing tiger puffer and a meatier red sea bream joined AquAdvantage® on the menu of genetically modified fish approved for human consumption.[142] These GM fish are developed by the Regional Fish Institute, a Kyoto-based startup backed by major industrial corporations such as Ebara, Ube Industry, Mitsubishi UFJ Capital and NTT

Docomo's venture capital branch.[143] The company promotes the use of CRISPR as a 'revolutionary tool', which 'promote evolutionary changes on aquaproducts from DNA level in super-short time' and 'significantly fast forward the current breeding process'.[144] The company claims to accelerate the course of natural history, bringing onto the market the fish that evolution would have given us soon anyway. As a testament to this view, the company branded one of its gene-edited fish '22-seiki fugu' – puffer fish from the twenty-second century.

It is rather uncertain whether natural evolution in the wild would have ever led to a 'red sea bream lacking the myostatin gene' that regulates muscle growth, leading to an 'edible part of about 1.2 times' and 'feed utilization efficiency improved by about 14%'; or a puffer fish that 'becomes 1.9 times heavier than the same species grown conventionally in the same farming period'.[145] These 'better fish' are already available on the market.[146] They did not have to undergo any safety testing, and will not need to be labelled. The company's business proposition is not only to generate genetically improved living organisms, but to franchise a 'smart aquaculture system' based on artificial intelligence. 'By combining the new fish variety (fast-growth & increased-filet) and the smart aquaculture facility', reads one of the company's promotional documents, 'the productivity per area is expected to be approximately 4 times higher than that of the normal setting'.[147] The company was included on *Forbes*'s 'Asia 100 to watch'. [148]

The search for 'better' and 'more efficient' bodies animates the global animal factory farm. In particular, CRISPR technologies open up new horizons of possibility for agribusiness, prompting proponents to herald the coming of 'livestock 2.0' and 'designer farm animals'.[149] Much research has been directed at identifying methods for suppressing myostatin – a protein identified as an inhibitor of muscle cell growth. Genomic studies on heavily muscled breeds of cattle and sheep suggest that an underactive MSTN-gene leads to a metabolism in which myostatin is below-average and muscle growth is faster. The gene became a target for early efforts at genome engineering in farm animals; and today cattle, sheep, goats, pigs and catfish have been genomically engineered to become highly effective meat factories.[150]

This approach is not without its challenges. Experimentation in pigs illustrates the technical difficulties and risks involved in genome editing. The first 'MSTN knockout pigs' experienced recessive leg weakness, an inability to stand, and early death.[151] Similarly, in 2020 the birth of Cosmo – the first CRISPRised calf produced by Alison Van Eenennaam's *All Boys* project – sparked a number of concerns. Cosmo's birth was widely celebrated as the first successful demonstration of a targeted knock-in of a large genetic sequence via embryo-mediated genome editing. In particular, the project involved copying the *SRY* gene – a long DNA sequence present on Y-chromosomes, which stimulates the development of 'male' phenotypic traits – onto the X-chromosomes. The final objective is to use Cosmo to create a new line of *All-Boys*-cattle. 'Cosmo's offspring that inherit this SRY gene', explained Van Eenennaam in an interview, 'will grow and look like males, regardless of whether they inherit a Y chromosome'.[152] According to the study, genome-engineered cows may increase the industry's efficiency since 'male cattle are about 15 per cent more efficient at converting feed into weight gain. They are more fuel-efficient than females. Additionally, they tend to be processed at a heavier weight'.[153] Yet, this daring experiment in sex-engineering was far from a complete success. In one arm of chromosome 17, the synthetic DNA failed to stick, and the cell filled the gap with random letters. On the other, seven copies of *SRY* had been inserted – two of them backwards, and along with the bacterial plasmid used as a vector for the SRY gene.[154] Once again, CRISPR revealed itself to be less precise than described, daunting the quest for total genetic control.

Female-only hens might soon join male-only cows on the molecular factory farms imagined by the livestock industry. Figures released by the Food and Agriculture Organization (FAO) show that each year, roughly 7 billion hens produce over 1,370 billion eggs to be marketed around the world. This growing population of egg-laying hens is constantly replenished by incubating billions of eggs. Once born, specialised 'chicken sexers' inspect the newborns. Females go on to live as egg-producing factories, and males go down a conveyor belt ending in high-speed spinning blades. Over 6 billion male chicks are killed every year in this way. Globally, the process costs over $6 billion annually. EggXYt, an Israeli biotech firm, has recently proposed genome editing to tackle this problem: splicing the chicken genome with DNA for luciferase enzymes, which occur naturally in some bacteria and produce a distinct bioluminescence. 'The chicken is made sex-detectable by inserting a biomarker onto the male sex chromosome', explains the CEO of the company. When illuminated with a fluorescent light, male embryos can be easily identified by their fluorescence and workers can dispose of them without waiting for the egg to hatch. This biomarker-based strategy promises to 'fundamentally change the economics of the poultry and egg industries' by preventing the incubation of male chicks, saving the industry considerable costs, while rebranding the egg industry as an ethical economic endeavour.[155]

Genome editing has been applied in other ways to engineer animal bodies that may increase the profitability of industrial animal farming. Frequent viral epidemics have been a major challenge facing the meat industry in recent decades – a problem exacerbated by the genetic uniformity of farmed animals and the crammed environmental conditions of so-called 'concentrated animal feeding operations'.[156] Conventional control methods and heavy use of antibiotics have proved unable to cope with the increased frequency and intensity of viral outbreaks. In this context, genome editing tools have been deployed in the hope of engineering disease-resistance into a wide range of farm animals, and thus support the continuation of intensive factory farming. Scientists have already developed cattle resistant to tuberculosis; poultry less susceptible to avian influenza; pigs resistant to African swine fever and porcine reproductive and respiratory syndrome viruses.[157]

Many of these genome-engineered animals have been presented as both an economic opportunity for the animal industry and as a way to mitigate the ecological devastation resulting from industrial animal farming. AquAdvantage® salmon and all-male cattle herds are all presented as artefacts of ecological modernisation which will lay down the steppingstones to a 'green' future. Yet, their material effect remains to accelerate production and further industrial expansion. Likewise, disease-resistant animals are presented as necessary to cope with increasingly virulent ecologies, but also as an opportunity for companies to increase the number of animals that can be safely kept in already-overcrowded facilities, thereby creating the condition for new pandemics to proliferate. This, in turn, is likely to call for new gene-editing interventions in an endless spiral.

A similarly contradictory dynamic is at play when we consider research projects aimed at engineering animals to withstand climate change, ecological degradation and rapidly shifting ecological conditions. An array of animals embodies this attempt at creating life-forms adapted to the ecological conditions imposed by the Great Acceleration. Take heat-resistant cattle: heat stress lowers both milk production and the ability of cattle to muscle up. As stressed by a recent study on the application of genomic tools for 'climate resilient dairy cattle production': 'Among all the domesticated production animals, dairy cows are most susceptible to heat stress as a result of the intensive long-term breeding done in them so as to improve their milk production, which has led to a higher metabolic heat generation in these animals.'[158] This has long constituted an obstacle to the expansion of cattle breeding in tropical regions, while climate change is likely to further shift the geographies of livestock production. In response, corporations such as AgResearch and Livestock Improvement have used genome editing to engineer 'heat-resistant' dairy cows. Holstein Friesian cattle are characterised by a heat-absorbing black-and-white coating: corporate scientists have worked to tweak their genome to lighten their coat colour 'to alleviate heat stress and provide adaptation to warmer summer temperatures'. Doing so, they argue, would promote 'genome editing as a promising new approach for the rapid adaptation of livestock to changing environmental conditions'.[159]

In March 2022, the US FDA approved the first climate-ready cattle for human consumption. These heat-resistant 'SLICK' cows are produced by Acceligen, a subsidiary of Recombinetics, Inc. and recipient of a $3.68 million grant from the Gates Foundation.[160] Not all cows are equally sensitive to heat stress. Some have evolved naturally to have short, slick-hair coats, which allow them to stay cooler in hot climates compared to their relatives with longer, coarser coats. However, these variants are not among the most productive in terms of milk or meat yield. This is why recent research has focused on engineering genes associated with the slick-trait to confer heat-tolerance to high-yielding breeds such as 'intensively managed lactating Holstein cows'.[161] 'As global temperatures continue to increase due to climate change, cattle experience heat stress more frequently and more intensely – even in traditionally temperate, non-tropical environments', stated the director of the Foundation for Food & Agriculture Research, one of the funders of the initiative: 'Adapting cattle to withstand the effects of heat stress is critical to ensuring global food security.'[162]

The FDA's approval opens the door to more widespread use of gene-edited farm animals. 'We expect that our decision will encourage other developers to bring animal biotechnology products forward', says the director of the FDA's Centre for Veterinary Medicine, thereby 'paving the way for animals containing low-risk [intentional genomic alterations] to more efficiently reach the marketplace'.[163] Acceligen's Naturally

Cool™ cattle are presented as the heralds of a new market in climate-ready animals.[164] They are also presented as a means to expand cattle farming in tropical and subtropical regions, where heat limits productivity. A socio-ecological contradiction underlies this geographical expansion. The livestock industry has grown exponentially in the last half-century, becoming a major contributor to global emissions and a direct cause of climate change. Today, the same industry is exploring ways to genetically engineer its living means of production to withstand warmer climates. If this engineered adaptation proceeds according to plans, preventing an impending industrial crisis, global emissions from cattle are likely to push global warming even further.

If colour engineering and coat engineering are intended to create animals better adapted to the Anthropocene, other alterations are aimed at curbing cattle contributions to climate change. Genome engineering, in other words, has been deployed both as a means of climate-adaptation and climate mitigation. In 2019, the New Zealand Ministry of Business, Innovation and Employment awarded a contract to AgResearch Limited to develop 'climate-smart dairy cattle'. The programme aims to alter the cows' gut microbiome to emit less methane.[165] Low-methane cattle has received support from big-tech lobbying groups such as the Information Technology and Innovation Foundation (ITIF). According to ITIF, altering the bovine genome provides a simple solution to increased emissions from animal farming: 'As the individual genes responsible for the presence of these [methane-producing] microbes are identified, it will become straightforward to use gene editing to knock out those most responsible for high methane production bacteria or increase the expression of others that favour low-methane species'.[166] Through a swift CRISPR-induced genomic alteration, the need to radically transform animal agriculture seemingly evaporates.

Public resistance to GM animals is broadly recognised as an important factor in preventing deregulation and therefore as an obstacle to establishing a 'free market' of these living commodities. For this reason, the 'climate-ready' animals described above might be particularly important in swaying public opinion in favour of deregulation. A recent article in *Wired* promises its readers: 'A More Humane Livestock Industry, brought to you by CRISPR'.[167] Genetically dehorned cattle feature heavily in such media accounts. The fact that cows have horns has long represented an inconvenience for factory farmers. Exposed to stressing conditions, farmed animals frequently horn one another and sometimes injure their human handlers. This represents a cost to producers, a threat to workers, but also a desperate expression of animal agency and resistance by creatures trapped in what Marx once called the 'animal prison system'. Mechanical dehorning has been long practised in industrial dairy production: a violent practice which reflects the systemic violence inherent to industrial animal farming. This is why Van Eenennaam's team has been recently celebrated for breeding genetically 'dehorned' cattle, which supposedly contributes to creating 'a more humane livestock industry'.[168] And yet, genetically dehorning cows so that they can no longer react to their oppressive circumstances does nothing to eliminate the inhumane conditions that make dehorning a necessity in the first place.

Which animals will ultimately populate the molecular factory farm remains to be seen. Even a cursory review of the emergent uses of genome editing in industrial farming suggests that what were once insurmountable biological barriers to capital accumulation, are now technoscientific variables to be manipulated. The process of real subsumption has reached the genetic threshold: capital not only transforms the production process and the environment in which it takes place, but it is increasingly able to develop artificial life-forms, conceived to support, accelerate and extend accumulation.

Marx once said that capital 'creates for itself an adequate technical foundation' through the mechanisation of production and a systematic reorganisation of labour processes guided by the search for speed, efficiency and increasing productivity.[169] Today, capital not only reconfigures how labour is performed in different spaces of production, it reconfigures the very texture of life and its metabolic unfolding. Not only does it re-design labour processes, but also the genotypical and phenotypical organisation of living bodies and the metabolic processes they rely upon. The real subsumption of life under capital offers new ways to speed up production, boost productivity and secure the conditions for further economic expansions. An unceasing race towards the generation of ever-more genetically enhanced living means of production is not only an increasingly attractive technical possibility; it is also a socioeconomic tendency fuelled by the same coercive laws of exploitation and competition that have long enforced technological dynamism throughout the Great Acceleration.[170]

6

Engineering Extinction Ecologies

Resurrection, Annihilation and Genetic Biocontrol

De-extinguished woolly mammoths roam the Siberian tundra, while coral reefs are kept on genetic life support in a warming ocean. Trojan rats scour pristine islands, keeping in check the excess vitality of their own kin, while genetically sterilised mosquitoes drive their own populations to extinction. In the bioengineered ecologies imagined by biotech capitalists and genomic technoscience, genome editing will resolve some of the most urgent crises facing the biosphere. Genetic adaptation is presented as a way of keeping desirable species from extinction, while genetic control is promoted as a solution to the ecological shifts embodied by invasive species and increasingly resilient pests. Bioengineering offers new ways of steering global ecologies in the increasingly volatile and uncertain conditions imposed on Earth by the Great Acceleration. The genomic gaze transforms entrenched socio-historical trends towards extinction and habitat loss into technical questions to be resolved through bioengineering. How can ecosystems be more fully engineered, from the genome to the globe? What role do biotechnological corporations and entrepreneurial scientists aspire to play in the labour process of governing the biosphere?

The search for genetic tools that would enable at once the planned eradication of unwanted pests *and* the resurrection of charismatic species reflect the growing environmental ambitions of molecular biology. It is part of a more general search for technological 'fixes' to design, alter and govern ecosystems. These techniques are increasingly attractive because they suggest the possibility of fabricating a genetically engineered, bespoke biodiversity adapted to capital's world ecology. In this vision, the biosphere is increasingly reduced to a single, planetary, cybernetic system, whose feedback loops can be managed through constant human interventions performed on the molecular scale. The gene becomes a controllable lever that enables not only 'mastery' over the bodily metabolism of living organisms, but also indirect control over entire ecosystems and, ultimately, the capacity to steward the biosphere.

In this chapter, we consider closely the genealogy of three paradigmatic products of this frenzied search for planetary stewardship. Each represents a technoscientific response to the ecological crisis, which relies on genome editing technologies to steer natural history. Each is based on interventions at the submicroscopic scale of the genome, designed to institute a complex chain of reactions that would indirectly affect biological systems at ever-larger scales. The first section reviews recent research on so-called de-extinction technologies and investigates the capital investments poured into 'resurrecting' the woolly mammoth. Proponents of the project hope that tinkering with the genome will eventually create a new life-form, whose niche-making can be harnessed to prevent the melting of the tundra. Manipulating the genome would unleash a cascade of biological and ecological effects, successively impacting on the physiology of an organism, the biopolitics of a population, the environmental *milieu* inhabited by that population and, ultimately, the climate of the entire Earth system. The second and third sections consider the mirror side: the use of genome editing to drive extinction. We dissect two projects that intend to mobilise strategies of 'genetic biocontrol', including emerging 'gene drive' technologies. The first of these analyses the quest for genetic means of biocontrol to eradicate rats from New Zealand and the Galapágos, while the other interrogates the use of genome editing to collapse mosquito populations and other insects in the name of combatting malaria and agricultural pests.

WELCOME TO PLEISTOCENE PARK

Genetic De-extinction and Mammoth Eco-Zombies

In the midst of alarms of an impending 'Sixth Mass Extinction', genome engineering has been hailed as a technique that could not only slow down the rapid erosion of biodiversity loss, but also reverse this long-term trend by bringing extinct species back to life.

The idea of de-extinction, of course, is not new. It has inhabited the realm of science fiction for decades, at the very least since Michael Crichton imagined a Jurassic Park filled with cloned dinosaurs. Even before that, German naturalists dreamed of restoring the wild ancestors of modern cattle who had roamed the central European plains until the early seventeenth century. In the 1930s, Lutz and Heinz Heck tried to realise the dream by using the very same principle that had led to the domestication of the aurochs (*Bos primigenius*). The idea was simple: if twentieth-century cattle were the final results of centuries of selective breeding, it might be possible to reverse domestication by the same means. It would be sufficient to individuate the fiercest and wildest-looking cows, and systematically breed them. The result of this 'back-breeding', which lasted only 12 years, was a new breed: Heck cattle. Of course, there is no way to know whether Heck cattle truly resemble the long-gone aurochs – but the animals have been a popular item of pastures and zoos throughout Europe.[1]

The idea of bringing back the aurochs has not been extinguished. Several European countries have entered a puzzling competition to do exactly that. One of the largest projects was launched in 2007 by the Taurus Foundation and Rewilding Europe. Rival projects exist in the Netherlands, Germany and Hungary.[2] Compared to traditional back-breeding, contemporary projects have a distinctive technological advantage. New genomic technologies have provided researchers with the opportunity of sequencing the auroch's genome, starting from organic remains retrieved by archaeologists. The auroch's genome is then used as a rough guide for back-breeders, an indicative model of what they are trying to approximate.[3] Will the result of these de-extinction experiments be a future filled with competing pseudo-aurochs, joining Heck cattle in the enclosed spaces of zoos, parks and museums of natural history? Supporters of these projects, however, often portray a different image of the future in which wild aurochs will once again roam the European plains, re-wilding them.[4]

De-extinguishing the auroch might seem far-fetched, but it is not an isolated case. In July 2013, a team of Spanish and French scientists almost reversed time: they brought a Pyrenean ibex back from extinction, only to watch it become extinct again exactly seven minutes and thirty-six seconds later. The method used to resurrect this wild mountain goat was the same used to create Dolly in 1996. It entailed extracting the nucleus of an adult cell – which had been conveniently collected from the last living member of the species, Celia, back in 2000 – and transplanting it into voided eggs from domestic goats. The 57 embryos were implanted in the wombs of as many goats. Only seven of them got pregnant and of those, only one was brought to term giving life to Delia, an almost perfect genetic copy of Celia. It asphyxiated rapidly due to a malformation of the lung.[5] The Pyrenean ibex went extinct, once again. For some, the experiment represented 'a material triumph of resurrection biology'.[6] It was, nonetheless, a gruesome endeavour that did little to reverse the historical trends that led to the extinction of the Pyrenean ibex, which was mainly driven by ruthless hunting practices and the gradual loss of habitat. Without politically resolving these underlying trends, it remains doubtful that de-extinction projects can resolve the biodiversity crisis, which currently threatens 1 million plant and animal species, nearly 11.5 per cent of global biodiversity.[7]

The dream of de-extinction was not buried once and for all with Delia's infant body. In September 2021, Colossal – a new biotech start-up – was launched by a motley crew of venture capitalists, wealthy conservationists and academic geneticists united by a single bizarre goal: resurrecting the woolly mammoth. George Church, one of the synthetic biologists that led the CRISPR-revolution, has been writing of genetically modified mammoths for years and has been at the helm of the venture since its beginning. In 2015, the project took more concrete form through a $15 million investment from various venture capital firms – including Climate Capital Collective, Breyer Capital and Draper Associates – and Thomas Tull, producer of the Hollywood feature movie *Jurassic World*.[8] Colossal has estimated that this first round of funding will be just enough to jump-start the project.[9]

Because the last mammoths disappeared around four thousand years ago it is unlikely that a well-preserved cell will ever be found, so cloning is not a feasible option. By contrast, Colossal's plan for a 'Mammoth 2.0' relies on genome editing technologies to reshape the DNA of living species and make it resemble that of a related extinct species.[10] As the company's website openly recognises, their 'landmark de-extinction project' will not actually *resurrect* an extinct species. Rather, any hypothetical offspring resulting from the project will be a genetically modified 'cold-resistant elephant with all of the core biological traits of the Woolly Mammoth'.[11]

Church first publicly raised the idea of hybridising existing with extinct species at a presentation at the National Geographic Society in March 2013.[12] The conference was organised by genetic conservation NGO

Revive & Restore – founded by Stuart Brand, famous New Age, eco-modernist editor of *Whole Earth Magazine* – and coincided with other highly mediatised events which helped 'de-extinction' grab global headlines.[13] A couple of weeks later, the cover of *National Geographic* already showcased a mammoth resolutely walking out of a laboratory test tube and onto the tundra.[14] The reality is rather more complex. Colossal will have to tread a rather long and tortuous route before unleashing its population of trademarked mammoths into the wild. The first step on the path to the imagined Pleistocene future would be to obtain a complete mammoth's genome sequence. Research teams from around the world have already deployed advanced sequencing technologies to analyse thousands of organic remains – dried bones, frozen skin and flesh – with the hope of recovering some of the genetic material characterising the ancient beast.[15]

This is extraordinarily difficult since, once an organism dies, its chromosomes shatter into pieces that get smaller over time. Eventually, the DNA strands become so fragmented that they are impossible to recompose and interpret. In 2015, a team of scientists led by Love Dalén analysed the frozen remains from one of the last surviving individuals roaming on Wrangel Island 4,300 years before our present day. The genome was highly degraded, having fragmented into pieces over the millennia. If the genome represents the book of life, time had shredded it into thousands of paper confetti. How could one reconstitute the ancient text? In order to reassemble the pieces into a coherent whole, scientists used the genomes of contemporary elephants as a rough guide, just like those confronted by a jigsaw puzzle may refer to the image printed on the box.[16] We now have something that may resemble the mammoth genome, but that is a scientific approximation that can hardly be verified.[17]

The second step is identifying the main genetic sequences that differentiate the ancient mammoth genome from those of its contemporary relatives. Once the main genetic differences between mammoths and elephants have been identified, the third step would be to use CRISPR in order to edit the elephant genome and make it resemble as much as possible the mammoth one. That would represent the most extensive editing venture ever attempted. Scientists have estimated that there are roughly 1.5 million nucleotide-level differences between mammoths and elephants.[18] Yet, Colossal projects to manufacture a rather unfaithful version of the mammoth's genome by focusing its editing efforts on selected genes. According to its CEO, the company will 'insert 60 genes that are unique to the woolly mammoth into the genome of an Asian elephant'. By focusing on these genes, the company aims to induce a few key phenotypic changes to elephant bodies such as 'increased adipose tissue (body fat), long hair and sebaceous gland development, domed cranium, shorter ears and tail, as well as cold-adapted haemoglobin which allows for a more efficient O_2 transfer in the cold'.[19]

The third step presents an even more complex challenge. If Colossal was to ever create something that resembles a mammoth genome *in vitro*, how could those DNA strands be turned into an actual living, breathing creature? Church's team plans to insert the edited genome into eggs from an Asian elephant.[20] The embryos could theoretically be incubated in living Asian elephants. But this would be at once impractical and problematic since the elephant is threatened with extinction itself. For that reason, Colossal plans to develop an artificial elephant womb – another gigantic task in and of itself. The closest thing the world has ever seen was an artificial lamb womb developed in 2017, a 'biobag' in which a premature lamb foetus was successfully incubated for four weeks.[21] These next steps would require additional investments of millions of dollars. Following this treacherous path, the company has set itself the ambitious goal to deliver the first mammoth calf by 2027. This would not be the end of the project. The ultimate hope is to bring back whole herds of these great beasts, release them in the Siberian tundra and let them proliferate there once again.

Colossal's giant ambitions resonate with long-standing efforts by scientists to reconstitute the long-gone ecosystem of the Pleistocene epoch in northeastern Russia. The project started in 1989 under the guide of Sergey Zimov, the director of the Northeast Science Station in the Republic of Sakha, and it has recently obtained financial support from the Russian government. 'My colleagues and I for the past decade', writes Zimov in an article published in *Science*, 'have been working to reconstitute the mammoth ecosystem in one modest parcel of the northern Siberian region of Yakutia. We call our project Pleistocene Park.' 160 square kilometres of Kolyma lowland have been already closed off in order to 'gather the surviving megafauna of the mammoth ecosystem', 'increase the herbivore density sufficiently to influence the vegetation and soil', and, finally, proceed to reintroduce the bison, the Siberian tiger and, hopefully, the mammoth. Pleistocene Park would constitute what Zimov calls a 'Northern Serengeti'.[22] Much as the actual Serengeti is often erroneously portrayed in Western culture as a place without people and outside of social time, Zimov imagines an enclosed time capsule in which an ancient ecosystem would be restored.[23]

De-extinction is often presented as a form of restitution, a symbolic gesture by which humanity may start atoning for its past sins and begin reversing the long historical trends that have led to the present ecological crisis.[24] This is only part of the story. Other motives also energise Colossal's great hunt. The first is, obviously, the hope of making money. Manufacturing mammoths is certainly a weird business proposition, but venture capitalists are convinced that they are not throwing money away either. Although the company openly recognises that they are unlikely to make profit from the mammoths themselves, the project represents a 'moon shoot' for the biotech industry. 'It's very similar to the Apollo program, which was a literal moonshot', Ben Lamm explained in a recent interview. Just like in the Apollo project, 'where a lot of the technologies that were created were able to be monetized over time', Lamm says, Colossal is 'very confident that there will be some pretty interesting breakthroughs in the world of genomics, multiplex editing [...], and in software databases for genetic reconciliation'.[25] In short, the mammoth project is presented as an incubator for developing new technologies that may promote potentially profitable ventures. Although the business proposition might be shaky, the project is part of a broader marketing strategy aimed at presenting synthetic biology as the shiniest high-tech tool in the quest to shape living organisms in profitable ways.

At the time of writing, Colossal is expanding. In March 2022, the company raised an additional $60 million through a second round of seed funding. Thomas Tull, a lead investor in the project, explains Colossal's mission as driven by the desire of 'developing breakthrough tools that are poised to have a material impact on science and biotechnology: from the eradication of diseases to the development of new drugs, CRISPR DNA sequencing, and even solving challenges around reproduction'.[26] 'Proving the technology with de-extinction is only the beginning', explains another investor. 'These same technologies will be able to solve a huge array of human problems. Synthetic biology will allow us to create new life-forms that can address massive problems, from oil and plastic cleanup to carbon sequestration and much more. Solving tissue rejection and artificial wombs will go on to help improve and extend life for all humans.'[27]

Whether Colossal succeeds at developing and then monetising these technologies remains to be seen. What is clear, considering the portfolios of the various venture capitals who have invested in Colossal, is that the company forms part of an investment strategy that financially bets on high-technological solutions to social issues. At One Ventures, the main investor in Colossal, sums up the company's mission as de-extinguishing lost species and 'genetically engineering other advanced biological solutions to combat climate change and pollution'.[28] The CEO of Animal Capital, has equally high ambitions: the 'Animal Capital team is thrilled to be one of the first and biggest investors in Colossal. Ben and George's mission goes so far beyond re-populating the earth with woolly mammoths; its core focus will be around fighting climate change and disease through technology'.[29]

How might genetically engineered elephants be understood as an 'advanced biological solution to combat climate change and pollution'? Evolutionary molecular biologist and author of *How to Clone a Mammoth*, Beth Shapiro of the UC Santa Cruz Genomics Institute, has emphasised that we should 'consider de-extinction as a means to create *ecological proxies* for species that are no longer alive, which may benefit ecosystems, for example by restoring critical interactions among species'.[30] The company's promise hinges on the mammoth-elephant chimaera being able to re-introduce a specific animal culture, a way of being in nature which in turn produces particular environments: 'It will walk like a Woolly Mammoth, look like one, sound like one, but most importantly it will be able to inhabit the same ecosystem previously abandoned by the Mammoth's extinction'.[31]

If reintroduced, the aspiration is that the elephant-mammoths will alter the ecosystem, knocking down trees, breaking down moss, fertilising with their droppings and re-establishing light-reflecting grasslands that might mitigate climate change. Researchers have argued that the revived woolly mammoths 'may act as ecosystem engineers'.[32] According to Shapiro, reintroducing the mammoth would spark a chain of events leading to ecosystem function restoration. 'Precise replication of an extinct species', Shapiro claims, 'is not necessary to achieve the conservation-oriented goals of de-extinction. In the majority of ongoing de-extinction projects, the goal is to create functional equivalents of species that once existed: ecological proxies that are capable of filling the extinct species' ecological niche'.[33] Sergey Zimov, the biologist currently directing the construction of Pleistocene Park, has conducted extensive climate modelling on the effects that the rewilding project may have on the local ecosystem. In collaboration with Colossal, they aim to reintroduce the mammoth 'so that we can benefit from their natural climate-combating geo-engineering capabilities'.[34]

It is interesting to pause on the idea of 'ecological proxies'. The figure of the 'proxy' harks back to mediaeval law, indicating 'the agency of one who acts instead of another', or a 'person who is deputed to act in place of someone else'.[35] The idea is that the living organisms resulting from Colossal experimentations are deputed to take the ecological role once played by the mammoths thousands of years ago. The ontological question becomes secondary. Colossal's vision suggests that it does not matter if the animals they create will really be mammoths, what matters is whether this animal will behave like a mammoth. Will they interact with the environment as mammoths used to? Will they shape for themselves an ecological niche that resembles the Pleistocene Earth?

From this functionalist perspective, each species constitutes a distinct force-vector with a certain impact on local and global environments. Local ecologies, and the global biosphere, are dynamic and complex systems, whose dynamics may be radically transformed by a single extinction (or de-extinction). One of the ultimate goals of Earth System Science is to elaborate models to quantify the web of subtle inter-species interactions that constantly reshape the world. This knowledge can then be utilised to inform biogeoengineering projects such as the one pursued by Colossal: utilising species as levers for the indirect regulation of entire ecosystems. If each species shapes its own ecological niche through its metabolic interactions with the environment, genetic modification constitutes an indirect way of constructing the world.[36] Mutant elephant-mammoths are imagined as living means of producing a new ecology. 'Never before has humanity been able to harness the power of this technology', declares Lamm in another interview, 'to rebuild ecosystems, heal our Earth and preserve its future through the repopulation of extinct animals'.[37] Who controls the gene controls the body, who controls the body controls the species, and who controls the species controls the world. Planet Earth becomes a never-ending engineering project pursued through strict control over biodiversity and a series of biopolitical decisions over which species must be extinguished, which must be protected, which must be re-introduced and de-extinguished.

Such bioengineering visions rely on a determinist assumption: animals have no social culture, but purely instinctual behaviours coded in their genes. The project assumes that introducing targeted genetic mutations can determine animal behaviours and lifestyles. As Heather Browning has pointed out, questions of animal socialisation are far from resolved: 'the ways in which these animals interact with their environment, find and extract food, find and make shelter, and interact with one another and other wild animals, are all unknown'.[38] How would these animals learn these behaviours, in order to niche-make like its mammoth relatives? 'You don't have a mother for a species that – if they are anything like elephants – has extraordinarily strong mother–infant bonds that last for a very long time', Browning tells the *New York Times*. In overlooking this, releasing gene-edited elephants onto the tundra risks producing nothing more than herds of 'functionally ineffectual eco-zombies'.[39]

Even if Colossal's quest for mammoth herds was never to succeed, the project is already operative and performative in the present. It promotes an emergent sociotechnical imaginary in which the combination of molecular biology, system science, and capital-intensive machinery form the ideological matrix for addressing the current biospheric crisis.[40] In this way, it serves to erase the complexity of extinction and its relation to capital's relentless expansion through the web of life.[41] The project bolsters a conception of nature that is becoming increasingly commonsensical, acquiring an air of timelessness when it is, in fact, profoundly historical. It fosters expectations that genetic biotechnology may soon lead to elaborate forms of biospheric control, whereby a class of systems-scientists – in collaboration with wealthy entrepreneurs and billionaire philanthropists – steward the world by employing different species as specialised ecosystem engineers. Finally, it displaces the focus from addressing the socio-political drivers of extinction. Why should we strive to reduce global emissions if technology will soon provide instruments to govern climate change and steward the global ecosystem? Should we address the socio-political drivers that are causing the extinction of millions of species if the species that we really care for may be promptly resurrected, maybe even improved? Is it really necessary to democratise human niche-making and ecology?

It is significant that Colossal's project hinges on the theological promise of a coming resurrection, promoting a narrative of salvation achieved by reaching through the mist of time to an innocent past; a time before humanity fulfilled its *telos* in the creation of its own geological era.[42] Salvation narratives, with their religious overtones, are secularised not only in political discourses but also in the sober statements of Enlightened science, where they infuse both promise of progress and threat of apocalypse. As Donna Haraway points out, 'Secular salvation history depends on the power of images and the temporality of ultimate threats and promises to contain the heteroglossia and flux of events.'[43]

In the salvation drama that infuses environmental technoscience in the Anthropocene, the mammoth is offered as a *katechonic* icon. The pressing, future-oriented temporality imposed by climate change – melting arctic ice, leading to increased methane emissions, leading to increased concentration of greenhouse gases, leading to further climatic change – is short-circuited by the salvific image of a mammoth resurrected to save humanity. It is a dazzling promise that is as much religious as it is scientific: an ecological theology that presents technoscience as a creative spirit – a force that will not only bring back the dead, but also offer new means of earthly salvation – and capital as the medium that enables the spirit to become embodied. Without financial backing, the mammoth chimaeras remain science fiction, trapped in the imaginations of scientists, venture capitalists, and amateurs. Only once they are infused with capital, do the ideas of technoscience become embodied, taking root in elephant wombs, and gaining critical mass in the world.

GENE-DRIVING RATS AWAY

Ecological Xenophobia and Genetic Annihilation Technologies

The woolly mammoth represents everything people love about the wilderness: it is exotic, majestic and absolutely distant. It gives the impression of being easy to control, and it does not present a direct threat to tax-paying urban dwellers. It is not surprising that it has rapidly become the charismatic icon for de-extinction, and a paradigmatic example of how genome editing technologies may not only profit corporations but also restore threatened global ecologies – or at least a technoscientific proxy for them. It is marketed as a symbol of that oft-imagined remote and fragile wilderness, which calls for our attention and protection.[44]

There is, nevertheless, *another* wilderness, composed of all those living organisms that stubbornly resist every attempt at human control. It is the wilderness of pests and vermin; whose excess of vitality overcomes artificial borders and threatens the landlord's mastery over space. This is a wilderness that few people love. And yet, it persists not only without human protection, but despite ceaseless attempts to eradicate it. It is the wilderness of flies and mosquitoes, bedbugs and cockroaches, mice and rats. As a paradigmatic pest, the rat is everything the mammoth is not: it is small, almost perversely banal and absolutely ubiquitous. The list of sins attributed to the rodent only got longer as modernity unfolded. Long considered an economically damaging pest due to its capacity to ravage fields, infiltrate granaries and ransack kitchen cupboards, in the eighteenth century rats began to be seen as vectors of disease. In the twentieth century, with the rise of ecological governance, rats started to be seen as an ecological villain: an invasive species proliferating without restraint, wreaking delicate ecosystems and driving less resilient life-forms to extinction.

Rats have always been small yet important actors in the making of capital's world ecology: nestled on explorers' ships, rats spread diseases and remade the ecology wherever the settlers arrived, plundering birds nests for eggs and 'outcompeting' the local fauna for food. Rats colonised the world, exploiting the global vectors that European colonisers established in their imperial quests. Today, rats rank at the top of the long list of so-called 'invasive species' that governments routinely compile. They are condemned as undesirable ecological migrants that mischievously cross borders they were never supposed to cross. They are object of widespread ecological xeno-phobia: a fear of the alien Other that calls for endless governmental interventions to suffocate the pernicious vitality of the intruder. They become the focus of extensive and highly organised death campaigns, issued by a multiplicity of private and public actors in order to bring an end to their exuberant proliferation. In many ways, the contemporary war against invasive species exemplifies the dark side of Earth-system thinking. Keystone species must be protected, or even de-extinguished, because their niche-making activities are valued as desirable from a system-wide perspective. Invasive species must be eradicated for the opposite reason, because they contribute to creating landscapes considered void of value.

State-organised rat eradication campaigns are nothing new. Already in the first half of the twentieth century, many countries systematically organised such campaigns in order to secure agricultural production and the health of the population.[45] Contemporary rat extermination campaigns add a new accusation to the list: rats disrupt delicate ecosystems. One in every four species currently faces extinction.[46] Yet, climate change is also associated with the proliferation of a small subset of species that appears well-adapted to life in the Anthropocene.[47] Rats are a good example. The rapid expansion of both urban areas and agricultural lands is advantageous to rats. Global warming is turning many parts of the world into ideal rat breeding grounds.[48] Growing rat populations, as currently seen in many parts of the world, must therefore also be understood as a symptom, and not only a cause, of ecological transformations. In the discourse of invasive species, these political-ecological complexities disappear from view. The urgent need to tackle the socio-economic causes of extinction and the proliferation of rats is abruptly turned into a call to arms against an alien enemy, which must be exterminated so that valuable lives may prosper.

Nowhere has this process been quite as pronounced as on islands. Not only do islands hold a disproportionate amount of the world's biodiversity, they also experience a disproportionate degree of biodiversity loss. Because of their unique biogeography, islands are often characterised by highly endemic biodiversity, meaning that they host species that are specific to the island in question. Partially because of this reason, islands have been long represented by Western literature as places outside of history, where nature could be observed in a pristine state. As Riley Taitingfong has shown, Western representations of island geographies – especially outside the European continent – have systematically over-emphasised their disconnection from the rest of the world.[49] The supposedly isolated nature of island ecologies is not a transhistorical truth, but rather a product of a particular social imaginary that is profoundly linked to the history of Western imperial expansion into these spaces. It is precisely this fantasy of the island as a remote speck of land in the midst of the vast ocean void – as a wild enclosure removed from the global flows of socio-natural history – which justified turning Caribbean islands into testing grounds for nuclear warfare in the twentieth century.[50] This imperial fantasy has been increasingly challenged as global logistics and international trade rapidly transform island ecologies. As a result, islands are today at the centre of global debates on invasive species, and what to do with them. Islands represent 'microcosms for the emerging biodiversity and socioecological landscapes of the Anthropocene'.[51]

The Galápagos islands illustrate this perfectly. Invasive mammals, including rodents, arrived there with seventeenth-century privateers and eighteenth-century whalers. By the time Charles Darwin arrived in the archipelago on the HMS *Beagle* in 1835, introduced species had already changed the local flora and fauna.[52] Today, Island Conservation, an NGO based in the USA, is planning the eradication of every rat and mouse in the archipelago starting from its experimental base in Floreana.[53] For decades the NGO has been experimenting with various eradication techniques, employing automatic rifles, helicopters, drones and a variety of pesticides. Campbell estimated that it would take '10 years, $26 million and 35 shipping containers full of poisoned cereal to clear Floreana of rats'.[54] The project also entailed dumping tonnes of brodifacoum – one of the most common rat poisons – from helicopters; in Campbell's own words: 'systematically paint the whole island'.[55] Painting an island with poison, rather predictably, turned out to affect animals other than rats. Poison spreads through ecosystems, endangering the wildlife in whose name the war on rats is waged. The dream of a techochemical fix to ongoing ecological transformations quickly revealed itself profoundly contradictory.[56]

Throughout the world, island ecologies have been exposed to similar large-scale military mobilisations.[57] Take New Zealand. Until the thirteenth century, there were no land-based mammals apart from some rare bat species. When European explorers and settlers arrived in the eighteenth and nineteenth centuries, they brought with them the widespread black rat (*Rattus rattus*) and brown rat (*Rattus norvegicus*). These mammals established colonies throughout the archipelago, adapting to their new environmental conditions. They learned to hunt the local fauna: feeding on local birds, seeds, snails, lizards, eggs, chicks, larvae and flowers. Iconic species, such as the kiwi and the giant kakapo parrot, became easy prey for the rats – as well as for the dogs, cats, stoats, weasels, and possums – brought to the islands by well-beaten shipping routes.[58]

The new predators are often singled out as the main drivers of biodiversity loss, causing 'considerable losses to both primary production and tourism'.[59] In fact, it is not easy to determine the impact of hunting, deforestation, pollution and habitat loss. Considering that these anthropogenic factors are currently leading to the extinction of roughly a quarter of the global fauna, it is safe to assume that they contribute to the specicides so often blamed on the exuberance of rodents. In the past fifty years, nevertheless, the black and brown rats have become the targets of extensive extermination projects, with varying degrees of success. The plan has so far relied mostly on chemical warfare combined with traditional mechanical traps. It has proven extremely demanding both in terms of ecological effects, economic costs and human labour.[60] New Zealand has eradicated circa 10 per cent of invasive mammal predators from its offshore island area, but the chemical *blitzkrieg* is starting to exhaust itself. As long as there is still a food source, colonies bounce back, and rats have now been found to be growing resistant to many toxins. According to a recent study, only two islands stand a chance of becoming rat free by 2025: 'Our results should be viewed as an examination of [...] potential outcome if transformative eradication advances are not made', declared one of the authors in an interview to the *New Zealand Herald*. 'Fortunately, universities, government researchers and private enterprise are already involved in exploring new and exciting transformative technologies to overcome limitations in the existing eradication toolbox.'[61]

Faced with the limitations of using poison as a means of ecological engineering, New Zealand has been on a quest to find novel pest eradication tools. CRISPR is one of the new technological fixes that the New Zealand's government hopes to mobilise in its extermination campaign. Since 2012, an ambitious initiative nicknamed 'Predator Free 2050' proposes to turn New Zealand into 'conservation country' by eradicating invasive predators such as rats, possums, and stoats.[62] To finance the effort, the government set up a crown-owned company, Predator Free 2050 Ltd, with the aim of 'developing new predator-control tools and techniques'.[63] Many have wondered if the power of genetic command and control, which has already shaped the life of countless animals enclosed in laboratory spaces around the world, could now be turned into a weapon of mass extermination. While experiments in 'genetic control' – generally defined as 'the release of organisms with genetic methods designed to disrupt the reproduction of invasive populations' – have been carried out for over half a century, CRISPR promises new and effective ways to curb the vitality of undesired populations.[64]

One decisive step in this direction was taken in 2014, when Kevin Esvelt, Andrea Smidler, Flaminia Catteruccia and George Church laid out the foundations for what would become known as 'CRISPR gene drives'.[65] Gene drive is a genome engineering technique designed to spread traits through a population at above-natural rates. It is sometimes called 'super-mendelian' inheritance since it biases selection in favour of the engineered trait, resulting in a more than 50 per cent chance that the offspring will inherit it.[66] Each gene-drive system works in a slightly different way, but most of them consist of two key components: the gene to be driven to dominance and the CRISPR-Cas9 assemblage. This gene-drive package is spliced into the animal genome. When an engineered organism mates with a wild one, their offspring gets one copy of DNA from each parent: a gene-drive version and a non-modified one. When the chromosomes from the parents line up in the egg, the CRISPR-Cas9 assemblage is activated. The genomic scissors recognise the copy of the natural gene in the opposite chromosome and cut it out. Once the natural gene is damaged, the cell's repair machinery intervenes. To repair the break, the cell uses the unbroken gene-drive chromosome as a template, copying it onto wild chromosomes. The embryo will develop carrying two copies of the gene-drive in each of its cells, to be inherited by future generations. As a result, the engineered trait spread quickly through the population. As a group of researchers at Department of Entomology of the University of California, Riverside, put it, gene drives have the 'remarkable intrinsic ability' to 'cheat evolution'.[67]

Multiple experiments have attempted to introduce gene drives in smaller organisms, from yeast to *Drosophila melanogaster*.[68] In 2018, the common laboratory mouse *Mus musculus* became the first gene-drive mammal. Although it was a proof-of-principle study without immediate practical applications, the authors argued that the GM mice would 'contribute valuable data to the ongoing debate about applications to combat invasive rodent populations in island communities'.[69] Andrea Crisanti, geneticist at Imperial College London, sums up the same idea when suggesting that with gene drive technology 'you can modify evolutionary trajectory. You can cause extinction'.[70] Gene drives bring the molecular power over life out of the laboratory and into the 'wilderness'. They open up a vast field of intervention. Genome editing technologies enable unprecedented control over genetic mutations, which can now be planned and realised, although with persistent off-target effects. Yet, without gene drives the passage from a single GM organism to a genetically engineered population remains laborious, expensive and time consuming. Most importantly, it can take place only under strictly controlled conditions, in which the sexuality of the animals can be supervised. A single genetically engineered animal released into the wild is unlikely to affect the genetic composition of an entire population, especially if the induced mutation impedes its reproduction.

This is why genetic engineering technologies have so far caused the biggest impact in scientific laboratories, industrial fermentation tanks, animal farms – controlled environments where breeding processes can be carefully organised. Gene drives represent a tentative automation of these breeding practices, ensuring that a particular genetic mutation spreads through a population. The promise is simple: once a single, experimental gene-edited organism is created and released, it will no longer be necessary to attend to artificial selection. Through automation, gene drives promise to accelerate the genetic engineering of entire living populations. Gene drives are often described as an unstoppable 'mutagenic chain reaction', spreading through wild populations at unprecedented speeds.[71] These descriptions play into and stimulate hypermodernist visions of total biospheric control, in which the Foucauldian 'power over life' is no longer administered at the relational scale of the landscape or of sexual encounters but enacted at the molecular level of the genome.

The reality is that gene drives do not work as well as they are supposed to. As soon as researchers began to test gene drives in caged populations in laboratories, living organisms rapidly developed resistance against them by accumulating mutations that prevented the drives from spreading. As a recent review of these experiments concludes, 'building a gene drive to manipulate or eradicate a population is like picking a fight with natural selection, and that fight might not be easy to win'.[72] Similarly, Charles Godfray of the University of Oxford has pointed out the persisting gap between what research in gene drives is trying to achieve and what has been achieved: 'The theory says that, in principle, if you release it once it would spread continent-wide. The reality is that would happen very slowly'.[73]

Research on gene drives has so far mostly relied on laboratory trials in caged populations but there are increasing demands for trials in wild populations. The idea has immediately sparked interest in the conservation world. 'Should we create a genetically modified rat so that its offspring is only male (or female)?' asks Island Conservation's website. In 2017, the organisation took a step in this direction by launching Genetic Biocontrol of Invasive Rodents (GBIRd), a partnership dedicated to the advancement of gene-drive research for invasive species management. The project mobilises an international consortium of public and private agencies and it is supported by a $6.4 million grant from Darpa, the research branch of the US military.[74] GBIRd's partnership is now searching for the perfect island laboratory for trialling these technologies in the field.[75]

Genetic control as a pest-eradication technology is garnering global attention. In 2019, researchers affiliated with, among others, Island Conservation, Australia's Health and Biosecurity, Commonwealth Scientific and Industrial Research Organisation (CSIRO), and the US Department of Agriculture published an article advocating for the use of gene drives to manage rodent populations.[76] Scientists based at the University of Edinburgh have theorised using gene drives to control the population of grey squirrels in the UK. The species, introduced in the nineteenth century, is considered invasive. The authors point out that 'current control methods such as shooting, trapping, and poisoning are inhumane, labour-intensive, expensive, and ineffective in dealing with the scope of the problem in most situations'. Instead, the article proposes the deployment of 'CRISPR-based gene drives' as 'a humane, efficient, species-specific and cost-effective method for controlling invasive species, including grey squirrels in the UK; filling a distinct void in the conservation toolbox'.[77] Their computer models show that the technology could potentially reduce the population by 60 per cent in a decade.

The forestry industry is particularly enthusiastic about the idea, as grey squirrels munch into their profitability. A report recently published by the UK Confederation of Forest Industries presents research on gene drives conducted by the Roslin Institute in collaboration with the European Squirrel Initiative, and endorses the implementation of Directed Inheritance Gender Bias (DIGB): 'an innovative genetic control strategy' which 'offers a genetic alternative "contraception" by skewing the sex ratio within the target population, leading to a population crash'.[78] One of the main reasons to implement the plan, suggests the Timber Union, is that the technology could be developed 'for less than the annual cost burden of grey squirrels'.[79] Loggers blame squirrels for eating into their profit margins, while portraying them as an invasive foreign species wreaking havoc in an otherwise peaceful British countryside. However, several environmental groups – including Animal Aid and the Interactive Centre for Scientific Research about Squirrels – are questioning policies of forest conservation by squirrel extermination. According to these critical voices, campaigns of squirrel extermination tend to be driven by political anxieties and economic interests, rather than by any real plan to address the main drivers of ecological collapse. The xenophobic discourse that dominates public discourse is deflected onto the forest, fuelling animosity against the foreign greys.[80]

Despite the enthusiasm expressed by the private and public actors who, for one reason or another, dream of eradicating a particular species from their backyard, gene drives have also generated controversy and debate. In 2015, the International Union for Conservation of Nature (IUCN) began discussions of the use of gene drives for conservation purposes, commissioning a report on the implications of gene drives for biodiversity conservation. The resulting document, entitled *Genetic Frontiers for Conservation*, sparked a spirited debate within the IUCN. Several members organisations protested against the report's ambiguity towards 'synthetic biology and engineered gene drives' which are described as replete with risks and uncertainties, but also as potentially 'beneficial to conservation and sustainable use of biodiversity'.[81] A counter-study conducted by ETC Group, a research and advocacy organisation, concluded that the integrity of the report was compromised by a number of authors' direct affiliation with groups that may have 'potential conflict of interest' such as Revive & Restore, Genetic Biocontrol of Invasive Rodents and Target Malaria.[82]

In 2016, it was the turn of the US National Academy of Sciences to consider the potentials and risks associated with emerging gene drive technologies. Their report titled *Gene Drives on the Horizon* highlights several potential uses for conservation, including for the eradication of rodents to protect island biodiversity. Just like the IUCN report of the previous year, it considers many of the potential risks associated with emerging forms of genetic control. It does, however, conclude that 'the potential benefits of gene drives for basic and applied research are significant and justify proceeding with laboratory research and highly controlled field trials'.[83]

In a surprising response article, a group of leading molecular biologists including George Church and Kevin Esvelt (one of the most established scientific authorities on gene drive technologies) warned against the conclusions adopted by the Academy of Sciences, arguing that field trials could result in 'unintended spread to additional populations' since 'even the least effective drive systems reported to date are likely to be highly invasive'.[84] Because of the fundamental interconnectedness of ecology, stresses Esvelt elsewhere: 'a release anywhere is likely a release everywhere'.[85] Releasing gene drive organisms poses several thorny political issues, which are unlikely to be resolved by future technological improvements. Since gene drives – at least in their present technological configuration – cannot be limited neither spatially nor temporally, how can a democratic decision concerning their deployment be made? Who has the authority to decide on a technology that may affect the future paths taken by natural history?

Reading the promotional documents for gene-driving extinction, one is easily struck by a sense of urgency; who, after all, is in favour of biodiversity loss? If rats – brought by humans – are responsible for the extinction of rare bird species, is there not a moral duty to kill them off? The political imagination is limited to genome editing or mass extinction. Such appeals to universal morality, however, often conceal vested interests and shroud in mystery the most important drivers of biodiversity loss. The ecological crisis becomes a vehicle to extend and expand new forms of power over life via technologies controlled by a handful of private companies and public institutions. In this topsy-turvy Anthropocene world, 'rewilding' stands for the release of genetically modified pseudo-mammoths, while 'conservation' means endless extermination campaigns. The protection of 'native species' is achieved by constantly monitoring 'invasive species' and organising cyclical counter-balancing interventions. Genetic control informs and enables a cybernetic form of ecological governance based on feedback loops. Ecological balancing is pursued by doubling down the interventionist stance of modern technoscience.

This hypermodernist approach to conservation is increasingly popular insofar as it fits with postmodern celebrations for the imminent death of nature. As Bruno Latour suggests, in a spirited praise of the eco-engineering promoted by the Breakthrough Institute, 'to breakthrough is to abandon the limit of limits'.[86] To solve the problems brought by capital's Great Acceleration, we need to abandon restraint and redesign life all-the-way-down. 'The environment', explains Latour, 'is exactly what should be even more managed'. It is not the moment of slowing down, rather we should 'intervene even more'.[87] Did we bring rats? We will counterbalance by gene-driving them away! Latour encourages his readers to abandon all precautionary principles, embrace 'the fabulous dissonance inherent in the modernist project' and, as we plan the next extermination, sing: 'Thank God, nature is going to die.'[88]

TROJAN MOSQUITOES

The Molecularisation of Interspecies Warfare

While debates over the political and ecological dilemmas posed by gene drives continue, genetic biocontrol is already making it out of the laboratory. The first field tests, however, are not taking place on Pacific islands and the living targets are at once smaller and more numerous than rats.

In July 2019, over 6,000 genetically engineered mosquitoes were released in a village in Burkina Faso. The mosquitoes were all male, and all sterile, a new strain developed by researchers at Imperial College London. The release was funded and organised by Target Malaria, an international research consortium mostly funded by the Gates Foundation ($75 million) and the Open Philanthropy Project ($17.5 million). It is the first step in an experimental process set to culminate with the release of mosquitoes equipped with a gene drive called X-shredder, designed to prevent the birth of female mosquitoes by disrupting the formation of X chromosomes in the sperm cells.[89] Only Y sperm cells would develop, shifting the sexual composition of the population. According to tests conducted in caged populations, the technology can generate 'fully fertile mosquito strains that produce >95% male offspring', thus 'providing the foundation for a new class of genetic vector control strategies'.[90]

The experiment has been presented as a humanitarian intervention that might eventually lead to a genetic solution to the major health risk associated with proliferating mosquito populations. According to the latest estimates from the WHO, there were 216 million new malaria cases and 445,000 deaths worldwide in 2016, with Africa accounting for 90 per cent of the cases and 91 per cent of the deaths.[91] The malaria-causing parasite has been described as 'one of the biggest killers in history'.[92]

In the last decade, a concerted global programme of malaria control lessened its global impact, reducing deaths by about 60 per cent between 2000 and 2015 through the provision of public health care, application of mosquito nets, and campaigns aimed at reducing the stagnant water basins central to mosquito reproduction.[93] The 2020 World Malaria Report, however, shows that global gains in combating malaria are slowing down, mostly due to 'insufficient funding', which 'have resulted in gaps in access to proven, WHO-recommended malaria control tools'.[94] Malaria continues to kill in over 87 countries. There, Target Malaria proposes to implement a different approach, directing funds towards the development of genetically modified mosquitoes designed to exterminate the air-borne malaria vectors.[95]

Genome engineering does not appear among the previously cited list of 'WHO-recommended malaria control tools'. In October 2020, the international organisation released a 'position statement' on the issue, pointing out that while genetically modified mosquitoes 'could be a valuable new tool in efforts to eliminate malaria', they also raise 'concerns about ethics, safety and governance and questions of affordability and cost–effectiveness, which must be addressed'. Ultimately, the report endorses 'a stepwise approach to testing GMMs', while stressing that 'community acceptance' for any release must be obtained.[96] There is a fundamental ambiguity at the heart of the WHO statement, which can appear as either an endorsement or a condemnation of Target Malaria's strategy.

The NGO claims to have conducted extensive 'public engagement' before each experimental release and that it has therefore met the concerns of the WHO. Yet, it remains controversial whether 'public engagement' translates automatically into 'community acceptance'. Target Malaria has tried to avoid getting bogged down by the issue by arguing that it would be 'logistically impossible' to obtain free, prior and informed consent from all those affected.[97] The statement appears rather dismissive of democracy. Democratic consultations may certainly be expensive and complex, but they are certainly not 'logistically impossible'. What is impossible is to establish if community acceptance exists without an explicit democratic vote on the issue. Short of that, we are left with an unresolved debate and two conflicting narratives.

On the one hand, Target Malaria has issued videos of local people who support the project, organised interviews for its promotional materials and facilitated some public meetings with international media sources. Charles Mugoya, chair of the National Biosafety Committee in Uganda, assured the *Washington Post* that there continues to be broad support for field trials.[98] On the other hand, there is a vocal and widespread opposition to the project, composed of local, national and continental civil groups. This front has emphasised that many of the locals engaged by Target Malaria oppose the project, while many more were never consulted. In a public statement, Mariann Bassey-Orovwuje, chair of the Alliance for Food Sovereignty in Africa, insisted, 'In Africa we are all potentially affected, and we do not want to be lab rats for this exterminator technology. Farmers have already marched in the streets of Burkina Faso to protest genetically engineered mosquitoes and we will march again if they ignore this UN decision. We are giving notice now that potentially affected West African communities have *not* given their consent or approval to this risky technology.'[99]

At the UN Conference for the Convention of Biological Diversity in Sharm-el Sheikh, Egypt, over 170 civil society organisations spoke up against the release. 'We don't want dangerous experiments in our country. We don't want corrupt politicians and scientists making decisions on our behalf', said Ali Tapsoba, president of Terre à Vie, an NGO that has described the experiment as a new form of medical colonialism.[100] The conference was followed by a statement signed by 43 civil society organisations – including the African Biodiversity Network, the Coalition pour la Protection du Patrimoine Génétique Africaine, the Community Alliance for Global Justice, the Eastern and Southern Africa Small Scale Farmers Forum, the Fellowship of Christian Councils and Churches in West Africa and the Indigenous Peoples of Africa Coordinating Committee – voiced clear opposition to the project. 'Experimenting with African lives', they write, 'to prepare the ground for this untested and extremely controversial technology, for which independent scientists have raised serious concerns, and for which more than 170 civil society organisations have called for a moratorium, is completely unacceptable.'[101]

The movement has emphasised that the rhetoric of humanitarianism legitimises forms of scientific experimentation with gene drive organisms that are considered too risky to be conducted in Europe and that remain strictly regulated by the EU's genetic engineering Directive 2001/18. Indeed, the only application for a GM insect field trial ever submitted in the EU – submitted by Oxitec to test a genetically modified olive fly – was rejected twice. Consulted by the Science and Technology Committee of the UK House of Lords, the British company complained that within the EU 'we cannot even get to the first hurdle of getting a genetically modified insect in a field cage'.[102] The reason why the EU has so far refused to approve the release of GM mosquitoes is the same that motivates opposition on the African continent: both the benefits and the risks remain unclear. A recent report by the African Centre for Biodiversity stressed that that the hype surrounding gene drives remains 'unsubstantiated', and that in focusing on a narrow high tech-fix it diverts resources away from tried and tested strategies.[103]

Such political conflicts over genetic biocontrol technologies are becoming increasingly common. Oxitec – a UK subsidiary of the US-based Precigen which has received over $18 million from the Gates Foundation to scale up its research on new genetic control techniques – has been one of the main actors in this genomic 'arms race' against both the mosquito and a long list of agricultural pests. While the company has so far been turned down by EU regulators, it has been able to release its GM mosquitoes in Brazil, Panama, Malaysia, the Cayman Islands and, most recently, the Florida Keys.[104] In April 2021, after more than a decade of regulatory back and forth and consultation with residents – many of whom were against the trials – the company placed mosquito eggs in several undisclosed locations around the Keys.[105] A number of hexagon-shaped boxes, emblazoned with the company logo, were placed on fenced private property for fear that disgruntled locals might sabotage the release.[106] All the eggs inside the carefully concealed boxes are male mosquitoes, which contain a modified gene intended to kill female offspring in the embryo stage. Now that they have been released, the company will be monitoring how far the engineered mosquitoes travel, and to what extent they mate with wild ones.[107]

The ultimate goal is to develop new products and to make profits. In November 2021, Oxitec announced 'the landmark commercial launch of its Friendly™ *Aedes aegypti* designed for use by homeowners, businesses, and communities to control the dengue-spreading *Aedes aegypti* mosquito'. According to the company, the commercialisation of Friendly™ 'represents the first time globally that the benefits of using biologically engineered mosquito control technology can be purchased directly by consumers'.[108] At a press conference for the occasion Grey Frandsen, CEO of Oxitec, commented 'we've placed the power of Friendly™ biology into a small, joyful box'.[109] Once delivered, customers need only to add water, place it in the garden and let legions of male Friendly™ disperse, inseminating females with many more prematurely dead offspring. 'While the Friendly™ *Aedes aegypti* males pursue and mate with female *Aedes aegypti*', proudly promises Oxitec website, 'customers need only reactivate the easy-to-use box once per month'.[110]

Oxitec's commercialisation of these Friendly™ autocidal insects represents the consolidation of a research trajectory which has been building momentum for over half a century. Scientists have been working on genetic control since at least the 1950s. The concept of 'genetic control' indicates emerging methods of reducing the population of targeted species by introducing damaging genetic mutations that disseminate by mating. This definition encompasses a number of well-established techniques, including: 'the use of chromosomally inherited genetic factors including rearrangements, mutations and transgenes, radiation- or chemically-induced dominant lethal mutations, as well as sexually transmitted symbionts'.[111]

The first application of 'sterile insect techniques' has been ongoing for over seventy years. Started in the 1950s, in Curaçao and Florida, the screw-worm fly (*Cochliomyia hominivorax*) eradication programme has developed into a large-scale international project, involving most of North and Central America.[112] In the last half a century, an estimated $1.3 billion have been spent to eradicate the insect. Today, 31 million *C. hominivorax* are raised each week in an industrial site located in Pacora, Panama. The flies are sterilised by ionising radiation and then released throughout southern Panama and up to 20 nautical miles within northern Columbia. In this 'buffer zone', designed to contain the fly from re-colonising Central and North America, wild and mutant insects mate but fail to produce viable offspring. As the population decreases, the ratio of fertile females to sterilised mutant males – which are continuously poured into the buffer zone – rapidly drops until the probability that a wild female will mate with one of the few remaining non-sterilised males is essentially zero. The local population is eradicated. Gradually, a new generation of flies pushes north from South America and a new eradication campaign begins.[113]

The international campaign of genetic control against the screw-worm remains exceptional in terms of scale. But it is not an isolated instance. Over seventy-five facilities around the world employ sterile insect techniques promoted by the Centre of Nuclear Techniques in Food and Agriculture of the International Atomic Energy Agency.[114] The insects mass-reared by these facilities are mostly agricultural pests, including fruit flies, moths, screw-worms and bollworms. However, repeated attempts to use radiation mutagenesis to genetically control the three major genera of mosquitoes – *Anopheles*, *Aedes* and *Culex* – have been mostly unsuccessful.[115] This is why, since the 1960s geneticists have focused on the development of 'selfish genes' for vector control.[116] This research has been driven by the fundamental aspiration of 'making "a better mosquito" that reduces the need for inundative releases'.[117] A recent review, authored by

Luke Alphey – in his dual role as a researcher at the University of Oxford and employee of Oxitec – affirms that traditional sterile insect techniques 'are obviously self-limiting as the lethal or sterile factor disappears rapidly from the target population, maintained only by periodic release of additional modified males. In contrast, gene drive systems are intended to spread themselves within the target population and are almost invariably self-sustaining'.[118]

While the idea of developing gene drives for pest control has a long history, it was not until the early 2010s that experimental research really began to pick up speed. In 2011, geneticists at Imperial College London succeeded in making an engineered gene dominant in *Anopheles gambiae*, reaching 85 per cent of the population.[119] In 2015 and 2016, studies demonstrated the feasibility of using CRISPR to suppress populations of two mosquito species (*Anopheles stephensi* and *Anopheles gambiae*).[120] In November 2018, researchers at Imperial College London – financially supported by Darpa and the Gates Foundation – succeeded in causing 'complete suppression' of a caged population of *Anopheles gambiae*.[121] By disrupting the gene *doublesex*, the team caused sterility in female offsprings and the population crashed within eleven generations. According to a recent review of these laboratory experiments, 'recently developed CRISPR-Cas9-based gene-drive systems are highly efficient in laboratory settings, offering the potential to reduce the prevalence of vector-borne diseases, crop pests and non-native invasive species'.[122]

What was previously only a hypothesis – that CRISPR could be used for population eradication – had been 'translated into a genetic tool able to suppress the reproductive capability of the mosquito population'.[123] According to Kevin Esvelt's statement during a recent National Academy of Sciences hearing: 'There is no societal precedent whatsoever for a widely accessible and inexpensive technology capable of altering the shared environment'.[124] Similarly, the most recent assessment of gene-drive technologies published in *Nature Communications* in July 2021 points out that 'despite aspirations that HGDs [homing-based gene drives] can solve world health issues, there are safety concerns due to the predicted ability of HGDs to persist indefinitely and invade non-target populations'.[125] Yet, these projects are hurtling forward, only partially slowed down by technical obstacles, civil resistance and persistent ecological unknowns.

By aiming to combat agricultural damage and disease through genetic interventions, these projects swiftly reframe complex socio-ecological problems into a war against the insect world. In the early twentieth century, Leland Howard, chief at the Bureau of Entomology of the US Department of Agriculture (USDA), summed up this interspecies conflict as nothing less than total war: 'The insect world is a menace to the dominance of man on this planet.' This war, according to Howard, 'will in all probability end in the elimination of one side or another.'[126] Imperial state bureaucracies have long engaged in world-making practices to eradicate mosquitoes and other insects. In the nineteenth and early twentieth century, this often meant engaging in large-scale ecosystem engineering with rather blunt tools: for example, Italian fascists in the twentieth century declared a war on the mosquito, which was fought by mobilising thousands of unemployed labourers to drain swamps. The workers died in great numbers, but this was considered equivalent to the sacrifice of soldiers in battle.[127] The consolidation of a chemical-industrial production model in the first decades of the twentieth century slightly changed this strategy. In particular, the use of DDT insecticide to support Allied troops in the tropics during the Second World War established the paradigm of entomopolitics for the remainder of that century: chemical warfare.[128]

In the process, the war on insects lubricated the circuits of accumulation for private chemical companies. Ecologically speaking, the consequences were disastrous; already in 1962, Rachel Carson's *Silent Spring* shone light on the deadly silence caused by the chemical paradigm.[129] The model continued unabated however, and blanket-spraying of insecticides remains a wide-spread practice. The agro-chemical system of industrial capitalism spells disaster for the insect world. A comprehensive literature review published in 2019 found that the main drivers of the 'worldwide decline of entomofauna' are 'in order of importance: (i) habitat loss and conversion to intensive agriculture and urbanisation; (ii) pollution, mainly that by synthetic pesticides and fertilisers; (iii) biological factors, including pathogens and introduced species; and (iv) climate change.'[130] Over 40 per cent of insect species are threatened with extinction as a result of the complex interplay of these four factors. Howard's war against the insect world has progressed steadily. And yet, that very war is also profoundly shaping global ecologies, creating complex and unpredictable chains of extinction events that might prompt a 'catastrophic collapse of nature's ecosystems'.[131]

It is important to stress that the interspecies war against the mosquito – and the entire insect world – is neither inevitable nor outside of history. It is a relatively recent phenomenon that has been profoundly shaped, and often magnified, by the development of capital's world ecology. When European powers began establishing a plantation economy in the Greater Caribbean, it stimulated ecological change that improved the breeding conditions of *Aedes aegypti* and *Anopheles quadrimaculatus*.[132] Dam-building in twentieth-century Egypt contributed to the spread of mosquitoes along the Nile.[133] Colonial interference with the physical topology of Bengal is recognised as a cause of the nineteenth-century malaria outbreak that ravaged the region.[134] After the 2009 financial crisis, unfinished skyscrapers in Jakarta turned into ideal incubators for mosquitoes.[135] Global commodity trade impacts malaria risk by driving deforestation and ecological change. This same commodity trade also fuels global warming, which is in turn likely to drive 'an overall global net increase in climate suitability [for mosquitoes] and a net increase in the population at risk'.[136]

Colonialism, industrial agriculture and global logistics have shaped the ecological conditions for mosquitoes in ways that have profound social repercussions. Mosquito-borne diseases are shaped by the social construction of space; where there is stagnant water, where forests are cut down, where dams are built, mosquitoes thrive. For this reason, the ecological history of mosquitoes and the diseases they carry is very much political. The struggle against malaria is a struggle for a different socio-economic system and a different human being in nature. By transposing the problem of malaria onto the molecular biology of the mosquito, programmes of genetic control depoliticise the issue. They represent a solution that bypasses the construction of space, providing a supposedly straightforward molecular alternative to a political rethinking of the social landscape.

Genetic biocontrol is often posited as a more precise technical solution than chemical pesticides; a solution that would avoid the dilemmas and unexpected consequences of twentieth-century chemical approaches to pest control. Yet, many questions remain surrounding the ecosystemic effects of introducing gene-drive in wild populations. The language of 'pests' obscures the many threads that tie mosquitoes to other beings in the web of life – relations that are only dimly known. Their eggs provide food for innumerable species of birds, frogs, and other animals.[137] In Camargue, a programme to curb mosquitoes through pesticides ended up 'affecting vertebrate populations following the suppression of prey species'.[138] In a Michigan forest, ecologists experimentally removed mosquito larvae from hundreds of small ponds. After the removal, the tiny ecosystems were radically altered.[139] Without mosquitoes, many plant species would lose a primary pollinator.[140]

Darpa's Safe Genes project – which funds most of the ongoing research on the new generation of gene drive technologies – exposes the ambiguities associated with this progressive molecularisation of a traditional military approach to nature. The 2016 *Worldwide Threat Assessment of the US Intelligence Community* describes genome editing technologies as a new category of 'weapons of mass destruction and proliferation', which threatens the stability of the international community. 'Research in genome editing', reads the report presented to the US Senate, 'probably increases the risk of the creation of potentially harmful biological agents or products. Given the broad distribution, low cost, and accelerated pace of development of this dual-use technology, its deliberate or unintentional misuse might lead to far-reaching economic and national security implications'.[141] Darpa has been particularly outspoken about the multiple risk posed by cheap gene editing tools. The 'rapid democratization of gene editing tools' is described as a source of increasing risks since 'the convergence of low cost and high availability means that applications for gene editing – both positive and negative – could arise from people or states operating outside of the traditional scientific community'.[142] According to the research branch of the US military, the integrity of the global genome is threatened by 'the accidental or intentional misuse of such technologies'.[143]

Gene drives are considered particularly unsettling from the point of view of national security insofar as 'traditional biosafety and biosecurity measures including physical biocontainment, research moratoria, self-governance, and regulation are not designed for technologies that are, in fact, explicitly intended for environmental release and are widely available to users who operate outside of conventional institutions'.[144] In response to these tendencies, the Safe Genes programme is 'studying control mechanisms for and countermeasures against gene drives' meant to 'protect warfighters and the homeland against intentional or accidental misuse of genome editing technologies'; 'prevent and/or reverse unwanted genetic changes in a given biological system'; 'control, counter, and even reverse the effects of genome editing – including gene drives – in biological systems across scales'.[145] Paradoxically, the threat posed by emerging genetic control technologies is precisely Darpa's rationale for funding further research on them.

The 'Gene Drive Files', a trove of emails uncovered by civil society investigators, reveal that the research branch of the US military has taken a leading role in promoting gene drive research. Since 2017, the agency has awarded around $65 million to different research projects, making it the single largest funder of gene drive research on the planet.[146] The fundamental rationale of the programme is stated clearly by the agency: 'guiding the development of an array of powerful, emergent genome editing technologies'.[147] As recently stressed by the coordinator of the Safe Genes programme Renee Wegrzyn, the research is not only defensive and reactive, it is also proactive in shaping the future of gene drive technologies. 'Our work flips the traditional understanding of dual-use research on its head', Wegrzyn said in an interview published on Darpa's official website. 'We set out to protect against misuse of genome editors, and by virtue of making progress in that mission, we're also laying the groundwork for safe, predictable, and potentially transformative applications of the technology'.[148]

In reality, the security risks created by gene drive technologies continue to be multiple and uncertain. The release of gene drive organisms has the potential to set off 'a cascade of eco-evolutionary dynamics', but what changes remain poorly understood.[149] An international coalition of biologists has warned in a public statement published in *Science*, that experiments involving the open release of modified organisms 'could have unpredictable consequences'.[150] Indeed, Oxitec's experiments with GM autocidal mosquitoes have *already* produced unintended consequences. In 2019, a team of independent researchers analysing an early trial raised alarms when noticing that some offspring of the GM mosquitoes did not die as planned. Rather, they reached sexual maturity, creating a new hybrid mosquito variant. According to the authors of the original study, this new breed is likely to engender 'a more robust population than the pre-release population due to hybrid vigor'.[151] Oxitec contested the study, prompting a scholarly discussion that remains unresolved. As population geneticist Jeffrey Powell recently stressed: 'the important thing is that something unanticipated happened. [...] When people develop transgenic lines or anything to release, almost all of their information comes from laboratory studies. Things don't always work out the way you expect.'[152]

One response to the scientific recognition that gene drives *will* produce unintended consequences has been to develop increasingly complex computer models in an attempt to anticipate what might be unknowable; to predict how the 'synthetic threads' of engineered genes will 'affect the web of life on Earth'.[153] The predictive accuracy of these computer models remains contested, and genetic biocontrol continues to be shrouded in social and ecological uncertainty. For this reason, ecological engineering using gene drives has met with fierce resistance. In 2016, over 170 civil society organisations called for a moratorium on both technical developments and experimental applications of gene drive organisms. Two years later, more than 250 organisations called for a moratorium on the release of such organisms; while a coalition of activists raised concerns that the development of gene drives for 'humanitarian' purposes such as malaria eradication might serve as a 'Trojan horse' for the development of new forms of genetic control to govern insect populations in agricultural fields, urban environments and touristic locations around the world.[154]

Research on new methods of genetic control, while often presented as a necessary risk to respond to a pressing global health issue, is also driven by the search for profitable applications.[155] Oxitec – the company that is driving much of the experimental research on new methods of genetic control against mosquitoes – is already working in this direction. So far, it has conducted field trials for two genetically modified autocidal insects: the agricultural 'pests' diamondback moth (*Plutella xylostella*) in New York and an engineered pink bollworm (*Pectinophora gossypiella*) in Arizona. The company has announced the development of a range of other genetically modified insects, including: the fall armyworm, the soybean looper, the medfly and the spotted-Wing drosophila. 'Our Friendly™ crop protection technology is simple', reports the company's marketing material, 'we have genetically engineered our male insects to carry a single self-limiting gene. After mating with targeted wild-type female pests, the male passes on to his progeny the self-limiting gene, which is designed to prevent the female offspring from surviving. [...] The self-limiting gene works by using the insect's own biology against itself'.[156]

Genetic biocontrol is increasingly attractive to agribusiness because it suggests the possibility of fabricating a bespoke biodiversity adapted to the requirements of industrial agriculture. It interests governments because of its 'dual use', which makes it both a potential source of insecurity and a means of securing that insecurity. The contours of a new biopolitical economy emerge slowly from these practices. Industrial production necessitates the annihilation of diverse life-worlds, and it relies on controlling the fundamental political distinction between valuable, noxious and worthless lives. These are all political and relational concepts. There is no 'natural pest': a living population can be worthless or noxious from a specifically situated perspective, while being a precious source of food, of services, of joy from another one. In today's political ecology, the distribution of different species within these three categories is mostly guided not so much by (bio)political decisions made by sovereign states, but by financial calculations performed in stock markets and corporate boardrooms. Market trends increasingly guide the separation between 'pest species' condemned to planned extinction, 'worthless species' to be let die, and 'valuable species' performing precious ecosystem services, and thus made to live.

These abstract biopolitical calculations do not affect only the specific species that become targets of genetic interventions. Genome editing technologies are increasingly becoming tools for large-scale operations of ecological engineering with the ambition to design and alter entire ecosystems. In this transition to bioengineering, the question is no longer simply to generate genetically engineered organisms, but rather to utilise genome-edited organisms as a lever for the reorganisation of whole ecological networks. Bioengineering enables a multiplicity of metabolic shifts, whose impact on the socio-ecological (re)construction of the biosphere remains uncertain. It offers new ways of taking possession of living bodies, genetically redesigning metabolic pathways and indirectly affecting global ecological processes at ever-larger scales. The overall metabolic shift induced by practices of genome engineering may well engender new ecological crises, further destabilising the socio-ecological metabolic cycles that have so far kept the biosphere in a livable state.[157] Or it could constitute a metabolic fix, which will patch over processes of ecological degradation, cement social structures of power and revitalise the global economy. This is, at least, the hyper-genomic dream shared by many molecular biologists and the billionaire philanthropists that bankroll their plans.

7

Pharmaceutical Lives

Humanised Mice, Pharma-Pigs, and the Molecularisation of Production

The making and selling of pharmaceutical commodities has been one of the most profitable businesses of the twentieth century. In the last decade, however, a number of reports have pointed towards an incipient profitability crisis connected to three main causes: patent expirations, a poor pipeline of new drugs, and pressure on prices due to political struggles for cheaper drugs.[1] The transition currently experienced by the pharmaceutical industry is also related to a more long-term tendency. It is difficult to keep expanding the market indefinitely – after all there are only so many drugs a person can take per day.

In response to these downward trends, Big Pharma has developed several counterstrategies. Firstly, the industry has reduced its investment in research and development of new drugs: 'after tax deductions only about 1.3 per cent of the money that the industry spends actually goes into basic research, the type of research that leads to new medications'.[2] There are now fewer products in the development pipeline, and the industry is increasingly focusing resources on ensuring that those make it through the regulatory process as quickly and economically as possible. Second, the industry is conducting a concerted global campaign to strengthen its patent privileges, resisting pressures to renounce cashing in royalties on life-saving drugs at a time of pandemic global crises. Third, given the threat of price controls, and protests against pharmaceutical patents growing, corporations are pressed to bring down production costs for existing pharmaceutical compounds. Consequently, considerable research funds have been shifted from the process of searching for new drugs to the quest for cheaper production strategies.[3]

Genome editing technologies play a prominent role in each of these strategical manoeuvres. As part of the effort to make testing processes faster and cheaper, pharmaceutical companies have invested considerable resources in developing new types of animal models. Genome editing furthers the quest for experimental bodies to partially substitute for expensive and time-consuming human testing. Genome editing also stimulates the accumulation of ever-growing patent portfolios by converting biological subjects into patentable, biotechnological artefacts. Finally, genome editing technologies play a prominent role in corporate strategies to bring down production costs and increase profit margins. In particular, they are contributing to a process of molecularisation of pharmaceutical production, which is taking assembly lines out of the factory and into the body of microbes, plants and animals.

An exploration of these three points informs this chapter. The first section focuses on the industrial production of millions of genetically modified 'model organisms', which are consumed every year by pharmaceutical and medical labs around the world. We consider questions posed by new forms of 'genetic humanisation' of animal models, and the way they challenge traditional boundaries between the human and the non-human. In the second section, we consider molecular 'pharming', a hybrid word that hints at the use of genetically modified farm animals as living production sites for pharmaceutical products. The final section is dedicated to the emergent xenotransplantation industry. In the last ten years, humanised pigs have been turned into a source of genetically designed hearts, kidneys, livers, corneas and skin tissues for human patients in need of an organ replacement. This may open up a new venue for capital accumulation based on the commercialisation of the chemicals, tissues and organs that compose the human anatomy. The body is increasingly transformed into a recombinant industrial product, purchasable piece by piece, protein by protein, organ by organ on the global market. The composition of the 'human brains, muscles, nerves, and hands' – which, according to Marx, represents the precondition and unchanging material base on which all labour-power ultimately rests – is itself commodified.[4]

CRAFTING EXPERIMENTAL BODIES

Humanised Mice at the Frontiers of Technoscience

'Experimental bodies', notes Ilana Löwy, 'are entities which can be substituted for patients' bodies in order to investigate diseases and look for treatments'.[5] The idea of using animals to stand in for human patients during inherently risky procedures – such as exploring physiology and the mechanics of disease, testing medical techniques, and trying out drugs – is not new. In the seventeenth century, William Harvey sacrificed thousands of mice to study blood circulation. In the eighteenth century, Robert Boyle's studies on respiration relied on close observations of mice and birds asphyxiating under a custom-made vacuum pump. In the nineteenth century, animals were used by surgeons keen to perfect their dexterity, by physiologists eager to observe the structure of internal organs, by microbiologists interested in understanding bacterial infections, by pharmacologists striving to test the effects of their concoctions, as well as by early geneticists investigating sexual reproduction and hereditary processes. All these experimental studies would not have been possible without expendable bodies that can be infected, injected, maimed and killed at low costs and with minimum social opposition.[6] Given the existence of cultural, political and economic constraints on human experimentation, medical science has long relied on the power of analogy, assuming that 'observations made in the laboratory using animals of one species will reveal mechanisms shared by numerous other species, and occasionally by all living organisms'.[7]

Experimental bodies have an additional advantage over human bodies: they can be subjected to procedures of standardisation and homogenisation. Bodies are infinitely variable, even within the same population. Each of us is sick in a unique way, depending on many factors including genetic endowment and socio-biological history. This poses an obvious obstacle to the reproducibility of scientific results. If each body is fundamentally unique, how can a scientist hope to generalise the results of an experiment performed on only a small subset of a given population? How can a scientist assure that the experiments performed on a particular set of bodies in London might be repeated on an unrelated set of bodies in Amsterdam? In the early twentieth century, the introduction of increasingly formalised techniques of biological standardisation offered a solution to this long-standing epistemological problem. Biological standardisation was first pursued by rigidly codifying the spaces in which experimental bodies conduct their lives. If it could be ensured that two separate sets of laboratory animals were kept in similar caged environments, while being fed similar food and stimuli, it could be assumed that their bodies would perform similarly in a given experiment. This, nevertheless, would offer only a partial solution to the problem. Genetic variations would still threaten the homogeneity of experimental results. Genetically uniform strains of mice constituted a second means of biological standardisation, which would rapidly turn the production of 'experimental bodies' into an industrial labour process.

The first standardised strains of laboratory mice were developed in the 1920s. While working in the laboratories of Harvard University, Clarence Little developed the basis of the method of production still in use today by inbreeding mice siblings over several generations. The most remarkable result of this activity was the establishment of the C57BL/6 strain. The strain started from a batch of common house mice, purchased from the amateur breeder Abbie Lathtrop.[8] Little noticed that, after forty generations of brother-sister mating, the strain of mice tended to stabilise. Each member of the population would then become an almost perfect genetic copy of the parents, creating a model-animal invariant over space and time. C57BL/6 was a population composed of nearly indistinguishable individuals, which appeared to Little as 'identical as newly-minted coins'.[9] The new genetically standardised mouse, also known as 'Black-6', constituted a perfect living platform for scientific experimentation; a standardised tool, which might be replicated in an endless supply of genetically consistent copies. Black-6 was a product *of* modern science *for* modern science: a living creature crafted to become a sort of Gold Standard of medical research. It remains today the universal equivalent that enables the circulation of comparable data across space and time.[10]

In 1929, Little founded Jackson Laboratory: an institution that remains to this day one of the world's largest suppliers of laboratory mice. Two billionaires played a key role in supporting the establishment of the first mouse factory. Roscoe Jackson, then president of the Hudson Motor Car Company, provided the initial seed capital. 'Being a good businessman', Little would later write, 'Jackson saw that [genetically standardised mice] added efficiency, accuracy, and repeatability to biological work'.[11] Subsequently, the Rockefeller Foundation became a main supporter of the laboratory's research in mammalian genetics. The rationale for such funding was suggested by Warren Weaver, then-director of the Foundation's Natural Science Division: 'Although a rational understanding of the behavior of man himself forms the underlying purpose of the program, man is obviously too precious and too complicated an organism to serve often as the experimental material.' Preliminary 'work with small mammals such as mice', Weaver concluded, would constitute 'a necessary step in the general progress toward knowledge of the genetics of that most important mammal, man'.[12]

Experimentation on mice could be the missing link between molecular biology and the eugenic ideals promoted by the Rockefellers. A few years after launching the lab, and while serving as President of the American Eugenics Society, Little suggested that 'many if not all of our major ills of today are dependent on the fact that we have not used our intellect in the making of men as we have in the production of machinery'.[13] Experimentation on mice, he hoped, would fill 'the wide and somewhat awesome gap between the [...] structure of the germ cells in *Drosophila* and the wise direction of a program of improvements in human biology'.[14] Jackson Laboratory would apply the insights deriving from recent research on molecular biology and genetics to the mass production of standardised rodents, testing its potential to direct the evolution of a mammalian species.

In 1932, Little started selling his standardised mice for 10 cents each, plus the cost of delivery and a money-back guarantee in case the animal died in transit. The business started to grow, generating enough money to expand both the breeding facilities and the number of standardised strains developed within their walls. In the following years, the parallel expansion of biomedical research and of the pharmaceutical industry turned millions of animals into living instruments for the development of medical knowledge. The production of experimental bodies gradually turned into a specialised industry, in which Jackson Laboratory was a dominant player. Speaking in front of the US Congress in 1937, on the eve of the first National Cancer Institute Act, Little argued for the necessity to shift research activities – and related funding – 'from working with the slow, unsatisfactory human material to material that is easy to handle, rapid breeding and conveniently controllable in the experimental laboratory'.[15] Not only could mice be utilised as experimental surrogates for human bodies, they had become better than the original.[16]

Around the same time, Little penned an article on *Scientific American* titled 'A New Deal for Mice', begging for public funding. 'Do you like mice?', he asked his readers, 'you don't. "Useless vermin", "disgusting little beasts", or something worse is what you are likely to think as you physically or mentally climb a chair'.[17] Thanks to his programme of genetic standardisation, Little assured, those nefarious pests would become 'the troops which literally by the tens of thousands occupy posts on the firing line of investigation [into the] nature and cure of cancer'.[18] Over the next twenty years, mice production soared. In 1939, Jackson Laboratory distributed over 100,000 mice across the globe. By 1953, the institute was selling almost three times as much. Standardised mice had become such a popular element of modern science that Little would go as far as suggesting to his lawyers that Jackson Laboratory might work towards 'arousing Disney's interest in [producing a] film to tell the story of our mouse (which might easily be a brother or some other relative of Mickey)'.[19]

The movie was never made, possibly on account of the fact that Black 6's peculiar laboratory lifestyle would hardly have been palatable to children. There was, nevertheless, public recognition of the growing importance of genetically standardised mice in twentieth-century technoscience. More than twenty Nobel Prizes have been awarded to scientists who either worked at Jackson Laboratory, or conducted experiments with the standardised mice developed there. Little's mice factory grew into a global conglomerate employing over 3,000 workers, shipping over three million experimental bodies a year, and running an operating budget of over half a billion dollars.[20] Little's programme of controlled evolution continued to be systematically applied to mice, guiding the development of a multiplicity of genetically distinct strains: a population fatter than the norm, another immunodeficient, another with a predisposition to lung cancer. Today, more than 11,000 strains of mice are available for order at Jackson Lab, each customised for a certain use. Black-6, however, remains unquestionably the most popular experimental body on the market, accounting for over half the rodents used in research laboratories around the US.[21]

Genetically each of today's Black-6 mice is a near carbon-copy of their common ancestors over a century ago. This fact has both scientific and economic value as it enables the comparability of experimental results across time. Stopping evolution is a complex labour process in itself. Generation after generation of random genetic mutations tend to add up, risking reshaping and transforming Black-6. During the first forty years of Jackson Laboratory, at least forty different sublines of Black-6 splintered off in this way. Today, guarding against genetic drift is one of the main occupations of molecular biologists employed at Jackson Labs, ensuring that Black-6 remains as it always was.[22] Naturally born mutants are promptly removed from the strain to preserve its contrived genetic purity. Every five generations, a batch of old-style Black-6 is thrown into the mix by defrosting, implanting and birthing cryopreserved embryos from decades ago: a strategic manoeuvre meant to stymie Black-6's natural genetic drift.[23] Genetic purity, an old eugenic ideal, turned out to demand the constant mobilisation of a frozen reserve-army of three million cryopreserved embryos kept in storage at −196 degrees Celsius.[24]

As of September 2021, there are almost 50,000 published research articles in PubMed that mention Black-6 mice.[25] Based on the number of bizarre experiments to which it has been subjected, we may know more about Black-6 than we will hopefully ever know about our own species. Each of these events has added value to the standardised mouse strain first established by Little over a century ago. As a recent manual of *Animal Biotechnology* sums up: 'The more that is known about a strain the more valuable it becomes.'[26] The mouse that never became a Disney's character got flown into space, toured around the world, poisoned with every possible intoxicant, exposed to nuclear radiations, administered electric shocks, sent scurrying through endless labyrinths and chosen to become the second mammal to have its genome fully sequenced.[27] Medical students got them sick and tried to cure them. Pharmaceutical corporations fed them their medicines and, whenever a new chemical cocktail performed as expected on their tiny experimental bodies, the products would be tried on people.

From time to time, things go horribly wrong. In November 1957, the pharmaceutical corporation Chemie Grünental launched a new sedative on the German market. Thalidomide was presented as a non-addictive alternative to opium-based narcotics. Safety tests conducted on standardised mice convinced the company that the drug was safe.[28] Based on these studies, the drug was licensed for over-the-counter sale. It was aggressively marketed as sedative, tranquilliser, and as a medication for colds, flu, nausea and morning sickness in pregnant people. By the mid-1950s, 14 pharmaceutical companies were marketing thalidomide in 46 countries under different trade names. By 1960, it was clear that it had dramatic side effects.[29] What remains to this day one of the most discussed iatrogenic accidents forced governments to review their pharmaceutical licensing policies. 'The House and the public', argued the British Minister of Health, 'suddenly woke up to the fact that any drug manufacturer could market any product however inadequately tested, however dangerous, without having to satisfy any independent body as to its efficacy and safety.'[30] In many countries, a key policy change was that drugs could no longer be approved purely on the basis of animal testing.[31]

The thalidomide case exposed a fundamental problem with the use of standardised mice as experimental objects for drug testing. It was only the most clamorous case of a long history replete with failures. The reason is not difficult to guess: the modern laboratory mouse is a very peculiar creature. First of all, it is a product of the ecological niche imposed upon it: the biomedical laboratory. As one of the most important historians of medicine of the past century suggests, 'if we may define the normal state of a living being in terms of a normal relationship of adjustment to environments, we must not forget that the laboratory itself constitutes a new environment in which life certainly establishes norms whose extrapolation does not work without risk when removed from the conditions to which these norms relate'.[32] There is an even more fundamental problem: a standardised laboratory mouse is still a mouse. Its metabolism is fundamentally different from human metabolism; as a result, cardiovascular studies based on rodents may be fatally misleading when applied to humans. Mouse cancer is not human cancer; oncological research based on animal experimentation often offers false hopes. Rodent immunity is not human immunity; what triggers a cytokine storm in one may be completely innocuous in the other.[33]

The creation of a global market in experimental bodies stimulated a growing debate in the 1960s and 1970s. Many scientists began noting the limitations of animal testing, especially as this became increasingly focused on a single standardised model organism.[34] How much can we still learn from Black-6? Some started to question the mouse's near absolute dominance over laboratory spaces, calling for a wider variety of model animals: 'armadillos for leprosy, prairie voles for autism, finches for language acquisition, leopard geckos and red-eared slider turtles for sex determination, sea slugs for learning and memory, and many more animals of choice for solving one problem or another'.[35] Others pushed for replacing animal testing with newly developed tests using human cells *in vitro* and computer-modelling techniques.[36] Yet – despite the philosophical critiques, political struggles, iatrogenic accidents and technological alternatives – the global market for laboratory mice has been booming, going from $2.7 billion in 2018 to an estimated $3.4 billion by 2023.[37] According to recent estimates, US biomedical research alone employs more than 111 million mice annually, while rodents constitute over 95 per cent of experimental bodies utilised by major laboratories.

The escalating popularity of mice experimentation has been largely driven by genetic engineering. In a paradoxical turn of events, the genetically standardised Black-6 mouse got reshaped in a thousand mutant forms. OncoMouse™, developed in 1984 by Philip Leder and Timothy Steward, was only the most notorious of Black-6 mutant offspring.[38] The birth of OncoMouse™ was a rather complex technoscientific journey, whose every twist and turn is painfully detailed in the section titled 'Production of Transgenic Mice' in the original patent application.[39] A new transgenic breed of mouse was born by the engineered sexual encounter between Black-6, an albino mouse and an activated oncogene. The new mouse would become a standard experimental body in oncological research and a living test for carcinogenic substances. As we saw in Chapter 2, DuPont – set on extracting the maximum possible rent from its newly minted proprietary life-form – demanded a hefty price for each OncoMouse™. Despite much contestation, these practices of rent-extraction persisted until the final expiration of the monopoly patent in 2005.[40] OncoMouse became once again a 'free living being', its freedom consisting in being freely available for exploitation by all pharmaceutical corporations and laboratories.

Meanwhile, thousands of other genetically engineered mice strains have been manufactured in specialised laboratories. A new industry has taken shape. As emphasised by Martin McGlynn, president and CEO of Stem Cells Corporation, 'mice are the backbone of biotechs, pharmaceuticals, and drug development'.[41] One of the leading centres in the search for artificially misfit rodent lives is located at Shanghai's Fudan University. In its 45,000 cages, mice affected by skin tumours live next door to mice that cannot move, while others systematically oversleep. Ultimately, researchers hope to create 100,000 strains of genome-edited mice, each sick in a unique way. 'We plan to mutate 70% of the mouse genome over the next 5 years', affirms Xiaohui Wu.[42] The centre specialises in a genome-editing technique known as 'knock-out'. By randomly disabling the rodents' genes, scientists are churning out hundreds of genetically modified rodents, assembly line style. Myostatin knockout mice present 'gross muscle hypertrophy'.[43] *Arhgef10* knockout mice 'exhibit social impairment'.[44] *Kirrel3*$^{-/-}$ mice have 'autistic-like behaviours, including social and communicative deficits, repetitive behaviours, and sensory abnormalities, as well as hyperactivity'.[45] Sorting through this growing population of knockout mice, research centres hope to individuate at least a few strains worthy of patenting and, hopefully, capable of generating rents on a global scale.

The rush to patent newly minted knock-out mice is likely to finish soon, simply because there will be no more genes to knock out. The mouse genome is composed of around 21,000 genes, and now the Knockout Mouse Project intends to knock out every single one of them, one by one, to reveal their function. 'It's a bit like trying to understand a car engine without a plan', explains David Adams of the Wellcome Trust Sanger Institute, 'piece by piece you pull parts out and then see how this contributes to the car not working.'[46] Mice, however, are not machines and this genetic disassembly can often lead to considerable pain.[47]

The project started in 2006, with contributions from over twenty institutes coordinated by the International Knockout Mouse Consortium, and largely funded through public grants coming from the EU, the US and Canada. By 2020, over $350 million have been spent and around 10,000 knock-out mice have been brought to life.[48] Subsequently, they are shipped to the International Mouse Phenotyping Consortium to undergo standardised physiological tests designed to reveal the function of the disabled gene. The resulting 'knockout mouse archive' is publicly available online: a list with over 20,000 entries spread over hundreds of pages. Selecting any one of the active entries opens a separate file indicating: the targeted gene, the phenotypic characteristics of the resulting knockout mouse, X-ray pictures, measurements, and a button to 'order the mouse from our member's centres'.[49]

These mice have both a use-value and a growing exchange-value. Their bodies are mass-produced, circulated and sold on a global scale. They are then employed by the global pharmaceutical industry, assisting the discovery, development and testing of new drugs and therapeutic products. They are used to define molecular mechanisms that contribute to disease, to validate therapeutic targets, to demonstrate the efficacy of novel drug candidates and to probe the function of specific genes.[50] Engineered mice are finding many new research applications. Yet, the process of production of mutant mice strains has remained substantially unchanged, at least until fairly recently. Mice factories have continued to rely on the laborious technique originally pioneered by Leder and Steward in the 1980s: genetically altering stem cells, injecting them into a mouse embryo, and then breeding multiple generations of animals. The whole process is labour-intensive and, as a result, engineered mice are expensive: going from a few hundred dollars for the most common strains to over 20,000 dollars for a custom mutant.[51]

Today, CRISPR is revolutionising how the pharmaceutical industry produces its favourite living means of production. It offers a relatively simple way to operate targeted genetic surgeries directly on a fertilised mouse egg, with a success rate going from 90 to 95 per cent.[52] The average production time has been reduced from two years to six months.[53] By making the engineering of mice far simpler and faster, CRISPR fosters a radical deskilling of the labour force employed in mice factories. 'When you made knockout mice before, you needed some skills', says Rudolf Jaenisch of the MIT in a recent interview. 'Now, you don't need them anymore. Any idiot can do it.'[54] As a result, it has been estimated that CRISPR has brought down the production cost of mice by 30 per cent, making the average cost of development of a new strain around $100,000.[55]

In the 1920s, inbreeding turned the everyday mouse into Black-6, a standardised model constantly escaping the pull of evolutionary forces by technoscientific means. In the 1980s, genetic engineering transmuted Black-6 into mutant forms, its genome functionally twisted on demand. Today, CRISPR is turning its body into an even more flexible surface. The instrumental value assigned to this reprogrammable experimental body has gradually increased. Genome editing techniques are enabling scientists to explore the function of specific genes by purposefully disrupting them. This expanding knowledge of gene functions offers a series of insights – particularly on the relation between genotype and phenotype – that are often generalised to all mammals, and therefore also to the human animal. CRISPR dramatically increased the value of the mouse as an exploratory probe, pushing further and further the expansion of the genomic frontier.[56]

However, some are beginning to question the future of the laboratory mouse. A better understanding of the complexities of genomics – which are displacing classic reductionist models – has exposed the inherent limitations and risks of translating biological insights across species.[57] Periodically, some event dramatically reminds us of the persistent gap between model animals and the human corporeality they stand in for.[58] Confronted by this long-standing problem, the solution is either to abandon the mouse, or to make it into a better model of the human condition. This is the context in which current research aimed at humanising the mouse has emerged.[59] This line of experimentation is paradoxically rooted in a growing awareness of the limitations of existing methods of animal experimentation, and in the conviction that genome editing technologies offer a convenient solution to the problem.

The humanised mouse embodies the modern dream of an experimental body which is at once human and non-human: non-human enough to be sliced, poisoned, probed and experimented with, but human enough to provide useful insights on how a human body would respond to such treatments. Humanised mice are currently produced using two techniques, which can be used in combination. The first is to engraft human cells onto rodent bodies. In this case, the initial step towards effective humanisation is to develop a severely immunodeficient mouse.[60] Disabling the animal's immune system enables the insertion of human cells, which would be otherwise immediately rejected. The genetically altered NOD SCID strain – produced by Jackson laboratory – is particularly valuable in this context. It was designed to have multiple immunodeficiencies, including defects in T, B and natural killer cells, enabling the engrafting of human cells *in vivo* without risks of immunitary rejection.[61]

The type of engrafted cells largely depends on the pharmaceutical products that need to be tested. For instance, cancerous human tissue can be transplanted onto an immunodeficient experimental body, turning the mouse into a living source of nutrients for the cancerous cells, which can grow and replicate undisturbed. Then, the cancerous human cells can be targeted with experimental chemical treatments to assess their efficacy.[62] Instead, if the humanised mouse is used for vaccine testing, researchers inject human haematopoietic stem cells. Researchers can then observe how the engrafted human cells respond to the immunitary prompt.[63] In 2021, dNovo transplanted human hair stem cells on a naked mouse in order to test its new treatment against baldness. The resulting mouse, nicknamed Spot, shows an all-too-human pattern of baldness: skin everywhere save for a melancholic tuft of hair on one side.[64] A month later, a team of scientists designed a mouse for experimenting pharmaceutical solutions to fecal incontinence by implanting 'a bioengineered innervated human internal anal sphincter in *RAG1*$^{-/-}$ mice'.[65] In all of these cases, the artificial living assemblage of animal and human cells is generally defined as a 'humanised mouse'. The rodent is no longer simply a metaphoric surrogate for the human, but rather a living surface biologically colonised by human cells.

Humanised mice are already used in medical research and pharmaceutical testing.[66] Nevertheless, they continue to have several limitations. Disabling the rodent immune system may protect the human cells, but it leaves the mouse wide open to viral infections and bacterial invasions, making it an extremely fragile and unstable host. Stringent biosecurity protocols are put in place to preserve these valuable animals. They are shipped in sanitised containers, conserved in expensive sterile caging systems, fed carefully hygienised food, and manipulated with disinfected instruments. Yet, the gravest threat to these engineered bodies comes from within them. The engrafted human cells are often able to mount an immune response that attacks the hosting body. Protecting the human cells from the mouse cells leads to the reverse problem: a condition in which the engrafted human cells become a deadly threat to the animal body upon which they depend.[67]

Recent research aims to both supplement and supplant this method of humanisation-through-engrafting. Instead of growing human cells within a mouse body, genome editing techniques can create a laboratory mouse with human genetic characteristics. Bioengineers have coined the term '*sapienisation*' to refer to 'this progressive conversion of mouse systems to resemble the *Homo sapiens* genome'.[68] The objective is simple: 'exchanging mouse genes for their human homologues', a recent review notices, 'has rendered the resultant mouse models more useful to test immunological, viral, and pharmacological outcomes'.[69] The new frontier of genome editing in rodents is making their brains deteriorate as if they had Alzheimer, their blood more vulnerable to leukaemia, and their livers more capable to metabolise drugs like people do.[70]

Humanised mice hold the promise of better pharmaceutical trials. For instance, bioengineers have recently worked on a new strain of mice, expressing many of the enzymes normally present in human embryos. This humanised mouse is promoted as 'a useful model [...] to assess embryonic toxicity of drugs in humans'.[71] Another recent project of sapienisation claims to have carried out 'the largest genetic humanization of the mouse ever attempted', thus creating 'a mouse with a unique type of genetically humanized immune system'.[72] The resulting HumAb Mice™ are advertised as 'highly valuable tools for drug discovery, either as a platform for testing or as a generator of therapeutic drugs'.[73] The biotech company Regeneron utilises 'a proprietary genetically engineered mouse platform endowed with a genetically humanized immune system to produce optimized fully human antibodies'.[74] These engineered mice have been turned into living means of production for human antibodies designed to boost the immune system.

All model animals occupy a paradoxical position in technoscience. Pharmaceutical companies must emphasise both the biological continuity between experimental animal bodies and human bodies (to legitimise their usefulness) *and* their ontological difference (to legitimise their suffering). Genome editing adds a layer of complexity, since this algebra of difference becomes increasingly contingent. A humanised mouse is more useful than a wild mouse insofar as it is physiologically similar to a human subject. Different companies compete to obtain the most humanised mouse. Yet, this race towards ever-more complete humanisation creates new anxieties. Can a humanised mouse become too human? This question has been at the heart of recent debates surrounding a phenomenon called 'off-target human chimerism'.[75] As pointed out by Hiromitsu Nakauchi, professor of genetics at Stanford University, there is always a risk that human cells will multiply not only in the targeted organ, but also in other parts of the mouse body.[76] In the US, Embryonic Stem Cell Research Oversight Committees have been set up to prevent the realisation of scenarios of this kind.[77] Similarly, the British Academy of Medical Sciences has recently issued a report regulating experimentation with animal–human chimaeras.[78]

Human–mouse chimaeras represent the culmination of a historical trajectory and the realisation of a modern dream. Throughout modernity, mouse bodies have been profoundly transformed by their enclosure in laboratory spaces and by their global circulation as living commodities and assets. They have been profoundly shaped by modern science's characteristic preoccupation with uniformity and standardisation, while themselves contributing to a scientific culture relying on comparable experimental data. Genetic engineering, and more recently genome editing, have turned the mouse into a technoscientific mutant designed on demand, produced to order, and shipped around the world to serve the peculiar needs of medical researchers, clinical toxicologists and pharmaceutical corporations. The mouse has become increasingly enmeshed in global socio-economic networks, which now shape its techno-natural history. What futures can we forecast in these experimental bodies, whose life is so thoroughly determined by modern technoscience and capital accumulation?

PHARMING FOR CAPITAL

Manufacturing the Human in Living Bioreactors

In 1982, the birth of OncoMouse™ attracted the attention of the world, promising a valuable living platform to test new therapies against cancer. Six years later, another transgenic mouse came to life and, while it did not attract the same amount of publicity, it sparked a new research trend that is slowly transforming the pharmaceutical industry. It was the first transgenic mouse to produce a human drug via its redesigned metabolism. The strategy used in that landmark experiment was not fundamentally different from the one used in the production of other transgenic mice. As the authors of the experiment laconically reported, 'DNA was microinjected into one-cell fertilised eggs and transgenic mice were derived'.[79] The technique was not revolutionary, but the application was unprecedented: the string of DNA inserted into the mouse genome was designed to stimulate the production of a human protein in the mammary glands of the resulting transgenic mice.

The protein in question was a so-called 'tissue plasminogen activator', a molecule involved in the breakdown of blood clots that is today often used in cases of pulmonary embolism, myocardial infarction and stroke. For the first time in history, the authors of the study reported, transgenic mice were successfully turned into a 'system for the cost-effective production of human therapeutic proteins in milk'. After a century of service as a living platform for *testing* the latest pharmaceutical compounds, the mouse had become a living facility for *producing* drugs. Admittedly, transgenic mice were able to produce only small amounts of the human protein. No attempts were made to scale up production to significant levels. Nevertheless, the authors immediately noticed that the experiment suggested a new system for producing 'foreign proteins in the milk of transgenic dairy animals'.[80]

No name was suggested at the time for this new mode of bioproduction. Today, it is generally known as 'pharming': a neologism that hints at the use of genetically modified farm animals as living pharmaceutical factories.[81] Advanced pharming practices represent a tectonic shift in the history of the pharmaceutical industry, from the chemical to the biological. The blockbuster drugs that made the fortune of Big Pharma in the twentieth century were mostly small molecules: aspirin, paracetamol, antihistamines, statins, as well as the active ingredient in Viagra. These drugs are quite easy to synthesise using chemical reactions. In comparison, proteins produced by human cells are far larger and complex molecules. Aspirin, for instance, is composed of twenty-one atoms; a single monoclonal antibody results from the combination of over twenty thousand atoms.[82]

Consequently, most biological proteins – such as antibodies, enzymes, hormones and clotting factors – cannot be chemically synthesised on an industrial scale. They are simply too large and geometrically complex.[83] Whenever a biological dysfunction prevents a human body from producing the proteins it needs, manufacturing a surrogate molecule requires harnessing the synthesising capability of living cells. In the early twentieth century, blood transfusions represented one of the first successful systems to transfer a large molecule from one living organism into another. A blood donor essentially acts as an external source for plasma proteins that the patient desperately needs. This sort of transaction is facilitated by the fact that a human body can produce blood quickly, and subtracting some of that blood does not usually compromise the donor's health. Many other molecules, however, are produced only in tiny amounts and cannot be easily extracted. For instance, some children struggle to gain height due to growth hormone insufficiency. This molecule is only produced in the brain and it cannot be easily and safely extracted from potential donors. Throughout the twentieth century, the only feasible alternative was to extract it from corpses.[84]

In the early 1980s, recombinant DNA enabled the industrial production of biological molecules of therapeutic value. Genentech pioneered the use of genetically engineered bacteria as living factories for the production of human growth hormone and human insulin. Many biological drugs continue to be produced in the same way: bacteria and yeast are spliced with human genes, injected into industrial vats and fed with sugars, left to multiply and, finally, they are destroyed to harvest the valuable molecules within their cells. This system has inherent limitations. Bacterial cells are extremely simple, at least compared to human cells. As a result, bacterial ribosomes are not always able to fold the complex proteins manufactured by mammalian cells. For instance, it has proven tremendously difficult to utilise genetically modified bacteria to produce complex monoclonal antibodies.[85]

Cell cultures have offered a partial solution to this fundamental problem. Since the early 1990s, the steel bioreactors originally developed for bacterial fermentation have been filled with isolated eukaryotic cells: hamster and mouse cell lines are today widely used as biofactories for therapeutic proteins and antibodies.[86] The cells are genetically modified to express human proteins and induced to proliferate in sterilised vats. Mammalian cells are able to produce most of the proteins that human cells produce, but they are much more delicate than bacteria. Cell cultures are prone to fail due to bacterial and viral infections, or simply because the environmental conditions in the vat are not perfectly set. A small variation in the composition of the broth in which the cells are grown can impede cellular growth. The oxygen and temperature conditions need to be laboriously calibrated, and it is often hard to produce the relevant protein at a sufficiently high concentration.[87]

Despite the limits imposed by the simplicity of bacterial ribosomes and the fragility of mammalian cells, the market for biopharmaceuticals – including recombinant proteins, monoclonal antibodies and vaccines – is growing rapidly. The recombinant protein industry already constitutes around 10 per cent of the drug market, driving 'continued and accelerated growth in the biopharma sector'.[88] A recent report estimated the value of the global biopharmaceutical market to $186 billion in 2017, a figure expected to double by 2025.[89] The active ingredient of the world's best selling drug, Abbvie's Humira, bringing in over $20 billion in sales in 2020 and $62.6 billion cumulatively between 2014 and 2017, is a monoclonal antibody produced by genetically modified Chinese hamster ovary cells.[90] Humira is prescribed for a myriad of conditions from arthritis to colitis, psoriasis, acne inversa and Crohn's disease. It is undoubtedly the most successful biopharmaceutical in history.[91] But it is not an isolated case: ten of the top 25 best-selling drugs are biopharmaceuticals, each generating over $1 billion in annual sales.[92]

Given the constantly expanding market, pharmaceutical companies are investing heavily in the search for scalable and economically efficient systems to expand biopharmaceutical production. As noted by Gary Stix, 'proteins are biotechnology's raw crude'.[93] Even so, the industry has struggled to obtain a steady supply of protein-based drugs from either genetically engineered bacteria or mammalian cell lines. Starting from the early 1990s, global demand for monoclonal antibodies often outstripped industrial capacity.[94] The development of cloning techniques, combined with genome editing, offers a way to replace industrial bioreactors with living, breeding organisms. Pharmaceutical-producing cultured cells are no longer confined in industrial steel vats, but rather they grow within the bodies of insects and mammals.

In the early 1990s, Genzyme was one of the first pharmaceutical corporations to invest in the search for animal bioreactors.[95] It launched GTC Biotherapeutics, a spin-off dedicated to realising the industrial vision presented in a patent application titled 'Isolation of Exogenous Recombinant Proteins from the Milk of Transgenic Mammals'.[96] The application, whose owner was promptly recruited by Genzyme, described an industrial process for 'producing large quantities of recombinant protein products in the milk of transgenically altered mammals'. The resulting patent granted monopoly ownership over the precise recombinant technique as well as over any 'transgenic mammal that produces the desired recombinant product in its milk'.[97] The company was set up to explore the possibility of turning this proprietary idea into an industrial production process for recombinant human proteins.

To demonstrate the potential of the new technology, GTC's first goal was to develop a herd of transgenic goats designed to produce a molecule called antithrombin: an anticoagulant produced by the human liver. The recombinant technique chosen – generally known as microinjection – had already been widely experimented in the production of transgenic mice. First, the company's scientists isolated the gene coding for human antithrombin, splicing it with a mouse promoter gene (a DNA sequence that controls when and how an associated gene is expressed). Then, they injected this chimeric DNA assemblage into fertilised goat eggs, which were subsequently implanted in the wombs of pseudo-pregnant goats. The resulting transgenic animals carried a genome composed of goat, human and mouse genes. The mouse promoter serves a key function, activating the expression of the human gene only in the mammary glands. The cells composing the goats' udder are induced to produce antithrombin: mixing a new ingredient in their milk. Subsequently, the protein is extracted from the milk, purified, packaged, stamped with a trademark name, and brought to market.[98]

In 2009, ATryn™ became the first drug produced in genetically modified mammals to be approved both by the European Medicines

Agency and the US FDA. Laboratories around the world are working on the next generation of genetically modified animals, capable of producing pharmaceutical commodities in various bodily fluids from goat's blood to boar's semen, camel's urine and chicken egg yolk.[99] In 2014, the US FDA approved the commercialisation of Ruconest, a drug produced in the milk of gene-edited rabbits owned by the Dutch company Pharming Group.[100] In 2015, it was the turn of a recombinant protein produced in the eggs of transgenic chickens developed by Synageva (later acquired by a subsidiary of AstraZeneca).[101] In 2019, a vaccine manufactured by genetically modified insect pupae has been submitted for the first time to the European Medicines Agency for regulatory approval.[102]

Algenex, a leading Spanish biotech company, is hoping to turn insects into scalable and cheap biofactories for vaccines. The insects are infected with a virus modified to carry the foreign DNA into the insect cells. Once the pupae are infected and begin producing the protein encoded in the viral DNA, their bodies are mechanically crushed. The resulting juice is purified and packaged.[103] Currently, Algenex is only seeking approval to market its vaccine as a veterinary commodity. However, the company hopes that this will soon open the door to the development of human vaccines. 'The first vaccine on the market with our technology', declared Claudia Jiménez as general manager of the insect-based pharma company, 'will be the key'. In 2020, the company started working on a protein-based vaccine against Covid-19, also to be produced by genetically modified moth pupae. 'Each insect', anticipates Jiménez, 'has up to 27 doses of the Covid-19 vaccine, and it doesn't cost anything to raise them'.[104]

Pharmaceutical companies are keen to project a future in which genetic modification will turn living organisms into a source of all sorts of valuable pharmaceuticals. However, pharming remains expensive, and the industry has so far struggled to make it sufficiently profitable. GTC Biotherapeutics, for instance, has not profited from its genetically engineered goats. 'We have incurred net losses in each year since our inception, [and] we expect our expenses will exceed the revenue we collect from sales of ATryn for at least the next several years', reads a filing to the US Securities and Exchange Commission.[105]

Despite the lack of commercial success, the search for new pharming strategies continues. To reduce the cost of developing new strains of animal bioreactors, pharming corporations have recently turned to cloning.[106] GTC has experimented with cloning technologies to increase the efficiency of its genetic engineering practices by duplicating the most productive goats. Similarly, the Roslin Institute – which achieved worldwide fame through the birth of Dolly – has been pushing its cloning technologies as an alternative way of producing transgenic pharm animals. Already in 1997, Roslin showcased this possibility by producing Polly, a sheep producing a human blood-clotting protein in its milk.[107] Going forward, new genome editing technologies may further bring down costs. According to several experts, CRISPR will soon become the 'chief means for generating genetically modified livestock to be used in producing biomedical products, which should greatly reduce the cost and increase the speed with which livestock may be engineered'.[108] However, since on average the development of a new drug takes well over a decade, it may be too early to assess the extent to which CRISPR will transform the pharming industry.

In a way, pharming appears to represent an idyllic return to nature for the pharmaceutical industry by which open barnyards in the countryside take the place of sterilised laboratories in metropolitan areas, goats replace cylindrical bioreactors made of steel, and workers get to wear overalls rather than laboratory coats. Indeed, some of the scientific pioneers of this perplexing neo-pastoralism have presented pharming exactly in terms of a victory of nature over technology, of biology over mechanics, of bodies over machines. In an article published in *Scientific American* under the title 'Transgenic Livestock as Drug Factories', the authors of a research programme that led to the creation of bioengineered pigs designed to produce human protein C in their milk, stress that mechanical bioreactors can scarcely approximate the productive capacity of a complex living organism. 'The mammary gland proves far superior to any cell-culture apparatus ever engineered for these tasks', the authors note, since it is 'optimised to maintain a high density of cells, to deliver to them an ample supply of nutrients and to channel the valuable proteins produced into an easily harvested form'.[109]

These emergent qualities cannot be easily replicated in an artificial metallic bioreactor. Moreover, both engineered bacteria and cultured cells must be cyclically destroyed to extract the recombinant proteins within their cells. They are single-use living means of production. Animal systems, instead, can be designed to excrete the desired proteins in their bodily fluids so that production can continue uninterrupted. Individual goats age and die, but the herd may be conserved as an endlessly self-reproducing productive facility. 'Despite all their efforts to improve industrial cell-culture facilities', the genetic engineers conclude, 'it turns out that a generation of bio-chemical engineers were unable to match the abilities of a tool for making proteins that nature had already honed'. [110] The article, following a well-worn trope in the biotech world, ends in a celebration of the wisdom of biology over mechanics.

Yet there is something uncanny in this triumph of nature driven by contemporary capitalist technoscience. While pharming displaces the production of pharmaceutical commodities from the factory to the farm, the latter – and everything it contains – is thoroughly transmuted in the process. In the visionary gaze of the authors of the previously cited study, pigs appear at first as 'experimental sows', and then as 'transgenic livestock bioreactors'; the mammary gland is 'a tool for making proteins'; while the whole process appears as a 'chimerical vision'.[111] These descriptions appear uncanny, in the proper Freudian sense, insofar as they reveal that beyond the façade of something that appears thoroughly ordinary, a mutation has occurred. For centuries, animal lives have been turned into instrumental tools for capital accumulation, and their metabolic activities harnessed as forces of production. There is nothing new about that. In thousands of industrial farms all around the world, goats are fed fodder which is metabolically transformed into milk. Milk is extracted, shipped to markets and turned into capital.

In traditional farms, the living organism remains a biological blackbox, whose hidden internal metabolism is mostly taken as given. Domestication has certainly transformed the bodies of cows, pigs and chickens, but only as a result of centuries of diffused universal labour and only by selecting for random mutations that could not be designed or planned. A single farmer could not design their own living means of production or corporealise a clever business idea. Only today, does the body become an open platform of production, whose metabolic system can be engineered. Corporations routinely design their living means of production based on pressures from fluctuating global markets. Business plans are incarnated into animal bodies, whose physiology comes to reflect abstract calculations performed by corporate strategists, synthetic biologists and metabolic engineers.

Farm labour is similarly transformed. Milking always referred to a relatively simple process of appropriation by which farmers extract the products of a body over which they exercise a purely external form of control. Farmers feed their goats, they appropriate their milk, they shape their environment, but they cannot design their internal metabolism. Pharmers utilise genetic power-knowledge to shape the metabolic processes of the animal lives they attend to. Metabolic engineering complements, rather than replaces, traditional methods of environmental control. Transgenic animals used as pharmaceutical factories must be subjected to even stricter disciplinary regimes to prevent infections. To this end, they must be kept in enclosed, contained and carefully sanitised environments. Farmers have been taking possession of animal products for centuries. Pharmers, on the other hand, penetrate the body to establish control over the entire metabolic process of production.

A MARKET FOR XENOTRANSPLANTATION

Harvesting Humanised Organs

The recent emergence of corporations dedicated to the industrial manufacturing of humanised organs for xenotransplantation represents one of the most striking new frontiers of accumulation opened by genome editing technologies. Obviously, transplantation is nothing new. The first successful kidney transplantation took place in 1954 and, starting from the 1980s, the operation has become increasingly commonplace. In the following years, surgeons successfully moved thousands of livers, hearts, lungs, pancreas and intestines from one body into another. As the success rate increased, demand increased. Today, 150,000 solid organs are transplanted every single year, but the World Health Organization (WHO) has estimated that this covers only about 10 per cent of global demand.[112]

In the US alone, 30,000 procedures are carried out annually, while the waiting list counts over 100,000 people. Seventeen people die every day from the lack of available organs for transplant. In the European Union, the situation is only marginally better: in 2019 more than 58,000 patients were waiting for an organ transplant.[113] There is undoubtedly a global shortage of donors. This has led several countries to launch campaigns promoting organ donations as well as regulations that put in place opt-out systems in which everybody is considered an organ donor, unless otherwise stated.[114] Neoliberal economists, on the other hand, have lobbied for a liberalisation of the international trade in organs, a proposal that would make the likelihood of receiving a transplant directly dependent on a patient's relative economic status.[115] Medical research in xenotransplantation has long offered the prospect of a technical solution to the problem, enabling the use of organs harvested from various mammals as a surrogate.

Surgeons have been attempting to put simian kidneys, livers and hearts into patients since the 1960s. They had little success. Patients generally died within a few months, mostly because the patient's immune system attacked the alien organ. But the idea of xenotransplantation persisted. In 1984, a terminally ill infant received a baboon's heart into her chest, which kept pumping for 21 days before giving in.[116] In the late 1980s, pigs started to be considered the most promising source of organs for future experiments, given the similarity between human and swine physiology. Several attempts were made throughout the 1990s, all leading to rapid death.[117] Then, a medical surgeon at the University of Alabama singled out α-gal as the main obstacle on the road to xenotransplantation. This sugar is present on the surface of all porcine cells, and is the main trigger causing adverse immunitary reactions in humans.[118] In the following years, major corporations invested millions of dollars to speed up the search for an effective way of developing α-gal-free pigs. Among others, Novartis began to heavily invest in xenotransplantation research, budgeting over $1 billion to support scientific research in the field.[119] Other companies, such as Genzyme (which would soon develop its pharma goats) and PPL Therapeutics (which played a prominent role in the pharming business), developed their own research programmes. In 1996, Salomon Brothers predicted that the global market for transgenic organs could reach $6 billion by 2010.[120]

The corporate excitement was short-lived. No effective methods to eliminate α-gal from porcine cells were found, and a new insurmountable problem arose: retroviruses. These pathogens, whose most notorious representative is HIV, integrate their genetic material into the genome of the cells they infect. There they lie dormant, conserving the potential to reactivate themselves in ways that remain unclear.[121] In the early 2000s, the sequencing of the pig genome revealed it to be sprinkled with dozens of dormant porcine endogenous retroviruses (PERVs). Studies conflicted as to whether they could become active once placed in a human host. While the ability of PERVs to infect isolated human cells *in vitro* has been demonstrated, no PERV infections have so far been reported in a living human subject.[122]

It is, however, a possibility that cannot be easily brushed off. The consequences of a porcine retrovirus being activated in a human host's body are unpredictable. Not only could it cause serious illness in the individual, it could also create the risk of a zoonotic spillover – not unlike the one that originated both the HIV and the Covid-19 pandemic.[123] In the early 2000s, corporate investments in xenotransplantation began to fade. In 2004, Novartis cut off its investments in xenotransplantation research. Most other companies followed suit. The challenges seemed too daunting, with no feasible solutions in view. *Fortune Magazine* captured the twilight of a corporate dream with the headline 'Novartis to Pigs: Keep Your Kidneys'.[124] For the next decade, xenotransplantation was consigned once again to the realm of science fiction.

Recent advances in genome editing technologies have reversed this trend. CRISPR enables scientists to edit the pig genome more quickly than ever before. Suddenly, the two main obstacles to a global market in animal organs appear surmountable. Genome editing technologies have been used to knock out the sugars that coat porcine cells. Several companies have already developed so-called gal-knockout pigs. 'Our first gal-knockout pig took three full years', says a transplant surgeon at Indiana University. 'Now we can make a new pig from scratch in 150 days.'[125] In December 2020, the US FDA approved one of these knockout pigs – developed by Revivicor and trademarked as GalSafe™ – as a marketable product both for human consumption and for biomedical applications such as xenotransplantation.[126] The company has, subsequently, worked to knock-out other protein antigens that could trigger the immune system.

Recently, the new genomically designed pigs have begun to be harvested for their organs. A number of trials have been carried out using baboons as living recipients. All of them died, although one survived nine months with a pig's heart beating in his chest.[127] The experiments were widely hailed as record-breaking achievements that rekindled the hope for a future in which companies will flood the market with humanised animal organs. Public authorities started accepting demands for new human trials. In 2021, surgeons at New York University transplanted kidneys harvested from gene-edited pigs into two legally dead people. The organs appeared to function normally, at least as long as the brain-dead recipients were sustained by ventilators.[128] On 7 January 2022, surgeons at the University of Maryland Medical Center received the green light to perform the first clinical trial on a living patient. A pig heart was placed in David Bennett, a 57-year-old volunteer. Ten days after the operation the patient seemed to recover. 'The new heart is still a rock star', said Bartley Griffith, who led the transplant team. 'It seems to be reasonably happy in its new host. It has more than exceeded our expectations.'[129] The possibility of an immune rejection, however, remained; as well as long-lasting concerns that the transplant may cause the activation of retroviruses buried in the pig's genome. The pig's heart, in fact, had not been voided of PERVs genetic sequences.[130]

In both landmark surgeries, the organs were provided by Revivicor, which has positioned itself as the main corporate player in xenotransplantation. The company, originally a spin-out of PPL Therapeutics, was purchased by United Therapeutics in 2011. Since then, it has built an advanced pharming facility in Alabama, and secured an international patent that covers its own brand of 'multi-transgenic pig for xenotransplantation', 'methods of making such animals', as well as any 'organs, organ fragments, tissues and cells, derived from these animals'.[131] The pig heart implanted in David Bennett's body was harvested from one of these patented pigs, whose genome had been subjected to ten modifications. First, Revivicor's scientists knocked out three genes, which normally enable pig cells to synthesise sugars known to trigger adverse immune responses. Then they knocked out the gene for a growth hormone receptor, reducing the chances that the pig heart will continue growing beyond the limited capacity of the human chest. Finally, they knocked in six human genes aimed at reshaping the porcine organ to more closely resemble the human heart. The resulting chimeric organ is now registered under the trademark Uheart™.[132]

'This has been a crazy, exciting week', reported Revivicor's chief executive David Ayares in an interview. The CEO is optimistic that the operation will open a new market in which chimeric organs will be routinely harvested from humanised pigs and sold as trademarked commodities. 'This is not a one-off', he pointed out in another interview. 'We're going to take this all the way through to human clinical trials, and hopefully have an unlimited supply of donor organs.'[133] To realise its business vision, Revivicor is building a second, much larger pharming facility in Virginia. It plans to supply hundreds of organs a year, which could be delivered quickly wherever needed using an onsite helipad. The company hopes to complete constructions of the new facility by the end of 2023, and then apply to the FDA for a full clinical trial.

When the media announced the first xenotransplantation, investors quickly saw how genetically modified pig bodies could be turned into sites of capital accumulation. 'As breakthroughs such as xenotransplantation become more common in health care, we are going to see massive amounts of money flow into this industry', one investor remarked: 'As venture capitalists, we can see this trend in its infancy and start to position ourselves right now'.[134] So far, Ayares has declined to answer questions concerning how much each pig costs, and what will be the price of the harvested organs. Meanwhile, Revivicor's mother-company United Therapeutics keeps soaring on the financial market. Its stocks went up almost 25 per cent in 2021, approaching a market capitalisation of $9.3 billion. Bennett's surgery was a milestone for the entire xenotransplantation industry. XVIVO, a Swedish biotechnology company who produced the 'heart perfusion device and proprietary solution' which preserved the heart prior to transplantation, saw their stocks soar by 15 per cent in one day, reaching a market capitalisation of $925 million.[135]

Twenty days after the transplant, however, the narrative surrounding xenotransplantation took a turn for the worse: a test indicated the presence of porcine cytomegalovirus DNA in Bennett's body. After six weeks, the patient developed a fever, while tests showed a marked rise in the quantities of viral DNA circulating in his body. After two months, he died. While investigations continue, cytomegalovirus has been indicated as the most likely cause of death: the viral infection in the heart causing an adverse immunitary reaction. Revivicor has so far declined to comment on the death, and has made no public statement about the virus. Mike Curtis, CEO of a competing company that is also breeding humanised pigs for xenotransplantation, took the opportunity for a subtle critique: 'It was surprising. That pig is supposed to be clean of all pig pathogens, and this is a significant one'.[136] The accident has reawakened anxieties towards the potential of future zoonotic pandemics being triggered by xenotransplantation practices. Porcine cytomegalovirus is considered to be highly species-specific: the infection slowly taking over the pig's heart without spreading to the surrounding human cells. Yet, this unexpected event indicates the possibility of other viruses lurking in porcine cells, posing new sorts of risks.

These concerns are unlikely to stop future investments in xenotransplantation. Griffith reaffirmed that experimentation on xenotransplantation will continue: 'This doesn't scare us about the future of the field. [...] It's just a learning point. Knowing it was there, we'll probably be able to avoid it in future.'[137] Many other companies are pursuing xenotransplantation as a business. The challenge of retroviruses has been recently tackled by eGenesis, a company co-founded by George Church of Harvard University. In 2016, Church's team isolated sixty-two PERVs hidden throughout the pig's genome. They utilised CRISPR to knockout all of them, generating a new strain of PERV-free patented pigs.[138] In March 2021, the company raised $125 million and announced to be preparing a human trial for its trademarked HuCo™ organs. Among the financiers are major investors in healthcare, including Farallon Capital Management and Bayer's venture capital branch Leaps.[139] Meanwhile, Miromatrix Medical, which aims to bioengineer pig kidneys and livers for human transplantation, went public with a $43 million offering.[140] NZeno has begun breeding miniature pigs, whose organs are designed to remain human-sized without the anti-growth-hormone modification used by Revivicor.[141] 'Big pork' is also increasingly interested in the emerging market for humanised porcine organs. Smithfield Foods, the world's largest pork producer, is running a biosciences unit to develop pig organs for xenotransplantation as part of a public–private consortium established through a $80 million grant from the US Department of Defense and over $214 million in federal funding.[142]

In the early 2000s, Nancy Scheper-Hughes described the growing 'global traffic in human organs' as resulting from 'a rapacious demand for scarce organs and tissues', which was bound to create 'a division of the world into organ givers and organ getters'. 'Global capitalism and advanced biotechnology', she wrote in an anthropological investigation of this dark global market, 'led to new medically incited "tastes" for human bodies, living and dead, for the skin and bones, flesh and blood, tissue, marrow, and genetic material of the other'.[143] While Scheper-Hughes pieced together an increasingly gothic portrayal of the sellers, buyers, brokers and networks that compose the global market in human organs, a number of international meetings convened to address the issue. In November 2000, the United Nations introduced the Palermo Protocols, which condemned organ trafficking and demanded its immediate dismantling.[144] In the midst of neoliberal globalisation, and despite some fringe demands for a free global market in organs, international law still imposes a limit to business.[145] Yet, the trade still goes on, burrowing innumerable invisible logistical channels connecting poor sellers in need of cash (mostly residing in the Global South) to sick buyers in need of organs (mostly residing in the Global North).

Revivicor's trademarked humanised pigs represent both a market response to insufficient organ donations and a technoscientific response to the legal limits imposed on the trade in human flesh. Humanised pigs have a major business advantage over all-too-human organ *sellers*. They are sufficiently humanised to be a source of hearts, kidney and corneas to quench the growing thirst for transplants; while remaining sufficiently inhuman to avoid turning the whole business into a criminal activity. They have also, at least from the perspective of companies such as Revivicor, a major advantage over human organ *donors*. An increase in organ donations – the objective of many public campaigns – would represent a non-market solution to this persistent problem. Clive Callender, a surgeon at Howard University College of Medicine in Washington, has been promoting for over two decades the National Minority Organ Tissue Transplant Education Program: a grass-roots effort emphasising community education and empowerment to 'increase organ and tissue donations within multiple ethnic minority groups (i.e. African Americans, Hispanics/Latinos, Asians, American Indians, and Alaska Natives)'.[146] The programme was funded by the National Institute of Health with $10 million between 1993 and 2008. This relatively small investment contributed to double minority donations percentages from 15 to 30 per cent.[147] Since, Callender has tried to obtain more funding to scale up to the national level: 'a model national donor education program […] is a worthy investment whose benefits greatly outweigh the underfunded support currently provided'.[148] Public investments in this direction would expand organ donations and promote social practices of solidarity and mutual aid.

Investments in corporate xenotransplantation ventures, instead, secures a further expansion of international trade and the creation of a new medical-industrial sector holding power over life. It opens frontiers of capital accumulation that have been so far foreclosed. It secures the further commodification of both human and animal bodies. It cements the neoliberal dictum that presents the expansion of commodity trade as the solution to each and every socio-political issue. The entire research endeavour is formulated within market terms and synchronised to the solipsistic dance of finance. Taken together, the contingent histories of humanised mice, pharma-goats and GalSafe™ pigs reveal the extent to which market pressures are increasingly turning health into an 'industry'. Framing issues of organ availability in the neoclassical tropes of 'scarcity' obscures much broader questions regarding the philosophy and practice of health. It also raises questions about the role of finance and patents in shaping what technologies get developed and how. Is the hype around xenotransplantation directing resources away from establishing more equitable health care systems and sustaining existing practices of organ donation? Might it further entrench corporate control in medicine?

8

Bioengineering the Human

Human Genetic Capital in a Neoliberal Environment

Bioengineering takes possession of the living vectors that make and unmake the world. It encompasses a set of techniques, elaborated in the last half a century, which offers new ways of genetically altering living bodies, redesigning metabolic pathways, and thereby indirectly affecting socio-ecological processes at ever-larger scales. Just like viruses replicate by redirecting protein synthesis in infected organisms, capital accumulation is increasingly conducted through the hijacking of living cells. As we have seen, this viral strategy of bioproduction is increasingly central to business operations in a number of sectors – including agriculture, aquaculture, livestock breeding, the pharma industry, the chemical industry, the textile industry, etc. Not only is capital claiming monopoly power over manufactured living artefacts, but life is increasingly turned into a means of manufacturing.

Humans are animals too. They are living organisms composed of trillions of cells, each relying on their genetic composition to direct catabolic and anabolic processes.[1] Like all living beings, humans are metabolic agents constantly interacting with their environmental circumstances, re-shaping their lived ecology in the process. 'Labour', writes Marx, 'is first of all, a process between man [*sic*] and nature, a process by which man, through his own actions, mediates, regulates and controls the metabolism between himself and nature. He confronts the materials of nature as a force of nature. He sets in motion the natural forces that belong to his own body, his arms, legs, head and hands, in order to appropriate the materials of nature in a form adapted to his own needs. Through this movement he acts upon external nature and changes it, and in this way he simultaneously changes his own nature.' This metabolic labour 'is the universal condition for the metabolic interaction between man and nature, the everlasting nature-imposed condition of human existence'.[2] Human beings participate in the construction of the world *together with* other living organisms, which are collectively characterised by common metabolic processes that involve the constant translation of DNA into mRNA into proteins. 'The concept of metabolism', writes Marx in one of his last works, 'suggest[s] a dialectical interaction between nature and society' and 'posits both the human beings and the non-human world as active, indeed interactive agencies'.[3]

Many of the genome engineering techniques already deployed to govern non-human lives are being redeployed in order to steer the human metabolism. This chapter follows molecular production as it penetrates beneath what Silvia Federici has called 'the periphery of the skin', and analyses the multifold social implications of this process.[4] The first section is dedicated to the rapid development of genetic vaccines during the Covid pandemic. This molecularisation of vaccine production enabled a faster pandemic response, while causing political controversies surrounding corporate control over these new productive technologies. The second section interrogates recent advancements in gene therapy. The transformation of genome editing into a therapeutic procedure is triggering novel political struggles concerning issues such as: the unequal access to health caused by skyrocketing rates of profit and multiple layers of rent extraction; the risks associated with direct manipulations of the human genome; and the porous boundary between gene therapy and genetic enhancement. The third section reviews recent experimentation in heritable human genome editing. It analyses the multiple political debates it has recently sparked and the post-humanist imaginaries it has inspired. It concludes by scrutinising the signs of a 'liberal eugenics' to come.

GENETIC VACCINES

Hacking The Human Cell

All life-forms are defined by a degree of biochemical autonomy: the metabolic capacity to produce the molecules and energy needed to sustain themselves. Viruses, by contrast, must colonise living cells to extract the raw materials necessary for nucleic acid and protein synthesis, for reproduction and replication. Without a host cell, viruses are non-metabolic entities: they do not consume energy, they do not transform matter, they do not interact with the surrounding environment. They simply drift as pollen in the wind. Only once they establish contact with living cells do they become active agents: injecting their genomes in the host cell, and inducing the cellular machinery to dedicate itself to assembling copies of the virus.[5] Given their ability to direct the metabolism of the cell by hijacking the cellular machinery, microbiologists often call viruses 'nature's genetic engineers'.[6] Viruses are parasitic agents, ontologically reliant on the multitude of life-forms that compose the biosphere. This is why molecular biologists tend to exclude viruses from the realm of the living: a paradoxical form of non-life, which can acquire 'a kind of borrowed life' only by systematically exploiting the metabolic potential characteristic of living cells.[7]

The borrowed lives of viruses greatly influence global ecologies. Viral infections can force sudden changes in the metabolic activities of the living organisms that inhabit a particular environment. This can shift the balance of organic molecules being consumed and released, alter biogeochemical cycles and thus shape global geographies in profound ways. Take, for instance, ocean ecologies. In the last thirty years, marine biologists have increasingly emphasised the key role played by viruses in influencing ocean productivity and planetary biogeochemical cycles.[8] Viral pandemics are the norm in the strange world of the ocean microbiome. Every day, 20 to 40 per cent of marine bacteria are killed by omnipresent viral infections, releasing organic nutrients in a process known as 'viral shunt'.[9] Viruses control marine microbial populations, while facilitating nutrient and carbon cycling. Reflecting on this phenomenon, a prominent marine biologist recently noted that 'even though viruses aren't alive, they are significantly altering the course of life every day in the environment'.[10]

Beyond the oceanic zone, viruses play a pivotal role in shaping the biosphere, profoundly affecting microbial food webs, prokaryotic biodiversity and global biogeochemical cycles.[11] An estimated 10 nonillion viruses exist on our planet, just about 100 million times more than there are stars in the universe. According to a recent calculation, if they were columned one on top of the other, they would stretch for 100 million lightyears.[12] They occupy every niche of the biosphere, proliferating in seawater, drifting through the atmosphere and lurking in the deep soil. Only an infinitesimally small fraction of them interact directly with human bodies. Even among the few that do infect human cells, the majority are unnoticeable: they only slightly alter the human metabolism and avoid triggering the immune system. They are ubiquitous, persistent and innocuous, although they do influence human societies in subtle and hardly understood ways.[13] Sometimes, however, a virus develops whose cycle of parasitic replication disrupts the host's bodily processes. The viral infection becomes clear and detectable, causing a dramatic change in the functioning of the cell; sometimes even causing metabolic shifts that threaten the whole organism. If such viruses replicate rapidly, they can spread exponentially: an infection can spark an epidemic, an epidemic can grow into a pandemic. The virus not only disrupts the metabolic activity of a single organism, it also affects the total socio-ecological metabolism.

The recent Covid pandemic is a dramatic example of how viruses continuously shape global ecologies as they emerge from the metabolic interaction of millions of living organisms. The first reported case of the novel coronavirus was officially notified by the WHO on 31 December 2019. By early March, the infection had already spread across 114 countries, with 118,000 reported cases and 4,291 deaths. At this point, the WHO officially declared the emergence of a global pandemic, which triggered a number of legal, political and economic responses. This, in turn, led to a chain of events that drastically shifted the global socio-ecological metabolism. In 2020, the negative impact on industrial production and the disruption of global supply chains translated itself into a 6 per cent reduction in global emissions – a slowing of fossil capital's carbon metabolism.[14] The crisis, however, did not have an equal impact on everyone everywhere. On the contrary, it is today widely recognized that it led to a dramatic increase in global inequality, leading the UN Secretary-General António Guterres to denounce the existence of an 'inequality pandemic'. 'COVID-19 has been likened to an X-ray', Guterres reported in a public lecture, 'revealing fractures in the fragile skeleton of the societies we have built'.[15] Recent estimates have shown that in 2020 the number of people living on less than $5.50 per day increased by over half a billion, while millions lost their jobs. Meanwhile, the stock market boomed. In the last ten months of 2020, billionaires accumulated over $3.9 trillion, with the ten richest people amassing over $540 billion.[16] The pandemic has exposed and exacerbated existing class, gender and racial inequalities.

In the first months of the pandemic, most governments responded by restricting freedom of movement, deploying two main legislative instruments: compulsory quarantines and tighter border controls. The United States represented a major exception to this trend. Then-President Donald Trump declined to impose strict restrictions on freedom of movement, and criticised state-level administrations that chose to do so. Instead, the White House launched Operation Warp Speed, a $18 billion public–private partnership devoted to fast-tracking the development of a vaccine.[17] Even in those countries that implemented more stringent lockdowns, these were mostly seen as temporary solutions aimed at slowing down the spread of the virus, while economic and human resources were mobilised in order to speed up the search for a vaccine. The development of a vaccine promised a solution to the pandemic crisis that would enable states to restore global logistics, freedom of circulation and economic growth.

The origins of this governmental rationality, according to Michel Foucault, could be traced back to the nineteenth century, when national vaccination programmes emerged as novel instruments of social mobilisation against epidemic disease. Until then, the measures taken by early modern states in order to control epidemics were mostly based either on 'exclusion' or on 'confinement'. Lepers would be banned from the city to preserve life within. In case of plagues, the infected would be confined in their homes in order to prevent the spread of the disease. In *Security, Territory, Population*, Foucault shows how the introduction of vaccines in the early nineteenth century enabled a new mode of government no longer focused on 'the problem of exclusion, as with leprosy, or of quarantine, as with the plague'.[18] As cities became more reliant on global trade and logistics, governmental models based on 'exclusion' and 'confinement' became increasingly difficult and costly to implement. Starting from the nineteenth century, vaccination campaigns enabled a new way of governing epidemics that would minimise disruptions to logistical transportation and global commerce. They secured the perpetuation of a world-market system, relying on a regulated but constant movement of people and goods.

In his writings, Foucault offers two important suggestions. First, vaccines have played a fundamental role in the history of modernity, enabling the management of epidemics within the context of a society based on freedom of trade and freedom of movement. Second, the origins of vaccination as a technique are not to be searched within the science of medicine.[19] What is missing from Foucault's account is a study of the process of social production of vaccines, both as an innovative medical instrument and as a valuable commodity. If vaccination did not emerge from within medical history, where did it come from? To answer this question, we have to look elsewhere. In the archives of London's Wellcome Collection, visitors can request to observe an etching titled *L'Origine de la Vaccine*, sketched by François Depeuille in 1800.[20] As the title suggests, it depicts the process that led to the publication of the seminal *Variolae Vaccinae* by Edward Jenner, and the subsequent development of the first vaccine against smallpox. Right at the centre of the picture, an imposing cow stands, gazing towards the observer. Around her, four characters engage in transactions. On one side, a milkmaid is examined by a surgeon; on the other, a merchant offers a packaged vaccine to a finely dressed man. On one side, a transaction of knowledge; on the other, a commercial exchange.

L'Origine de la Vaccine captures the series of social transactions that enabled the production of the first vaccine, and subsequently of most other vaccines. Firstly, we are presented with a circulation of knowledge from the people to the medical profession: in 1768, an unnamed milkmaid suggested to Jenner that people in her profession, who were routinely exposed to the relatively innocuous cowpox virus, were seldom affected by smallpox epidemics. Thirty years later, the English surgeon tested this popular knowledge by extracting the cowpox virus from the hand of a milkmaid named Sarah Nelmes and inoculating it into the arm of a nine-year-old. The child was then repeatedly exposed to the smallpox virus, without developing the disease. It was the first scientific proof that it was possible to induce immunitary protection from smallpox by exposing the body to a similar, but innocuous, viral agent.[21] This method, in fact, was already practised by unlicensed farmers such as Benjamin Jesty, whose pioneering experiments in vaccine inoculations would be later recognised by the *Edinburgh Medical & Surgical Journal*.[22] It was a partial acknowledgment of the fundamental role of popular knowledge in the construction of medical expertise. Yet, neither Benjamin Jesty, nor Sarah Nelmes, nor the anonymous milkmaid would ever be called to share the significant monetary rewards that Jenner later received as the publicly sanctioned inventor of the vaccine.[23]

At the end of the eighteenth century, informal popular practices were converted into sanctioned scientific knowledge. Gradually, medical science began to grasp the principles that make vaccination an effective antiviral strategy: introducing a weakened version of a particular virus triggers the infected cell to produce antibodies. That knowledge was then transformed into an increasingly organised system of commodity production. Several private companies began commercialising cowpox as a vaccination against the smallpox virus. In the 1870s, vaccine farms were established both in Europe and in the United States. By the end of the 1890s, the marketing of glycerinated calves' lymph as a vaccine against smallpox was well established.[24] The production process required spreading the cowpox virus among cattle, harvesting lymph from the infected calves, purifying it, and selling the final product.[25] Cows were absolutely central to early vaccine production. They were living sites of viral fermentation in which the cowpox virus was injected and made to replicate, to be then extracted and turned into packaged vaccines. Indeed, the term *vaccination* simply indicated the extraction of viral matter from cows, whose Latin name is *vacca*.

Despite the planning of increasingly intensive vaccine farms, the *vacca* was not an easily scalable means of vaccine production. For a short time, ferrets and mice became the preferred animal bodies for cultivating selected viruses. Meanwhile, the sciences of immunology and vaccinology emerged, and new manufacturing processes were established. In 1931, Ernest Goodpasture introduced a new method of cultivating viruses in chicken eggs by injecting a live virus into the allantoic fluid of a fertilised egg, where it is left to replicate. After about three days, the virus-containing fluid is harvested and purified. The resulting viral matter is then inactivated and packaged as a vaccine.[26] This enabled the creation of easily scalable industrial processes for manufacturing large quantities of vaccines against smallpox, influenza, yellow fever, typhus and many other viral infections. Today, chickens continue to play a key role in vaccine production. Eighty-two per cent of the 170 million doses of flu vaccine distributed in the US in 2020 were egg-based. With each egg producing roughly one dose, the US flu-vaccine market alone requires 140 million eggs annually. The production is outsourced to a handful of specialised manufacturing companies, which breed chickens in secret locations governed by strict biosecurity protocols. Chicken vaccine factories are one of the main weapons in governmental strategies against viral epidemics, but they are themselves vulnerable to viruses. 'A pandemic of H5N1', says Leo Poon, head of public health laboratory sciences at Hong Kong University, 'can kill chickens substantially and there will be a huge drop in egg supply, and you will have a problem getting enough eggs to make the vaccines'.[27]

Another fundamental problem faces vaccine manufacturers: avian cells, such as those in chicken eggs, are not human cells. Once injected in chicken eggs, human viruses can mutate to better infect the avian cells. If this happens, the mutated virus no longer matches the ones circulating in the wild, and the vaccine may end up incapable of mounting an effective resistance against the wild viral strain.[28] As a result, vaccinologists have increasingly turned towards alternative production strategies. The polio vaccine developed by Jonas Salk in 1955 was the first licensed product developed using cell culture technologies.[29] The same technology is now widely used to produce vaccines for smallpox, polio, hepatitis, influenza, rotavirus, rubella and chickenpox. First, a cell culture is established by extracting living tissue from a selected subject. Madin-Darby Canine Kidney Cells – tissue derived from an adult Cocker Spaniel in 1958 – are currently the most used in vaccine production. These cells are injected with weakened viruses and placed in large stainless-steel vats, where the virus can replicate rapidly. Finally, the incubated viral matter is extracted, purified and packaged for commercialisation. Cell-based bioreactor systems are faster and cheaper than egg-based systems. They cut costs and increase profits, while shortening reaction time against rapidly mutating pandemic threats.[30]

Genetic engineering has enabled an alternative method for turning isolated cell lines into industrial sites of bioproduction. Instead of using living cells as sites of viral replication, their metabolic capacity is harnessed to manufacture viral proteins. In the late 1970s, it was established that the HBsAg proteins, which compose the outer envelope of the hepatitis virus, stimulate an immune response. The finding sparked a whole new category of vaccines. Subunit vaccines no longer rely on weakened viruses, but rather on the injection of isolated viral proteins – inactive bits and pieces of the whole virus.[31] Genetic engineering offers a cheap and scalable way to produce such proteins. To produce the first recombinant hepatitis-B vaccine, molecular biologists spliced the gene for HBsAg into a yeast genome. The resulting recombinant yeast cells began manufacturing the surface proteins of the virus as part of their modified metabolism. Recombinant vaccines have become increasingly attractive in the last twenty years. As they do not require the cultivation of legions of viruses in infected cells, they present less safety issues and accelerate production.[32]

From the earliest days of the pandemic response, researchers worked to develop a recombinant vaccine against Covid-19. 'This technology', reported Yves Balmer, head of microbial development at Swiss chemical and medical conglomerate Lonza, 'does not require the culture of virulent organisms and their subsequent attenuation/inactivation, which can create safety concerns'.[33] The production process, nonetheless, remains complex. It requires identifying a vital protein that would trigger a sufficient immunitary reaction, isolating the genetic sequence for that particular viral protein, splicing it into a living cell, thereby creating a stable cell-line expressing that genetic sequence. Then, the protein must be purified and mixed with chemical additives that help stimulate the immunitary response. By early 2022, more than twenty recombinant vaccines have entered clinical trials. Most of them are based on a fragment of the SARS-CoV-2's spike protein, which is manufactured using different types of genetically modified cells: Novavax and Sanofi/GSK use cells from the fall armyworm; Clover and Medigen rely on hamster ovary cells; Medicago harnesses the metabolic capacities of a tobacco plant. Over two years after the first reported outbreak of Covid-19, however, recombinant vaccines have not yet reached the global market.[34]

In the meanwhile, a new category of vaccines has emerged. The first vaccines to be developed against Covid-19, in fact, were neither traditional inactivated viruses, nor recombinant protein-based vaccines. They were genetic vaccines, which represent a significant departure from established vaccination technologies. Almost every vaccine ever sold before 2020 was based on one fundamental principle: injecting viral matter into the human body prompts the creation of specific antibodies that can mount a defence against subsequent viral infections. Genetic vaccines are based on a rather different principle. Instead of delivering viral matter directly into the human body, genetic vaccines induce cells to produce viral proteins. They deliver genetic instructions for specific viral proteins into the cell, as either DNA or mRNA molecules. 'RNA vaccinology', molecular biologists at Imperial College London explain, 'works by outsourcing the production of the vaccine protein antigen to the cells of the human body, based on the information in the RNA sequence'.[35] Through mRNA and DNA technologies, pharmaceutical firms relocate the production of pharmaceutical proteins from the factory into the human body, harnessing the metabolic capacities of human cells. Genetic vaccines turn the cell into a living factory of viral proteins.[36]

This approach has enabled pharmaceutical companies to develop a vaccine at unprecedented speed. Conventional vaccine development takes on average 8–14 years. By contrast, a genetic vaccine against Covid-19 was being tested less than two months after the first reported outbreak. Three days after the WHO first announced the emergence of 2019-nCoV, Chinese authorities employed Illumina sequencers to derive the full sequence of the coronavirus genome and shared it online. Moderna immediately began working with the US National Institutes of Health (NIH) to develop a genetic vaccine. 'They never had the virus on site at all', points out the chief executive of Illumina, 'they really just used the sequence, and they viewed it as a software problem'.[37] Relying on the power of genomic abstraction, and the global circulation of genomic data, researchers could approach the vaccine problem as, first and foremost, a question of genetic (de)coding: how to infer the correct mRNA sequence coding for the spike protein? And only secondarily as a material process of production and molecular logistics: how to synthesise sufficient quantities of spike-coding mRNA molecules? How to safely deliver them into human cells?

Vaccines were developed at record speed, but the scientific research underpinning vaccine production unfolded over decades, with the contribution of hundreds of scientists and supported by significant public funding. Genetic vaccines emerged from collective scientific research in at least three areas: the discovery of mRNA in the 1960s and subsequent efforts to synthesise it; the search for efficient ways of delivering genetic material into human cells; and research in virology searching for new means of inducing immune responses using artificial proteins. These three strands of research began to intertwine around 1987, when Robert Malone performed a landmark experiment at the Salk Institute for Biological Studies. He bathed human cells in a solution made of synthetic mRNA and lipid molecules. The human cells absorbed the artificial RNA and began producing the encoded protein. If synthetic mRNA could be used to direct protein synthesis, he jotted down in the experiment report, then it was possible to 'treat RNA as a drug'.[38] It was the first time that synthetic mRNA was used as a molecular means of production, effectively inducing human cells to assemble a desired protein. The experiment suggested the enticing possibility that human cells may be turned into programmable molecular factories, assembling drugs directly within the targeted body.

In the following twenty years, hundreds of researchers and billions of dollars were deployed to realise this intuition.[39] In 2020, this collective endeavour bore fruit, informing the production of some of the most profitable drugs in the history of the pharmaceutical industry. Once the genetic sequence of the Covid-spike protein was publicised by Chinese researchers, corporations synthesised the corresponding mRNA molecule in their labs, using methods developed in the 1990s by Katalin Karikó and Drew Weissman.[40] Those synthetic molecules were then enclosed in a protective lipid layer, facilitating their absorption within cells – a method elaborated by Peter Cullis in the early 2000s.[41] The mRNA-containing lipid bubbles were then injected into volunteers. Cells absorbed the compound, and began producing proteins that resembled the spikes of the coronavirus. Finally, this pseudo-viral matter triggered an immunitary response, training the body to defend itself against the attacks of coronaviruses.

The rapid deployment of Covid-vaccines has been presented as a triumph of corporate science. In a now infamous quip, the UK prime minister Boris Johnson went as far as suggesting that 'the reason we have the vaccine success is because of capitalism, because of greed'.[42] Nothing could be further from the truth. Even a brief account of the two centuries of research that led to the development of vaccine technologies (in general) and genetic vaccines (in particular) imposes a different perspective on the present. Many scientists recognise that genetic vaccines were developed on the basis of a diffused circulation of knowledge performed by countless people across several countries: 'You really can't claim the credit', pointed out Cullis in a recent interview, 'we are talking hundreds, probably thousands of people who have been working together'.[43] Similarly, Karikó, generally credited as one of the central figures in the field of mRNA research, insisted that 'everyone just incrementally added something – including me'.[44]

The final months of product development carried out by Moderna and Pfizer/BioNTech were heavily subsidised by public funds estimated to exceed $8 billion.[45] Pfizer frequently points out that it declined the federal funds offered by Operation Warp Speed. Nevertheless, its partner company BioNTech *did* receive substantial support from the German government in developing their joint vaccine. More generally, the development of the vaccine would not have been possible without freely available public research. The final products, nevertheless, were patented and the profits privatised. The commodification of genetic vaccines represents an enclosure of both popular knowledge and of public research, which turns medicine into a process of capital accumulation driven by an insatiable thirst for profits.

Based on financial statements produced by Moderna, BioNTech and Pfizer, a recent study has estimated that the three corporations have cashed in over $34 billion during 2021 – almost $100 million per day.[46] Thanks to their patent monopolies, which prevent others from developing equivalents, the three corporations have been able to impose astronomical profit margins on the vaccines. A recent study by a collective of engineers from the University of London has estimated that the mRNA vaccines produced by Pfizer/BioNTech and Moderna could be mass produced for as little as $1.18 to $2.85 a dose. Yet, the same vaccines are sold at an average price of $16.25 per dose.[47] This monopolistic pricing practice reflects the increasingly concentrated power wielded by corporations in global production.

Patents and monopoly pricing have not only exacerbated existing wealth inequalities, it has also slowed down vaccine manufacturing and distribution, creating the conditions for the perpetuation of the pandemic. Because of monopoly patents, the vaccine rollout remains limited and geographically distorted. On the eve of 2022, less than 4 per cent of people in low-income countries were fully vaccinated and, based on current vaccination rates, it may take another 57 years to achieve full coverage.[48] If the vaccine rollout remains localised in a small cluster of countries able to pay the steep monopoly prices set by private suppliers, new variants are likely to emerge and regular booster shots will be necessary. As a result, the vaccines are expected to keep generating significant profits for years to come. 'We believe that a durable demand for our Covid-19 vaccine, similar to that of the flu vaccine is a likely outcome', suggested Pfizer's CEO Albert Bourla, commenting on his company's projections of $26 billion in Covid-vaccine revenues for 2022.[49] Some have put the blame for what WHO director General Tedros Ghebreyesus has called the 'catastrophic moral failure' related to unequal vaccine distribution on cumbersome bureaucracies and vaccine hesitancy. But the root of the problem lies in a dysfunctional system of production monopolised by private corporations driven by profit incentives.[50]

In response, more than a hundred sovereign states have been calling for a waiver on vaccine patents to increase the global supply, drive down prices and speed up vaccination campaigns in the Global South. Yet, the three corporations that profit from the vaccine monopoly have refused to give up on their patents, claiming that this concession would not increase supplies due to the limited productive capacities of low-income countries. One question immediately arises: if one were to take seriously the companies' claim that a waiver would not increase competition on their products, why would they deny such a request? 'The real nonsense', argued Oxfam's Health Policy Manager, 'is claiming the experience and expertise to develop and manufacture life-saving medicines and vaccines does not exist in developing countries. This is just a false excuse that pharmaceutical companies are hiding behind to protect their astronomical profits.'[51]

In an attempt to demonstrate that it would be possible to scale up the production of equivalent vaccines in many parts of the world, the WHO launched a mRNA tech-transfer hub in South Africa. They asked Moderna, Pfizer and BioNTech to help teach researchers how to make the Covid-19 vaccines. The companies refused to grant their support, and the WHO went ahead without their help. In February 2022, Afrigen – the institute at the core of the WHO's hub – announced that it had successfully reproduced the Moderna vaccine: 'We didn't have help from the major COVID-vaccine producers', declared a leading scientist at Afrigen, 'so we did it ourselves to show the world that it can be done, and be done here, on the African continent'.[52] The WHO expects that a copycat of the Moderna vaccine will be ready for trials by the end of 2022. What will happen at that point remains uncertain. Moderna has so far declined to respond to WHO's requests to confirm that the company will not enforce its patents in case the copycat becomes available for distribution.[53]

Political struggles over the new molecular means of production will not dissipate with the end of the Covid pandemic. Rather, they are set to become increasingly central in the twenty-first century. Through genetic interventions, steering metabolic activity and directing protein-synthesis, molecular biology is introducing ever-new ways to turn living bodies into molecular factories. Genome editing technologies, as a result, take an economic, social and political dimension. This epochal transition echoes Marx's reflections on the growing importance of science as a productive force, and of technology as a new site of enclosure. In the sixth and seventh notebooks of the *Grundrisse*, Marx depicted a future in which capital accumulation would no longer be based only on 'the theft of labour time', but would increasingly rely on the appropriation of dispersed knowledges incessantly produced throughout society.[54] Marx emphasised not only that social knowledge becomes a powerful force of production, which enables the introduction of new forms of machinery, but also that 'the accumulation of knowledge and of skill, of the general productive forces of the social brain, is thus absorbed into capital, as opposed to labour, and hence appears as an attribute of capital'.[55]

Today, the advent of genome engineering and synthetic biology has a similar effect. On the one hand, these are scientific practices that enable the transformation of the living cell into a versatile, embodied and diffused site of production. Technoscientific knowledge is not only reified into machines, it is increasingly embodied in the flesh of the living.[56] On the other hand, these new means of molecular production are enclosed, privatised and translated into an apparatus for the expanded accumulation of capital. The sources of genetic knowledge are dispersed and popular, but they now appear as an attribute of capital, as corporate property. The political struggle for democratic control over genetic biotechnologies is therefore also a demand for a direct control over the body as a site of (re)production, over its metabolic activities as a form of labour, and over the ways in which new means of control over metabolic forces are set to shift the complex political ecologies in which new pandemics are already brewing.

CRISPR CLINICS

The Politics of Gene Therapy and Somatic Engineering

Since the 1970s, sociologists have been debating the expansion of medical authority over ever-increasing aspects of everyday life. This progressive medicalisation of society has been defined by the International Epidemiological Associations as an ongoing tendency 'by which conditions, processes, or emotional states traditionally considered nonmedical are redefined and treated as medical issues'.[57] As the list of illnesses compiled by health professionals constantly grows, the social relationship between doctor and patient becomes increasingly ubiquitous. The number of days each person spends as a patient under medical surveillance steadily increases. In parallel, the number of drug prescriptions is constantly on the rise. From the early 1980s to the early 2000s, global prescription drug sales tripled to nearly $400 billion. In the last twenty years, the global pharmaceutical market tripled again, reaching an estimated value of $1.27 trillion.[58] Medicalisation, in other words, has led to pharmaceuticalisation – that is, 'the translation [...] of human conditions, capabilities and capacities into opportunities for pharmaceutical intervention'.[59] Painkillers, antibiotics, antihistamines, contraceptives and statins are ubiquitous. Anxiety and depression, high blood pressure and cholesterol, asthma and allergies, stomach acidity and obesity are increasingly prevalent conditions kept in check by constant shopping trips to the local pharmacy.

Whether small chemicals or recombinant proteins, most drugs are specifically designed to avoid inducing genetic mutations. Testing procedures always include a careful assessment of the potential impact that the medication may have on the genome. Drugs that are considered mutagenic are normally excluded from the market, as any potential positive effect of the drug must be weighted against the increased risk of cancerous mutations. As we have seen, mRNA vaccines steer the cellular metabolism to stimulate the production of viral proteins *without* modifying the genome. The small quantities of synthetic mRNA do not reach the cell nucleus, they do not impact the DNA contained therein, and they are broken down by the organism in a few days.[60]

Gene therapy represents a radical departure from this approach. It constitutes an emerging field of medical intervention, which aims to govern processes of molecular production taking place in human cells by inducing intentional mutations in the genome. In biology textbooks, metabolic activity is often presented as a unidirectional chain: DNA informs the synthesis of mRNA that, in turn, informs protein synthesis. Gene therapy represents a technoscientific intervention that influences all levels of metabolic activity: the manipulation of the DNA substrate impacts the production of mRNA, which influences protein-synthesis. As a result, genetic interventions are theoretically able to refashion the entire metabolic process in profound and long-lasting ways.[61] The genome becomes a biological lever whose manipulation can influence the (dys) functioning of living bodies.

Genomic science is producing mountains of data concerning the 'function' of specific genes and the correlation of specific genes with particular diseases. Many bodily dysfunctions involve rare genetic mutations, which disable the production of certain proteins, or induce the production of misshapen proteins that end up being inoperative or even toxic. Frequent injections of custom-made recombinant proteins can partially supplement bodily metabolic activities of protein-synthesis, but the promise of gene therapy is to enable a new form of medical intervention at the genetic 'basis' of disease. 'Correcting' the human genome has been a driving dream of molecular biology since its inception in the early 1930s. It was only with the introduction of recombinant DNA techniques in the 1970s, however, that this idea became a feasible research strategy. Theodore Friedmann and Richard Roblin coined the concept of 'gene therapy' in a seminal 1972 article.[62] Given the constraints of early genetic engineering techniques, their proposed strategy was rather straightforward. When a bodily dysfunction was found to result from the metabolic effects of a genetic mutation, it may be possible to induce the production of the missing proteins by delivering a functional copy of the gene into the affected cells. Prashant Mali, a bioengineer at the University of California, San Diego, has called this approach gene therapy 'version 1.0'.[63]

One of the first attempts to realise Friedmann's and Roblin's vision took place in September 1990, when researchers at the National Institutes of Health (NIH) conducted an experimental operation on a four-year-old girl affected by severe combined immunodeficiency.[64] This condition is linked with a rare genetic mutation by which cells fail to synthesise an essential enzyme called adenosine deaminase. Given the central role played by these enzymes in the immune system, the patient experienced a high susceptibility to infections. The medical trial pioneered an approach to gene therapy that remains the most used. First, researchers drew the patient's blood. Then they infected the blood cells with a culture of genetically engineered viruses spliced with the human genes stimulating the production of adenosine deaminase. The viruses went on doing what they do best, injecting their modified genome into the infected cells and hijacking protein synthesis. Given that their viral genome had been modified, the result was not the creation of viral factories spewing out viruses, but a set of blood cells induced to manufacture the desired enzymes. In essence, says Charles Gersbach, director of the Center for Advanced Genomic Technologies at Duke University, the approach was 'taking advantage of the viral shell as a Trojan horse to deliver therapeutic gene cargo'.[65] Finally, the researchers injected the genetically engineered cells into the patient's body, and the cells went on secreting the desired enzymes directly into the bloodstream. The experiment was not a definite cure, but it proved that gene therapy could induce the production of desired proteins in a human body.

This early technique, also known as 'gene trapping', informed a series of medical trials in the early 1990s which yielded a few therapeutic successes, many setbacks and some dramatic adverse events.[66] In 1995, the NIH issued a warning, stressing risks from insufficient knowledge of genetic dynamics and uncertainties concerning the effects of the most widely used viral vectors.[67] In 1999, a further gene therapy trial was conducted on an eighteen year-old patient affected by a mild form of ornithine transcarbamylase deficiency: a bodily dysfunction that causes toxic levels of ammonia to build up in the blood. Standard treatment for the condition involves a restricted diet and the constant intake of protein supplements. Jesse Gelsinger was already being treated in this way, but gene therapy promised to reverse his pathology by inserting a gene promoting the synthesis of ornithine transcarbamylase into his liver cells. Within days, the adenoviruses used to deliver the gene sparked a severe immunitary reaction: a cytokine storm that blazed through Gelsinger's body and killed him within hours.[68] Gelsinger's death hit the entire field, leading the FDA to suspend clinical trials.[69] Gene therapy returned to being a theoretical hypothesis, rather than an active field of experimentation.

James Wilson, the geneticist who headed Jesse Gelsinger's operation, was banned from conducting further human trials and was later called to write 'Lessons Learned' as part of an investigation into the case.[70] The report offers a detailed analysis of the technical errors that led to Gelsinger's death. It also points out a number of political and economic issues that influenced the outcome of early gene therapy trials, and remain deeply relevant today. First of all, Wilson suggests that the hype surrounding genomic science may lead researchers to systematically overestimate their understanding of the human metabolism: 'We were drawn into the simplicity of the concept', Wilson sums up in an interview: 'you just put the gene in'.[71] Given this tendency, Wilson argues that clinical scientists should embrace a strict precautionary principle, asking themselves: 'If the worst case scenario played itself out – not the potential or likely, but the worst – would that be acceptable?' If he had posed himself this question, Wilson says, he would not have performed the trial.

The report points out that market pressures may lead researchers to discard this precautionary question, pushing experimentation in dangerous directions in view of financial and professional gains. 'One of the most troubling allegations that surfaced following the OTCD gene therapy trial', Wilson writes, 'was that decisions were influenced by the potential for personal financial gain, especially as it related to my affiliation with a gene therapy biotechnology company called Genovo, Inc. These allegations emerged at a time when more global concerns had been rising regarding financial conflicts of interest in other clinical trials conducted in the United States. [...] Upon reflection, I realize my initial reaction to these allegations oversimplified what is a more complex issue and that concerns raised about the potential for financial conflicts of interest in my role as sponsor of the IND were indeed legitimate.' In Wilson's view, scientists' participation in financial ventures provides incentives to push research forward, even when confronted with considerable dangers and risks. 'Both Penn and I owned stock in Genovo', he points out, 'and it is possible that a success in the OTCD gene therapy trial could enhance the value of Genovo (and other gene therapy companies) through encouraging proof-of-concept clinical results.'[72] Wilson highlights that many scientists' decisions can be influenced by market pressures: since the financial rewards of technological advancements are largely privatised (while the negative effects are not), the potential benefits are systematically over-estimated (while the social and ecological risks are not).

Ten years after their publication, Wilson's reflections on the risks of gene therapy experiments in a neoliberal social context appear even more relevant. Recent developments in molecular biology have restored the credibility of gene therapy and renewed its appeal as a frontier for capital investment. In 2013, two parallel events made noise in the biotech industry and contributed to reviving the quest for gene therapy. On one side, CRISPR was hailed as a more precise mutagenic tool, which could reduce the risks associated with gene therapy. On the other, the European Commission approved the commercialisation of the first gene therapy product on the European market: Glybera, marketed by the Dutch corporation uniQure as a medication for lipoprotein lipase deficiency.[73] Developed by researchers at the University of British Columbia, Glybera promised to open a new field of medical possibilities. Hundreds of newspaper headlines captured its launch, stressing its record-breaking price tag of $1.2 million for a single injection. 'It's not a crazy price', commented the chief scientific officer at UniQure in an interview. 'People say it's the most expensive drug in the world and what have you, but in the end, all of these products, even priced at $1 million, are going to be generally cheaper than replacement therapy [...] Pricing shouldn't be a political decision. It should be a rational decision based on merits and values.'[74] The drug's impact on society was disappointing at best: a single insurance company agreed to cover the price for the drug, a single injection was sold and, after four years, uniQure withdrew Glybera from the market.[75]

In the following months, a global political struggle developed. The Canadian researchers who originally developed the technology protested against uniQure's pricing strategy. Michael Hayden denounced that 'as the clinician scientists, we lost control of the project completely. They decided that they were going to charge a million dollars a shot. In 2012, in Europe, this was outrageous.'[76] Meanwhile, an international group of biohackers affirmed they had created a knockoff version of the million-dollar drug in a warehouse in Florida for $7,000. While the pirate gene therapy was little more than a provocation, it did spark debates. Hayden, interviewed as one of the scientist that pioneered research on the drug, supported the message conveyed by the biohacking collective: 'The right to access medicine is a social-justice issue', he said commenting on the pirate drug, 'any way to provide potential benefits to patients is entirely meaningful, and I would never stand in the way'.[77] Meanwhile, Canada's National Research Council relaunched Hayden's research on developing an alternative to Glybera to stay in public hands.[78]

The Glybera case shows that gene therapy represents an emerging means of capital accumulation by which public research is systematically privatised. This trend has continued in the last five years. In August 2017, the FDA approved the very first gene therapy drug for sale in the United States. Developed at the University of Pennsylvania, Kymriah genetically reprogrammes a patient's own white blood cells to recognise and target growing cancers. Swiss pharmaceutical corporation Novartis acquired the rights to distribute the drug, pricing it at almost $500,000. The announcement was met with surprise in the medical community given that early estimates indicated that the manufacturing costs would account for only 5 per cent of that price.[79] An analysis published in *Health Affairs* estimated that Novartis will reap an 84 per cent profit on the drug and concluded that 'the drug is overpriced. Novartis took a drug developed with significant investment from US taxpayers and is on course to make substantially more than its already healthy performance'.[80]

A few months later Spark Therapeutics launched Luxturna, a gene therapy for Leber's congenital eye disease priced at $1 million. Zynteglo, a gene therapy product targeting a rare genetic blood disorder, was marketed at $1.7 million. Zolgensma, a gene therapy for spinal muscular atrophy developed by a Novartis subsidiary, costs over $2 million per dose. Many of these gene therapy products have been developed in response to rare pathological conditions. Significant profits can be made in these niche markets by exploiting the lack of alternatives to set high prices. Average profit margins of over 80 per cent are reported in the rare disease industry, well above the 16 per cent registered in the rest of the pharmaceutical industry. This makes drugs aimed at rare diseases an increasingly attractive business option, and precipitated a strategic shift by major pharmaceutical corporations away from so-called block-buster drugs with low profit-margins and towards so-called niche-buster drugs. The strategy relies on establishing lucrative monopolies on niche markets, enforcing high prices, and relying on insurance companies to cover the bills.[81]

The number of clinical trials exploring new techniques of gene therapy is growing exponentially worldwide. The most recent global review, compiled in July 2020, reported over 2,100 trials currently ongoing, the highest number ever registered.[82] As of August 2019, just 22 gene therapy products have been approved by drug regulatory agencies from various countries. From now on, the US Food and Drugs Administrations estimates that it will be approving ten to twenty new gene therapies every year.[83]

* * *

Gene therapy's ongoing renaissance has been boosted by the introduction of new genome editing techniques. The new approach, often described as 'gene therapy 2.0', is no longer based on the delivery of additional synthetic genes.[84] Instead, it uses genome editing technologies to 'edit' a pathological genetic mutation. When a pathogenic mutation in the genome is singled out through sequencing, the question is no longer: Can we introduce a substitute gene? But rather: can we 'correct' the mutation?[85]

The first clinical trial with CRISPR gene therapy took place in October 2016 at Sichuan University in Chengdu, China. The experiment focused on developing genome-edited cells that would trigger an immunitary reaction against cancerous growth. The Chinese team extracted blood from a volunteer, employed CRISPR to disable the gene for a protein that can shield cancer cells from being detected by the immune system, and then reinjected the engineered cells into the patient's body.[86] A few months later, researchers at the University of Pennsylvania headed by Carl June obtained the green light to conduct an equivalent trial. Interviewed by *Nature*, June pointed out that the dawn of CRISPR genome editing is likely 'to trigger Sputnik 2.0, a biomedical duel on progress between China and the United States'.[87] In fact, this genome editing race has been ongoing for at least a decade. By early 2020, three quarters of all clinical trials ever performed took place either in the US or in China.[88] A growing number of them are testing CRISPR-Cas9 as a therapeutic tool in human volunteers.[89]

Most of these trials are still in the early stages, making it impossible to draw firm conclusions about the efficacy and safety of CRISPR therapies. One of the most promising fields of experimentation involves sickle cell disease and beta-thalassemia. Both diseases affect millions of people worldwide and are linked with a single-gene mutation, making them good candidates for gene therapy. Sickle cell anaemia is linked with a mutation in the gene instructing the production of haemoglobin, an essential protein for the functioning of blood cells. The mutation causes red blood cells to become stiff, assuming a sickle shape that makes them pile up, causing blockages and damaging vital organs. Beta thalassemia occurs when a different genetic mutation causes a reduction in the production of haemoglobin.[90] Vertex Pharmaceuticals and CRISPR Therapeutics, two US-based biotechnology companies, recently announced the first results from clinical trials performed on beta-thalassemia and sickle cell patients treated with CTX001, a CRISPR-based therapy. In both cases, CRISPR is directed to knock out a gene that, shortly after birth, suppresses the production of foetal haemoglobin. The genetic intervention reactivates the production of this protein, which contributes to alleviating the dysfunctions caused by an insufficient or defective synthesis of adult haemoglobin. The approach involves the extraction of hematopoietic stem cells from the patients, which are then genome edited and injected back into the body. The two companies have recently announced that they plan to publish the results of the trials by the end of 2022, and then file for patents and permits.[91]

Another corporate partnership between Allergan and Editas Medicine has launched trials to treat Leber congenital amaurosis – the same eye condition targeted by Spark Therapeutics' Luxturna. Researchers injected into the eye of volunteers a virus containing the CRISPR assemblage. Aptly named EDIT-101, this was the first trial to attempt CRISPR genome editing inside a human body. Early results will be reported only later in 2022, but the attempt has already been hailed as a landmark event in the history of medical science.[92] These are only two examples. Several companies are heavily investing in CRISPR gene therapies, including Novartis, Amgen, Sangamo Therapeutics, Spark Therapeutics, CRISPR Therapeutics, Merck, Editas Medicine. Many of them are startups, often created by scientists conducting their research in the public sector. Big Pharma is starting to make inroads in this area as well. According to a recent study by the management consulting firm McKinsey, sixteen of the twenty largest pharmaceutical companies have recently acquired assets in the gene therapy sector.[93] In 2020, nearly $20 billion in funding flowed into biotech companies developing cell-, gene- and tissue-based therapies.[94]

As we have seen, most gene therapy products approved so far have extremely high price tags, which do not reflect manufacturing costs but rather the high profit margins imposed by corporations in monopolistic positions. It is not clear to what extent future gene therapies will reverse this trend. What is certain is that the granting of exclusive patents on many CRISPR applications to the Broad Institute is already creating additional layers of rent extraction, which could inflate the price of future CRISPR-based gene therapies. After the Broad Institute obtained control of several patents on CRISPR technologies in 2017, it promptly issued licences to a newly formed private corporation called Editas Medicine.[95] In this way, the company obtained an exclusive right to exploit CRISPR for medical purposes. Other companies will have to obtain permits from Editas to use CRISPR for gene therapy, adding significant royalties to the costs faced by manufacturing companies. For instance, Editas signed a $737 million deal with Juno Therapeutics and a $90 million deal with Allergan to licence their use of CRISPR tools.[96] Commenting on the likely impact of these recent deals on future drug prices, Jim Kozubek has pointed out that 'investors in Editas demand a cut on each CRISPR application. As investors engage in layers of transactional deals along the top of the food chain, the costs of gene therapies go up while the financiers may shift blame for a lack of patient coverage to insurance companies.'[97]

Addressing increasingly impactful systems of rent extraction, and the unequal social environment that may result from them, is only one of the pressing political questions raised by recent advancements in human genome editing. Three additional issues have attracted considerable attention in recent years.

Firstly, the scientific community is still assessing the long-term risks associated with genome engineering in human bodies. Several clinical trials have been deferred due to fear of off-target effects, generation of unexpected chromosomal alteration and adverse immunitary responses.[98] A study published in 2021 has shown 'that when CRISPR-Cas9 is used to edit the genome, cells with cancer-associated mutations are likely to be selected to survive; and this is more widespread than scientists previously understood'.[99] Human cells have developed powerful mechanisms to preserve the integrity of the genome, avoiding the accumulation of dangerous mutations that could disrupt the bodily metabolism. The p53 protein – often dubbed 'the guardian of the genome' – plays a prominent role in preventing cancer.[100] Within each cell, p53 continuously monitors the genome and, whenever it identifies dangerous mutations, it triggers cellular suicide. The problem is that the targeted-mutations intentionally induced by CRISPR genome editing are often enough to trigger p53, which kills off many of the affected cells. Cells in which the guardian of the genome is absent or inoperative are much more likely to be successfully genome-edited, but they are also most at risk of developing dangerous mutations. This finding suggests that genome editing still holds considerable unknown risks. As the leader of the study points out, 'we may be selecting for cells that carry mutations in key cancer driver genes when using CRISPR-Cas9 editing, and that could be potentially dangerous'.[101]

Another centre-point for debate is the porous border between gene therapy and genetic enhancement. Can gene therapy products be turned into means of genetic enhancement? Accusations of 'gene doping' have already entered the legal realm during a court case in Germany over ten years ago. The trial led to the conviction of Thomas Springstein, a track coach of the German national team, who received a 16-month suspended jail sentence for supplying doping products to athletes. In an email recovered by investigators, Springstein reached out to a Dutch doctor complaining: 'Repoxygen is hard to get. Please give me new instructions soon so that I can order the product before Christmas.'[102] Repoxygen is a gene therapy product developed by UK-based company Oxford BioMedica for the treatment of anaemia. It uses a virus to deliver an EPO gene into human cells, boosting the production of erythropoietin and thereby increasing the oxygen-carrying capacity of blood cells. This makes it a promising drug for anaemic patients, but also for oxygen-hungry endurance athletes. While it is unclear whether Springstein ever obtained Repoxygen, this was the first documented attempt at 'gene doping'. In light of these trends, the World Anti-Doping Agency (WADA) has decided to explicitly ban from professional competitions any athlete guilty of 'nontherapeutic use of cells, genes, genetic elements, or modulation of gene expression, having the capacity to enhance performance'.[103]

Recent developments in gene therapy have also rekindled long-standing debates concerning medicalisation and pharmaceuticalisation. Gene therapy products have so far focused mostly on single-gene diseases. Molecular biologists, nevertheless, are investing considerable energy and resources to expand the range of medical(ised) conditions to be targeted by future gene therapy products. One area that has received increasing media hype in recent years concerns products promoting weight loss and tackling the so-called 'obesity pandemic'.[104] The pathologisation of obesity is a recent event. The American Medical Association officially reclassified obesity from a subjective condition to a 'disease' in 2013, following a spirited debate both within and outside the association. In fact, the decision was in direct conflict with the recommendation of the expert Council that the Association had convened to investigate the merits and risks of reclassifying obesity as a disease. In its final report, the Council suggested *against* the pathologisation of obesity since this would promote further stigmatisation and unnecessary medical treatments. 'The medicalization of obesity', argues the report, 'could detract from collective social solutions to environmental forces that shape people's behaviors and impact a number of conditions beyond just obesity. Thus, public efforts to enhance the built environment to make healthy eating and physical activity choices easier may receive less attention, despite providing substantial health benefits at every body weight.'[105] The final decision ignored these concerns, stressing the potential benefits of increased public funding for new surgical and pharmaceutical treatments. In a single stroke, one-third of the US adult population was redefined as suffering from a newly coined disease, requiring pharmaceutical or surgical interventions.[106]

Molecular biologists are now moving the first steps towards the goal of developing new gene therapy products for the expanding 'obesity market'.[107] One of the first studies in this area was published in 2015 by researchers at the Broad Institute. The article reports experiments in humanised mice, which suggest that editing a gene called FTO may induce cells to dissipate more calories as heat. 'Knowing the causal variant underlying the obesity association', one of the authors of the study concludes, 'may allow somatic genome editing as a therapeutic avenue'.[108] In another set of experiments performed on humanised mice, a team led by Jee Young Chun targeted a gene that plays a role in the metabolism of fatty acids, claiming 'a 20% reduction of body weight and improved insulin resistance and inflammation after just six weeks of treatment'.[109] The study similarly concluded that this finding provides the basis for developing new gene therapy products.[110] These early experiments might not lead to gene therapies against obesity, at least not any time soon. However, they are indicative of an ongoing tendency. The medicalisation of conditions such as obesity is likely to lead not only to a further pharmaceuticalisation of everyday life, but also to increasing genetic interventions. As argued by the Council of the American Medical Association, this risks erasing the social, political and environmental drivers of many recently medicalised conditions. Surgical, pharmaceutical and genomic fixes are presented as technological solutions to social problems that demand better nutrition, less stressful lifestyles, shorter working hours, appropriate sporting facilities, and more green spaces. Political struggles are turned into personal problems to be solved by gaining access to expensive gene-editing procedures controlled by corporate actors.

As we have seen, many scientists and medical professionals have already recognised the many contradictions that beset this approach. A new genomic politics is emerging, concerning issues such as the unequal access to health caused by skyrocketing rates of profit and multiple layers of rent extraction; the known and unknown risks associated with direct manipulations of the human genome; and the difficulty of preventing gene therapy products from being turned into means of further pharmaceuticalisation of everyday life. The list is set to grow longer. As we will see in the next section, the role of genome politics in twenty-first century neoliberal societies is only starting to take shape.

NEOLIBERAL EUGENICS

Human Capital in the Genetic Supermarket

Thirty years after the start of Human Genome Project, the human genome has turned from an object of cartographic 'exploration' into a target of technological modifications. At the heart of this quest is the modernist expectation that life can be improved by rational planning, intelligent design and technological control. Applied to human beings, gene therapy reflects this aspiration to 'correct genetic defects', 'repair the genome's spelling mistakes', and 'fix errors in the natural genetic ciphering'.[111] While gene therapy is altering the social practice of medicine, it also implies considerable risks. Each blood cell in a patient affected by sickle cell anaemia contains the same disease-causing mutation because they all derive from the same cell, the initial zygote. What began as a genetic mutation in a single fertilised egg cell has been copied into billions of descendant cells. Somatic therapy requires targeting a 'genetic mistake' in each target cell, or at least in a significant subset of them. The genetic intervention needs to be performed countless times. Every time practitioners use CRISPR to induce mutations, there is a danger that the molecular 'scissors' will cause unplanned metabolic shifts – both off and on target.[112] Yet, two parents affected by sickle cell anaemia – even if they had undergone successful gene therapy – would pass on the mutation to their children. Any and all changes made to the genome die with the patient.

Experiments are now under way to engineer the human germline; to introduce heritable changes to the human genome. In this way, the genetic intervention is performed on the zygote immediately after *in vitro* fertilisation. Nearly half a century ago, *in vitro* fertilisation technologies brought the human embryo 'out of the darkness of the womb into the light of the laboratory'.[113] Today, germline editing could allow scientists to 'correct' supposedly 'faulty' genetic scripts before birth, by performing a single targeted mutation directly at the zygotic source from which all other cells derive. The edited DNA would be present in every cell, and it would be passed down generation after generation.[114] With CRISPR, modern technoscience's long-standing dream of 'rationalising' the human finally appears within grasp.

Jennifer Doudna summarises the technical appeal of germline editing in the following way, summoning the scriptural metaphors for life to make her case: 'Imagine trying to correct an error in a news article after the newspapers have been printed and delivered, as opposed to when the article is still just a text file on the editor's computer'.[115] Similarly, according to George Church, 'using one altered germ cell rather than a billion somatic cells is very likely to be a billion times less risky because each of the billion cells has an independent chance to add to the risk of initiating cancer'.[116] Germline editing therefore represents an enticing proposition. The intervention needs to be done only once, lowering the risk of off-target effects. Intervening directly at the embryonic stage also makes it thinkable to alter multiple traits at once, enabling new forms of 'human design'.

While germline engineering is technically less challenging than somatic editing, it is ethically and politically more controversial. Any changes to germ cells will be inherited by offspring, potentially altering the gene pool of *Homo sapiens*. For this reason, germline editing has intensified long-standing debates about the social and political implications of technoscientific efforts to control human evolutionary patterns. It is not the first time that the public is confronted by sustained debates surrounding 'designer humans', nor is it the first time that the spectre of eugenics haunts biology-laboratories. We could rather say that each major advance in the history of molecular biology revived the fundamental question of engineering evolution. By the late nineteenth century, a new understanding of hereditary patterns based on Darwin's and Mendel's studies led many to promote using this new form of knowledge to direct human evolution. By preventing individuals harbouring 'bad genes' from reproducing ('negative eugenics') or by favouring the reproduction of individuals bearing 'good genes' ('positive eugenics'), it was thought that the 'human stock' could be rationally improved.[117] In the 1930s, the first experiments in artificial mutagenesis performed by Hermann Muller inspired the budding eugenic movement, suggesting that technoscientific progress would soon offer more sophisticated means to rationally direct human evolution.[118] In the 1970s, recombinant DNA techniques revived expectations that genetic control would end once and for all the 'reproductive roulette' and enable scientifically directed human evolution.[119] The advent of genome editing technologies has, once again, stimulated a flurry of utopian and dystopian imaginations. In this section, we do not intend to picture either utopian or dystopian futures. Rather, we ask: What worldviews are inscribed in demands to engineer the human germline? What social relations do these worldviews presuppose and further?

When CRISPR-Cas9 gained worldwide attention for its capacity to edit the germline of a number of living species, this immediately implied that genome engineering could be applied to humans. Two biotech companies – OvaScience and Intrexon – announced a $1.5 million joint venture to explore the possibility of germline editing in human egg cells already in December 2013. In public presentations to investors, as part of a successful effort to raise $132 million in capital investments, Ovascience suggested that the company's main goal was to use new genome editing technologies to correct 'mutations before we generate your child', offering a service to 'individuals who aren't just interested in using IVF to have children but have healthier children as well, if there is a genetic disease in their family'.[120] These claims could be dismissed as unrealistic marketing strategies. Yet, by early 2014 Jennifer Doudna was already suggesting that the 'steady march of CRISPR research' had arrived 'right to *Homo sapiens*' evolutionary front door'.[121] The question was no longer *if* CRISPR-mediated targeted mutagenesis would be applied to human beings, but how and when.

Doudna's narrative was ambiguous from the start. On the one hand, she proclaimed the political neutrality of CRISPR technologies and funded a number of companies that offer genome editing tools as relatively cheap commodities, freely available on the global market. On the other hand, she stressed the dangers associated with these technologies if only 'somebody like Hitler had access' to them.[122] This ambiguity has persisted ever since. Doudna has been vocal in calling for caution in what she often presents as an almost inevitable run towards human germline editing. Yet, she has also encouraged a 'prudent path *forward* for genomic engineering and germline gene modification'.[123] This was, in fact, the title of a white paper she published in March 2015 – together with a group of high-profile scientists and gene editing pioneers including Paul Berg, George Church and David Baltimore – following the Innovative Genomics Institute's (IGI) Forum on Bioethics. The Forum, which explicitly mimicked the 1975 Asilomar Conference on Recombinant DNA, renewed a well-established tradition in the history of science: a small group of self-appointed experts – many based in prestigious institutions with significant financial stakes in the development of a particular technology – holding a closed meeting to discuss ways to self-regulate.

Three weeks after the white paper's publication, a group at Sun Yat-sen University attempted to edit the gene responsible for beta thalassemia in eighty-six nonviable human embryos.[124] The results seemed to undermine the celebration of CRISPR as a precise word processor for genetic codes. Less than half the embryos were edited, and those presented a number of unintended cuts and off-target effects. The resulting article, published in *Protein & Cell*, concluded that CRISPR is insufficiently precise to safely pursue germline editing in the human species. The experiment sparked a global debate, which is still ongoing. Calls for a global moratorium on germline editing became increasingly widespread, partially because the study demonstrated that experiments in germline editing were both possible *and* risky. Speaking at National Public Radio, George Daley, dean of Harvard Medical School, warned that 'we should brace for a wave of these papers, and I worry that if one is published with a more positive spin, it might prompt some IVF clinics to start practising it, which in my opinion would be grossly premature and dangerous', and suggested that the published data 'reinforce the wisdom of the calls for a moratorium on any clinical practice of embryo gene editing'.[125]

Later that year, the US National Academies of Sciences and Medicine, the British Royal Society and the Chinese Academy of Sciences launched the first International Summit on Human Gene Editing. Nearly five hundred scientists, ethicists, legal experts and advocacy groups from over twenty countries came together in Washington to discuss how to regulate genome editing technologies.[126] The conference did not suggest a global moratorium on future research in germline editing. Instead, it produced a set of guidelines for future experiments, inviting researchers to refrain from pursuing any research plan that involved bringing into the world gene-edited human beings.[127] Yet, in the absence of a moratorium, research pushed ahead. In August 2017, a research group based at Oregon Health and Science University used CRISPR to induce targeted mutagenesis of the *MYBPC3* gene in human embryos. The goal was to 'demonstrate the proof-of-principle that heterozygous gene mutations can be corrected in human gametes or early embryos'.[128]

When the media bomb finally exploded, it was not entirely unexpected. Early experimentation had demonstrated that inducing targeted mutagenesis in human embryos was likely to have dangerous unexpected effects, but also that it was no longer a technical impossibility. In other words, by this point in time, the main obstacle for any laboratory wishing to make the unprecedented attempt at birthing a genome-edited human was no longer technical and scientific, but ethical and regulatory. These social lines of defence soon proved insufficient to prevent the first documented attempt to engineer the human germline.

On 26 November 2018, He Jiankui of Southern University of Science and Technology in Shenzhen went on stage at the Second International

Human Genome Editing Summit, organised by the US National Academies of Science and Medicine, the Royal Society of the United Kingdom, and the Academy of Sciences of Hong Kong. Speaking in front of a crowded room, He announced that the first genome-edited humans had been born just a few weeks before. The first 'CRISPR-babies', a neologism created by the international press, were two twins: Lulu and Nana. He's team claimed to have conferred HIV-resistance to the twins by altering the *CCR5* gene, the main receptor that allows the virus to enter healthy cells.[129] Several scientists present at the conference immediately questioned He's work, given the risks associated with the use of CRISPR-Cas9 in human embryos and the existence of safer options to prevent HIV transmission from parents to newborn babies. In the following weeks, these initial criticisms turned into a global outcry, prompting condemnations by commentators all around the world.[130]

The Chinese government quickly stepped in, placing He under house arrest, raiding his laboratories, and seizing all the existing documentation. On 30 December, the People's Court of Nanshan District of Shenzhen revealed that a third gene-edited baby had been born and ruled that the scientists had violated Chinese law 'in pursuit of fame and profit'.[131] He Jiankui was condemned to three years in prison and a fine of $430,000 for coordinating the experiment.[132] The *MIT Technology Review* published excerpts from a research paper that He's team had prepared before their arrest. Experts invited to comment upon the article cast doubt on the methodology and the outcomes of the experiment. Fyodor Urnov of the Innovative Genomics Institute doubted the article's claim to have successfully reproduced the *CCR5* mutation often associated with HIV resistance, calling it 'a blatant misrepresentation of the actual data'. Other experts protested that the search for off-target effects had been incomplete, making it impossible to establish if the embryos harboured unintended mutations that could influence the twins' development.[133]

Within one year of his announcement, He Jiankui had been condemned for carrying out an experiment deemed to be unethical, dangerous, unnecessary, and poorly executed. He was almost universally depicted as a 'rogue scientist', suggesting that he worked alone and against scientific norms. It is, nevertheless, important to stress that the first attempt to introduce targeted genetic mutations into the human germline did not take place in a social vacuum, but within a context that provided economic incentives to conduct such an experiment. He Jiankui was not an obscure biohacker working in isolation. He was a successful scientist, deeply embedded both in global networks of research and in the biotech industry. In September 2017, a Chinese documentary described him as 'the new top shot in the Gene World'.[134] A recent review of He's available funds affirms that 'the basic research laboratory stage of He's experiment was almost entirely financed with public funds provided by He's university, the region of Shenzhen, and the Chinese government'.[135] On the other hand, a report by the Health Commission of China concluded that He 'funded his own research'.[136]

He had access to considerable private funds, which may have been mobilised in order to support the project with minimum institutional oversight. By 2018, He had founded three biotech corporations: Direct Genomics, launched with a $6 million start-up fund from the city of Shenzhen and over $40 million from private investors; Vienomics Biotech, sustained by direct private investments of over $12 millions; and Shinzen Nanke Biotechnology, with a capital of about $10 million. In addition, he held managing positions in another nine biotech companies working across different fields, including genomic sequencing, cancer screening, and software development.[137] While it is difficult to assess the total worth of He's financial investments, Jing-Bao Nie has estimated it to be 'at least a few billion yuan (more than half a billion US dollars)'.[138]

There have also been indications that He's experiment in germline editing may be connected to larger economic and financial projects. Jon Cohen, writing in *Science*, uncovered evidence suggesting that He was involved in sustained discussions with John Zhang, head of New York's *New Hope Fertility Center*, 'to discuss opening a clinic together in China'.[139] According to the report, in the same month when Lulu and Nana were born, the two travelled together to Hainan, 'where they discussed their vision with provincial officials'. Zhang has since confirmed that he considered the possibility of setting up a fertility clinic together with He.[140] The Putian Medical Group, a company which runs 80 per cent of private hospitals in China, has also been accused of being implicated in He's project. The Chinese Clinical Trial Registry attests that one of the clinics associated with the company facilitated the approval of the necessary ethics clearance.[141]

Given his embeddedness in the academic, financial and industrial world, it is difficult to uphold the popular image of He Jiankui as a 'rogue' scientist. This dominant, rather soothing narrative isolates this historical event as disconnected from the political and economic context in which it took place. It is difficult not to recall the warnings sounded by James Wilson in 'Lessons Learned', the memoir in which he reflected upon the structural conditions that influenced the tragic gene therapy trial of 1999.[142] Is this a second warning of the risks of a global genomic-industrial complex, whose financial interests shape the directions taken by medical research?

What is certain is that the so-called 'He Jiankui Affair' was not an isolated accident. Rather, it exposes deeper trends within a global scientific culture shaped by the commercialisation and financialisation of genome editing technologies. It has also sparked a global debate concerning the shortcomings of existing regulatory frameworks.[143] The absence of a global legal framework could enable corporations interested in offering germline editing to prospective wealthy parents to simply set up shop in countries with the most permissive legislation. In December 2018, William Hurlburt – a professor at Stanford University who has been repeatedly in contact with He Jiankui – reported that a fertility clinic in Dubai had asked if He was willing to collaborate with them to develop 'CRISPR gene editing for Embryology Lab Application'. Hurlburt decided to publicise this inquiry to stress the pressing need for stringent global regulation. 'There is a real risk of the commercialization of this', he said in a recent interview. 'There probably are fertility clinics eager to offer these services, and people naive enough to want them.'[144]

The road to global regulation, however, is paved with political questions: What kind of rules and norms should be put in place? Should germline editing be banned altogether? Should it be subject to a temporary moratorium, waiting for further advancement in genome editing technologies that would reduce the frequency of off-target effects? If the technology was judged sufficiently safe, how should it be regulated? Should it be permitted only to prevent the hereditary transmission of monogenic diseases, such as sickle cell anaemia? Should it also be allowed for prophylactic purposes, providing resistance to particular diseases as attempted by He Jiankui? Would genetic enhancement be admissable, for instance if used to induce a genetic predisposition to develop a more muscular body? Should it be provided directly by public institutions? Or, should it be made freely available on the market, giving parents the option to choose what genetic interventions they would like for their future children? Who would own the molecular means of human enhancement?

The fact is: no clear global consensus exists on any of these questions. Rather, we are witnessing the messy contours of an impending battle.

The opposition to germline editing is widespread and diverse. Most geneticists have warned that genome editing technologies are too imprecise at this stage, making any attempt extremely dangerous. Yet, many seem to imply that further research should be sponsored to make these technologies more precise. Many organised religions resist any form of experimentation that necessitates the instrumental use of human embryos. Yet, according to a recent poll, most US citizens appear to hold a mixed view, supporting germline editing if used in order to treat a serious disease, but opposing it if used for purposes of enhancement.[145] While this is a position embraced by many bioethicists, it is also increasingly clear that it conceals even more political questions and possible lines of division. As stressed by the US President's Council on Bioethics already in 2003, 'both *enhancement* and *therapy* are bound up with, and absolutely dependent on, the inherently complicated idea of health and the always-controversial idea of normality'. As a result, suggests the Council, 'relying on the distinction between therapy and enhancement to do the work of moral judgement will not succeed'.[146] People that formally agree on the use of germline editing in order to eradicate 'genetic disease' may actually hold very different views on which genetic mutations should be considered legitimate targets. Who draws the boundaries between 'genetic difference' and 'genetic disease'? What principles determine which mutations are considered as 'errors' in the 'genetic script'?

Down syndrome and dwarfism are paradigmatic cases that problematise the seemingly clear boundary between variation and disease. Many people living with disabilities have stressed that genome editing will enable new forms of discrimination.[147] Writing about CRISPR, disability justice scholar Rosemary Garland-Thompson has argued that attempts to genetically define and 'cure' disability effectively constitute a form of 'velvet eugenics', which 'standardizes human variation in the interest of individual, market-driven liberty at the expense of social justice and the robust diversity and inclusion upon which modern egalitarian social orders depend'. From this situated point of view, the risk of future germline editing range from 'producing medical harm to abrogating consent, intensifying genetic discrimination, increasing social inequality, promoting conditional parental acceptance, turning people into products, fostering a commercial medical industrial complex, and encouraging rogue scientific and medical practice'.[148]

There is no doubt that eugenics is alive and well. In fact, the eugenic movement has made a major comeback, enjoying a new wave of popularity. The union of neoliberalism and genome engineering technologies has birthed an increasingly powerful neo-eugenic movement, striving for a speedy deregulation of germline editing. This, according to the liberal eugenicists, would enable market forces to shape the human genome in the most efficient way possible. The concept of 'liberal eugenics' is not ours. It is a self-appointed badge of honour, worn with pride by a growing cohort of mostly white, male academics employed by prestigious institutions in the Global North. Ethicist Nicholas Agar launched the term in an often-quoted 1998 article published in *Public Affairs Quarterly*, followed by a monograph succinctly titled *Liberal Eugenics: In Defence of Human Enhancement*, and passionate follow-ups such as 'How to Defend Genetic Enhancement' and 'Why We Should Defend Gene Editing as Eugenics'.[149]

The argument presented in these works is singularly forthcoming in its basic thesis. 'I have argued', writes Agar, 'that the addition of the word "liberal" to "eugenics" transforms an evil doctrine into a morally acceptable one'.[150] The thrust of his argument comes from the conviction that twentieth-century eugenic programmes – pushed by authoritarian states such as Nazi Germany but pioneered in liberal democracies such as the United States, Canada, Sweden and Denmark – were 'evil' because they were guided by 'hopelessly wrong' understandings of genetics, which were enshrined into state-imposed reproductive policies.[151] State-led eugenic programmes attempted to steer evolutionary history in a predetermined direction, using individuals as instrumental means to 'enhance' collective entities such as the race and the nation. Liberal eugenics, according to Agar, would be different since it would require minimal state intervention: public authorities would be limited to setting up a free market in genetic enhancement services, establishing its operative rules, and policing its functioning. 'On the liberal approach to human improvements', writes Agar, 'the state would not presume to make any eugenic choice. Rather it would foster the development of a wide range of technologies of enhancement [...] Parents' particular conceptions of the good life would guide them in their selection of enhancements for their children.'[152]

The idea of delegating the organisation of a decentralised system for improving the human genetic stock to the 'invisible hand' of the market is not completely new. In 1969, Robert Sinsheimer – then-head of the Biology Division at Caltech – suggested as much. 'A new eugenics has arisen', he wrote in an article titled 'The Prospect of Designed Genetic Change', 'based upon the dramatic increase in our understanding of the biochemistry of heredity and our comprehension of the craft and means of evolution'. Commenting on the possibilities opened by early research on recombinant DNA, Sinsheimer affirmed that the 'horizons of the new eugenics are in principle boundless'. Its dawn was for him 'a cosmic event', 'a turning point in the whole evolution of life'. Yet, he recognised that 'the ethical dilemma remains. What are the best qualities, and who shall choose?' His answer was obliquely suggested rather than openly stated. He was confident that 'intelligent genetic intervention' could soon be deployed on an unprecedented scale: to improve the lot of the many 'losers of the chromosomal lottery', including 'the 50,000,000 "normal" Americans with an IQ of less than 90'. Sinsheimer was also convinced that this 'new eugenics could, at least in principle, be implemented on a quite individual basis, in one generation, and subject to no existing social restrictions'.[153]

Five years later, libertarian philosopher Robert Nozick offered a more detailed description of how 'the issue of genetic engineering' could be managed on the basis of a liberal market system and a minimal, night-watchman state. In *Anarchy, State and Utopia*, a book that continues to be a sort of right-wing libertarian manifesto in political philosophy, Nozick imagined how genetic technologies could be turned into molecular machines churning out corporate profits and consumer freedoms. 'Many biologists', he writes, 'worry over what sort(s) of person there is to be and who will control this process. They do not tend to think, perhaps because it diminishes the importance of their role, of a system in which they run a "genetic supermarket", meeting the individual specifications (within certain moral limits) of prospective parents. [...] This supermarket system has the great virtue that it involves no centralized decision fixing the future human type(s).'[154] Nozick presents 'the supermarket' as a neoliberal paradigm whose spatial architecture offers an ideal model for the organisation of a future society; one in which genetic engineering contributes to the selection of human types. Such a market model would decentralise decision-making processes over the relative value of genetic traits and, thereby, the steering of species-history through 'supply and demand'.

Many (neo)liberal theorists have since embraced Nozick's vision of a free market in genetic enhancement services. Contemporary supporters of liberal eugenics recuperate Francis Galton's idea that modern biological science offers precious tools for a eugenic 'science of improving stock'.[155] Yet, they distance themselves from the authoritarian methods once used to achieve this ideal. The market is presented as a different means to achieve the same eugenic ends: a space of freedom, whose invisible hand will shape the social body. In advanced neoliberal societies, we are told, the accumulation of genetic stock will not be guided by a single conception of the good, but by a collection of free individual choices. Genetic consumers will shape future generations by purchasing the genetic enhancement interventions that best reflect their values. Julian Savulescu – director of the Uehiro Centre for Practical Ethics at Oxford University – has argued that in the future genetic market parents will have 'a moral obligation' to act as rational economic actors, 'select[ing] the child, of the possible children they could have, who is expected to have the best life'.[156]

'Liberal eugenics' can be seen as a paradigmatic expression of neoliberal rationality in the age of genomic capital. In his intertwined genealogies of biopolitics and neoliberalism, Foucault suggested that 'the political problem of the use of genetics arises in terms of the formation, growth, accumulation, and improvement of human capital'. In other words, if we want to 'grasp the political pertinence of the present development of genetics', its 'implications at the level of actuality itself' and 'the real problems that it raises', we must look at how this new form of power-knowledge is integrated within existing social structures. The deployment of genetic biotechnologies does not take place in a social vacuum. Rather, it is profoundly influenced by long-standing social tendencies towards the reduction of people into 'human capital'. This mutation, according to Foucault, is an essential part of the neoliberal programme for 'the rationalization of a society and an economy' on the basis 'of enterprise-units'.[157]

The individual worker becomes 'a sort of enterprise': with their mental and physical abilities representing a form of capital on the basis of which they receive a stream of income. As a result of this social arrangement, the neoliberal subject faces constant pressures to improve and accumulate their human capital. 'How is human capital made up?' asks Foucault. On the one hand, people cultivate a set of acquired skills, which can be improved by investing in education, training, etc. On the other, they have a set of innate physical abilities, some of which are genetic. But how can one invest in genetic capital? 'As soon as a society poses itself the problem of the improvement of its human capital in general', writes Foucault, 'it is inevitable that the problem of the control, screening, and improvement of the human capital of individuals, as a function of unions and consequent reproduction, will become actual, or at any rate, called for'. The social construction of the neoliberal subject as an individual enterprise managing its own human capital drives the search for ever-new ways of genetic enhancement. In the end, according to Foucault, neoliberal society forces each of us to confront 'the costly choice of the formation of a genetic human capital', fueling a radical social transformation with 'racist effects'.[158]

The ink spent justifying a 'market eugenics' does not tell us much about the likelihood that this vision will be realised in the future. Bioethical debates on 'liberal eugenics' do, however, tell us something about the present. The less we believe in the power of social cooperation and of political action, the more technoscience appears as the only viable source of social change. Neoliberal society is rapidly approaching the point where large-scale genetic interventions to mitigate climate change appear more realistic than reducing carbon emissions; just as human enhancement through genetic means is presented as more feasible than providing free and public healthcare, education and nutrition to anyone, anywhere. 'Unfit for Life: Genetically Enhance Humanity or Face Extinction', the title of a recent public lecture held by Julien Savulescu at Oxford University sums up almost too well the logic of this neoliberal politics of despair.[159]

The impossibility to imagine a future beyond neoliberal market society creates a present in which mass genome editing appears as the only possible source of socio-ecological transformation, and therefore as something worthy of massive public expenditures and private investments. This, in turn, reinforces the cultural belief that genetic interventionism may fix the ecological and social crises proliferating in the present: not by creating a society more adapted to the organic needs of human and non-human ecologies, but by genetically adapting human and non-human ecologies to life under hyper-modern capital.

Conclusion

The narrative of industrial modernisation often departs from factories shrouded in the brown fog of nineteenth-century Manchester, 'where mountains of iron turnings are carted away to the foundry in the evening, only to reappear the next morning in the workshops as solid masses of iron'.[1] Mechanisation has, without a doubt, largely defined the industrial epoch. Modernisation, however, has another technoscientific facet. Departing from Gregor Mendel's monastery in nineteenth-century Moravia, we followed the first tentative steps towards the establishment of a technoscience of heredity. When Mendel started his research on pea plants, he did not operate in a monastic social vacuum, nor was he moved simply by scientific curiosity. He was part of a broader cultural context which strived to understand *how* particular breeds of sheep or crops emerged, with the goal of building a power-knowledge that would enable the conscious direction of these natural histories. Much as Newton's *Principia Mathematica* offered technical solutions to the social questions posed by seventeenth-century societies, this emergent biological research was never disconnected from social production. The answers given were technological and scientific, but the questions posed were social and political.

The birth of our hyper-genomic age is indebted to a quintessentially modern quest to grasp and control evolutionary processes. We have followed the drive to genetic command and control since its slow beginnings in the nineteenth century and through its own great acceleration after the Second World War. This collective search proceeded at the pace set by larger socio-economic processes of industrialisation and globalisation. The search for hereditary particles and for technological means of governing mutations mirrored a growing interest in engineering the living means of production employed in agriculture, in industry, in medicine. It also promised new ways of controlling the human animal as a breeding living organism. Early twentieth-century practitioners of this nascent discipline used chemical bombardment and nuclear irradiation to accelerate the rate of genetic mutations in living organisms. Artificial mutagenesis, a term coined by Hermann Muller in the 1930s, was hailed as 'the first deliberate, successful scientific interference with the process of heredity by external agencies'.[2] Genes started to be imagined as molecular levers, whose manipulation could provide control over evolutionary patterns: over 'those forces, far-reaching, orderly, but elusive, that make and unmake our living worlds'.[3] Atomic energy promised not only new ways of killing, but also new ways of shaping life. Control, however, quickly revealed itself to be fragmented and partial. Radiation could accelerate the rate at which random genetic mutations occur, but the price was high: most mutant organisms were neither useful nor fit for life.

With the informatic turn of the 1950s and 1960s, molecular biology spurred an epistemic shift which transformed life-forms into living *systems*. Deciphering the molecular structure of DNA further illuminated the nature of heredity, and exposed new ways to control and direct it. A major revision of the fundamental conception of biological processes was pushed through. Scientists increasingly thought of living organisms and their physiological activity in terms of genetic information, code, structure, sequence, transcription, translation and feedback. Life became a one-way street: DNA directs the production of RNA, RNA directs the production of proteins, and proteins ultimately combine to create the entire organism. In this cybernetic view of the organism, the gene was singled out as the regulator which could be used to steward metabolic processes at ever-larger scales: much like the throttle-lever controls the speed of Watt's steam engine, the genomic-lever is set to control living beings reified as biological systems. The living cell increasingly resembled a Fordist factory: a hidden abode of production endlessly churning out proteins on the basis of genetic blueprints. Molecular biology increasingly aspired to apply the engineering ideals that were revolutionising factory production to rationalise biology, to redesign living bodies to make them better, faster and more resilient. The application of 'principles of machine production' made thinkable, for the first time, the industrialisation of living metabolisms; their adaptation and rationalisation according to the needs of industry.

In the 1970s, these abstract aspirations were embodied in the recombinant bacteria mobilised by Genentech and other early biotech companies. Genetic engineering emerged as a new form of industrial labour. Mastering restriction and ligase enzymes, laboratory workers spliced together the genetic fabric of thousands of unrelated organisms 'creating a wide variety of novel genetic combinations'. Biotech companies could now steer the metabolic activities of living organisms. Bacterial bodies were turned into 'industrial bioreactors' converting feed into marketable human proteins. Crops were turned into 'chemical factories' synthesising bacterial toxins. Yeast were spliced with viral genes to churn out recombinant vaccines. Transgenic mice assisted the discovery, development and testing of new drugs and therapeutics products.

These chimeric bodies sparked ontological debates and furthered a new juridical, political and economic imaginary. A series of neoliberal juridical reforms originating in the US and later replicated internationally, re-classified these bodies as 'living artefacts': a peculiar composition of organic matter produced through modern manufacture. Genetically engineered organisms morphed into patentable objects and financial assets. Sequencing and splicing became routes to patenting; genomics became a strategy of capital accumulation. The neoliberal counter-revolution dovetailed with a corporate rush to claim monopoly rights on transgenic bacteria, recombinant mammals, bioengineered crops and isolated genes extracted from the cells of plants and animals.

Biodiversity hotspots suddenly appeared as repositories of valuable genetic sequences, which could be collected and recombined to generate new recombinant life-forms. The extraction of genetic matter from microorganisms, plants and animals was turned into an industry; and the biosphere mutated into a vast genomic mine waiting to be bio-prospected and excavated for deoxyribonucleic acids. Genomics not only offered new ways of producing and a new range of commodities. It also offered new ways of abstracting from the materiality of life, turning wet organic matter into a digital code represented by an endless series of A, T, C, G. This genomic abstraction increasingly serves as a vehicle for the financialisation of life: sequence, patent, enclose, repeat.

Once living assets entered the casino of the world financial market, the calculations of investors and the solipsistic dance of financial indexes became a material force capable of steering the course of natural history. The search for generous returns on investment becomes inscribed in the flesh of living organisms: in the cancerous body of each OncoMouse™, in each and every fast-growing AquAdvantage® salmon, in every seed of herbicide-resistant RoundUp Ready® soy. By directing the business strategies of hundreds of biotech firms around the world, Nasdaq's biotech index has become an evolutionary force that can reshape the physiology of a mouse, encourage the generation of new strains of bacteria, condemn or promote the splicing of potatoes with *Bacillus thuringiensis*. The financialisation of life entails two complementary aspects. Life-forms have been transformed into financial assets; and finance into a life-shaping force.

Capital shaped and steered the trajectory taken by biological research. Private foundations – such as the Rockefeller Foundation and, more recently, the Gates Foundation – have been active for decades in distributing funding, setting up research programmes, encouraging particular lines of experimentation, and lobbying for regulatory changes both at the national and the international level. Corporations, exposed to stringent competition in the world market, have invested heavily in turning emerging socio-ecological problems into frontiers of accumulation. How the genome is understood and acted upon, moreover, has been profoundly shaped by the historical doxa and cultural zeitgeists that dominated industrial societies. In the 1950s, molecular biologists projected onto the cellular world a mirror-image of their own time, turning the molecular frontier into a Fordist society bent on commodity production, producing copies upon copies of the items described by its industrial blueprints. In the 1990s, when neoliberal society became increasingly dominated by corporations driven by an abstract tendency to grow and self-replicate, molecular biology offered a rather congruous vision of nature: a world of automata driven by a selfish software code, whose single purpose is to secure its own conservation and endless self-replication.

As the opposition between capital and labour disappeared from view, the individual increasingly appeared as a lonely corporation born to compete for its survival, facing constant pressures to mobilise and enhance its genetic capital. Science became an increasingly competitive social environment, pushing a new class of entrepreneurial scientists to pursue lines of research that would grant financial rewards. Meanwhile, competition between nation states spurred public investments towards the construction of an expansive infrastructure for the new genomic age: a 'molecular railroad' intended to facilitate capital accumulation across multiple economic sectors. Sequencing went from a laborious, artisanal process requiring significant expertise to automated 'assembly line' operations carried out in gargantuan corporate establishments such as BGI.

In the 2020s, as the world spins into an ecological crisis that exposes the ungraspable complexity of planetary biogeochemical cycles, synthetic biologists think of the genome as an excessive, noisy and confusing software. Even the most simple organism appears to resist increasingly desperate efforts to simulate, model, understand and control it. Paradoxically, this only furthers aspirations to civilise the genomic frontier, simplify living organisms, and reduce them to organic systems that would be easier to interpret, control and manipulate. 'Natural biological systems are selected to continue to exist and evolve', write the authors of 'Refactoring Bacteriophage T7'. They are 'complicated systems that are difficult to understand and manipulate'. This is why, the authors conclude, genome engineering has the potential to improve currently existing living organisms: 'the genomes encoding natural biological systems can be systematically redesigned and built anew in service of scientific understanding or human intention'.[4] The audacity of the new generation of molecular biologists is rooted in the confidence that new genome editing technologies will finally endow them with the power to rewrite the genome, shape new life-forms, generate functional ecosystems, spin new synthetic threads in the web of life, and steer planetary biogeochemical cycles.

Molecular biologists promise 'a new era of genetic command of control' in which 'the genome would become as malleable as a piece of literary prose at the mercy of an editor's red pen'.[5] It is an aspiration that is rather reminiscent of Muller's rhapsodic pledge to 'place the process of evolution in our hands'.[6] In this new hyper-genomic era, which is rapidly unfolding under our own eyes, genome editing and synthetic biology are enabling new technoscientific modalities of manufacturing life, in a double sense: life becomes at once an object of manufacture and a subject which manufactures. Genome engineering not only midwives a new generation of proprietary life-forms; it also metamorphoses these life-forms into living machines, whose metabolism is engineered to transform production in different industries. Genome engineering gradually becomes a means of expanding, accelerating and securing the accumulation of capital on a planetary scale. It is the pulsing heart of an emerging biopolitical economy: a strategy of accumulation and a mode of government which relies on a technoscientific knowledge of biological processes and new means of manipulating them.

The newly birthed biopolitical economy is already troubled by roars of battle: Who will get to shape the genomic composition of countless biological organisms and thereby radically transform planetary ecologies? Who will own and control the new molecular means of production?

Who will master the complex technological assemblages – composed of machines, knowledges, ideas, laws, databanks, genomic arks, frozen embryos, proteins, enzymes and mountains of deoxyribonucleic acid – that are mobilised to produce mutant lives and mutant ecologies? The ongoing 'battle for CRISPR' is only the tip of the iceberg. Venture capitalists, technocrats, entrepreneurial scientists and multinational conglomerates are staking their competing claims: clashing for ownership over the core technologies of the new genomic age. They are spinning their narratives, turning the history of science into a means of claiming ownership over original inventions, born off the minds of special individuals worthy of monopolising the profits created by their single-handed efforts. The sanctioned historiographies of the genomic age are but the history of 'great scientists' and 'great investors'. A handful of companies are privatising the benefits – and socialising the substantial costs – of this newly found biotechnological dynamism. What perspectives may emerge, instead, from a history of technoscience from below: one that revindicates these technologies as emerging from endless social processes of co-production, mobilising a multitude of nameless lives across time and space?

Meanwhile, the molecular means of production are firmly in the hands of corporations, private foundations and the emergent class of scientists-entrepreneurs, who embody both the social authority of science and the political-economic power of the venture capital they mobilise. In the contemporary biopolitical economy, market competition guides restriction enzymes in their molecular journeys and suggest the double-stranded breaks that best serve corporate strategies. The targets struck by targeted mutagenesis are not random – that is the whole point – but who decides where to strike? Genome engineering is a technoscientific tactic pursued within laboratory walls, but the overall strategy is largely defined in corporate boardrooms and financial markets. The politico-economic objective pursued by genomic interventions, in other words, is never a purely technical affair. The Promethean enthusiasm towards a world in which 'evolution would finally be under human control' conceals that it is increasingly the 'invisible hand' of the market and the 'inhuman power' of capital that guide the newly engineered paths of natural history.[7] Technoscience breeds new mutant bodies, but it is capital that selects which of them are abandoned as curious experiments recorded for posterity in the pages of scientific journals, and which are scaled up and made to proliferate in the world.

Throughout the chapters, we have analysed some of the ways in which these genomic interventions are *expanding*, *accelerating* and *securing* the accumulation of capital on a global scale. In each case, the genetic intervention is intended to produce 'better' life-forms, which can increase the efficiency of existing bio-production system. Indeed, the main promise of genomic engineering is exactly this promise of producing a 'better life'. What 'better' means remains, however, very much a political question: a question that is largely defined by the needs of industry and the interests of those who control the new means of molecular production.

Capital strives, first and foremost, to *accelerate* the biological processes mobilised in production processes. In agriculture and forestry, a great deal of experimentation is focused on finding new ways to accelerate the growth rate of crops and trees; in pharmacology, metabolic engineering increasingly aims at accelerating microbial processes of protein synthesis; in aquaculture and the livestock industry, researchers tirelessly pursue the age-old goal of 'getting animals ready for their fate in less time'. Capital's characteristic need for speed is embodied in fast-growing salmons, tiger puffers and red sea breams; in fast-growing Myostatin Knock-Out Sheep and Cattle; in genetically modified eucalyptus trees that 'produce 20% more wood' and that are 'ready for harvest in five and a half years instead of seven'; in genetically modified bacterial strains, whose metabolism is redesigned to accelerate fermentation processes; in GM tobacco plants with a shorter 'photosynthetic recovery time', which are less vulnerable to bright light but grow 20 per cent faster; in 'photosynthetically efficient' crops.[8] Countless genomic interventions are directed towards a metabolic acceleration meant to speed up both cellular processes of protein synthesis and socio-economic processes of commodity production.

Genome engineering can also assist the *expansion* of capital accumulation by opening up new frontiers of production, by enabling the spatial concentration of living bodies, by creating whole new markets in areas traditionally outside the realm of commodity trade. A striking example of this trend is represented by growing corporate investments in the xenotransplantation business. While grassroot campaigns aimed at encouraging voluntary organ donations represent a non-market solution to the persistent shortage in transplantable organs, Revivicor's newly trademarked Uheart™ constitutes a tentative market response to the problem of insufficient organ donations. The humanisation of animal bodies via advanced genome editing techniques is conceived as the technological key that may open frontiers of accumulation that have so far been foreclosed by the criminalisation of human organ trafficking.

New frontiers of production may also be opened by adjusting living bodies to perform 'better' in otherwise highly stressful environments. Recombinetics' Naturally Cool™ cattle is described by company representatives as being adapted to 'withstand the stress caused by tropical production conditions by only changing a single base pair in prolactin receptor'.[9] The new brand of GM cattle, which has already been approved by the US FDA, promises to facilitate the expansion of cattle breeding in tropical regions. While this genomic intervention has been oft-publicised as ecological and humanitarian, it is likely to accelerate the ongoing expansion of intensive beef farming in the Amazon region. The environmental impact of industrial livestock-breeding is already profound. Indeed, 'cattle ranching is the largest driver of deforestation in the Brazilian Amazon'.[10] The introduction of increasingly efficient living means of meat-production is likely to intensify – rather than reverse – this trend. Naturally Cool™ has already been hailed for its potential of turning Brazil into 'another viable source in addition to the US for Angus beef and add billions of dollars in revenue to the industry'.[11]

Genome editing can facilitate the expansion of capital accumulation into new geographical areas, but it can also facilitate the intensification of production *in loco.* Take, for instance, genetically modified disease-resistant pigs, chickens and cows. These genetic interventions are often presented as an expression of a rather surprising ethical turn taken by giant multinational conglomerates that profit from industrial breeding in concentrated animal feeding operations. Yet, they also facilitate the accumulation of animal capital by reducing the losses caused by recurrent epidemics. Pandemic threats have imposed limits to the density of animal bodies that can be safely kept in each facility. Disease resistance opens the door to a further concentration of bodies in increasingly confined and unsanitary spaces. This would not only create the perfect conditions for new pandemics to proliferate – which will in turn call for new gene editing interventions in an endless spiral – it would also magnify the significant ecological impact of the livestock industry. As recent estimates already attribute 12 to 18 per cent of global greenhouse emissions to the livestock sector, anything that contributes to its further expansion is likely to accelerate global warming and ecological crisis.[12]

This leads directly to the third dimension of biopolitical economy. As the Great Acceleration in industrial impacts on planetary biogeochemical cycles threatens to undermine the ecological conditions for both capital accumulation and human well-being, genome engineering is increasingly evoked as a *deus ex machina*: a technoscientific fix by which seemingly unsolvable social issues will be abruptly resolved. Genetic bioengineering is presented, first of all, as a means of climate adaptation. Consider, for instance: heat-resistant cattle, drought-resilient crops, GM rice designed to prosper in high salinity environments, CRISPR crops 'well-suited for use in future conditions where temperatures and other climatic conditions near equatorial regions render farmlands less fertile', genetically modified semi-dwarf bananas 'more resistant to lodging as a result of intense winds, typhoons, and storms, anticipated to increase in severity as a result of climate change'.[13] Agribusiness is searching for new ways of securing its living means of production in a warmer and increasingly unpredictable climate. Global warming is also set to magnify the frequency, severity and incidence of a multiplicity of pathogens threatening both plants and animals.[14] In response, agriculturalists are knocking-out genes in an attempt to reduce the 'susceptibility loci' normally exploited by pathogens. We now have wheat, tomatoes and grapes resistant to powdery mildew, cucumbers that can withstand yellow mosaic virus, rice with added resilience to bacterial leaf blight, bananas resistant to BSV and grapefruits resistant to citrus canker.[15]

Genome editing is also morphing into a means of climate mitigation and a tool for increasingly audacious plans of planetary biogeoengineering. The basic idea is: since all life-forms are entwined together in the web of life, it is possible to steer that living web by tugging at one thread – or weaving in a new synthetic one. The genome is desperately grasped as an adjustable lever whose manipulation will enable technoscience to direct cellular metabolisms, govern the evolution of species and populations and, finally, steward biogeochemical processes at ever larger scales: from the gene to the body, from the body to the species, from the species to the world.

AquAdvantage® salmons, EnviroPigs™ and methane-light cattle are presented as savvy business propositions, but *also* as solutions to problems of phosphorus pollution, declining wild fisheries and increasing CO_2

emissions. We are presented with a quintessential neoclassical economist approach to nature, animated by the assumption that the contradictions of industrial agriculture can be resolved through the search for genetically designed marginal efficiencies. Genome editing is no longer simply a means to accelerate, expand and secure the accumulation of capital in specific branches of industry. Rather, it envisions transforming the planet into a never-ending biogeoengineering project, pursued through a strict control over biodiversity and endless biopolitical decisions over which species must be extinguished, which must be genetically protected, which must be de-extinguished, which must be cloned, which must be mutated. Since each species shapes its own ecological niche, through its metabolic interactions with the surrounding environment, genetic engineering constitutes an indirect way of constructing the world through the industrial production of living artefacts.

Synthetic biologists are formulating methods to scale up bio-geo-engineering 'from the test tube to planet Earth', reflecting the discipline's increasingly ambitious plans for planetary design.[16] The traditional aspiration to build 'better organisms' is giving way to intoxicating plans to build a 'better Earth'. Genetically modified bacteria, the old workhorses of molecular biology, are being reconstructed as means of 'earth terraformation' that 'could be used to safely prevent declines in some stressed ecosystems and help improving carbon sequestration'. These projects promote a form of geoengineering that would be relatively cheap and far-reaching 'thanks to the intrinsic growth of the synthetic organisms. This makes a big difference in relation to standard engineering schemes, where artifacts need to be fully constructed from scratch. Instead, once a designed population is released, appropriate conditions will allow the living machines to make copies of themselves and expand to the desired spatial and temporal scales, or even spread an engineered device.'[17] Such schemes aspire to transform sprawling industrial wastelands into functional habitats for engineered microbes designed to facilitate carbon sequestration. What kind of ecological shifts will be engendered remains uncertain.

* * *

Genomic biotechnologies have escalated a centuries-long race towards more 'efficient' lives and metabolic processes. Throughout our historical analysis, we have charted how some of these virtualities are already material in the present. It is impossible to determine where this race will ultimately lead. Much of it is presented as an *arms* race; between 'plants and pathogens', between humans and mosquitoes, between competing corporations or rival nation-states. Already in the 1990s, the Human Genome Project was presented as a biological 'Apollo project' part of a molecular 'space race'. Today, commentators remark that CRISPR has the potential 'to trigger Sputnik 2.0, a biomedical duel on progress between China and the United States'.[18] In whichever state's arbitrary territory we happen to find ourselves, we are bombarded with messages about the competitive advantage that will be generated by the adoption of new genomic biotechnologies.

War keeps recurring, both as a metaphoric trope and as a historical reality. Not only is there a war on mosquitoes, plant pathogens, rats; scientists have also raised concerns that these techniques might engender new forms of biological warfare.[19] Are we faced with a 'colonial war on nature', as Erwin Chargaff once argued in his critique of early genetic engineering? Perhaps. Genomic engineering certainly resonates with centuries-old practices of colonisation. Imperialism can take the form of plunder, appropriation and violence: seizing wealth, maiming and killing. Modern colonial ventures, however, were not simply a succession of isolated acts of plunder. They took upon themselves the task of refashioning colonised societies; engineering the socio-ecological landscape to make it legible, functional and profitable for the colonists.

A similar transition seems to be at play in the ongoing transition from traditional biopiracy to emerging forms of biocolonialism. The history of biopiracy has been, firstly, a history of acts of appropriation by which colonial powers seized life-forms, extracting them from their local ecosystems and making them circulate in profitable ways through imperial geographies. This history firmly persists in our own present. Today, however, these appropriations prepare a reorganisation of the molecular geographies of earthly life. Contemporary biocolonialism takes upon itself the task of redesigning living organisms, bioengineering their molecular landscape to render it legible, functional and profitable for the patenting corporation. Genome editing engenders a different form of life for the targeted bacteria, plant or animal; for their sanctioned owners; and for the the workers employed to labour with and through their bodies. AquAdvantage® salmon illustrates this vividly. Its genome has been redesigned to facilitate a particular process of production; its whole life-course and metabolic interactions have been engineered in order to facilitate its speedy arrival in the marketplace. Capital no longer simply appropriates the animal and plants mobilised in farms and factories; the animal's entire life and physiology is redesigned as part of an industrial labour process.

In a sense, one could say that this is nothing new: throughout modernity people have seen and treated animals as machines. This view is a product of modern times; as Marx observed, 'Descartes, in defining animals as mere machines, saw with the eyes of the period of manufacture'.[20] Yet, animals have routinely made people aware that they are *not* machines by misbehaving, by using their horns, by not performing as expected. Those flashes of animal agency and resistance explode modernist fantasies of total control over life. Living organisms return to being mesmerising, kindred beings of which we know very little. Are these moments of lucidity becoming increasingly rare, the more technoscience reduce life-forms into proprietary, designed biological 'systems'; the more technoscience aims to render life more legible, predictable, and controllable?

In all likelihood, the contemporary affirmation that genetic control is now firmly 'in our hands' will once again be disappointed, as it has been so many times in the past. It might be tempting to look back with condescension to the newspapers – and the Nobel Prize winners – that in the 1930s affirmed that X-ray mutagenesis would grant control over the invisible forces of evolution. With the lucidity of hindsight, the elated declarations of the 1970s that recombinant DNA would allow the complete breakdown of species barriers appear not only premature but misguided. Will future scientists look back in the same spirit to the newspaper titles – and the Nobel Prize winners – of our present time? CRISPR is hailed as the ultimate means of control over living processes: a 'word processor' that will allow seamless editing of human genomes, end disease, feed the world and solve climate change. Yet, CRISPR might soon become obsolete: 'Move over CRISPR, the retrons are coming', announces Harvard's Wyss Institute for Biologically Inspired Engineering, capturing the fact that today's revolutionary tool is tomorrow's technological fossil.[21]

Fantasies of total control are continually thwarted by their own contradictions. 'Every technical object contains its own negativity', Paul Virilio once pointed out.[22] In very much the same way, genome editing generates its own contingent and systemic negativities. CRISPR carries 'off-target effects', 'unexpected mutations' and 'genetic drifts' as its inescapable doppelgängers. Genome editing also re-invents the biological accident: security agencies now regularly warn their governments of the potential 'biohazards' that could be sparked by synthetic and engineered pathogens escaping from their confined laboratory environments. Such a possibility is not science-fictional; lab leaks have occurred in the past and could repeat themselves in the future. As a recent review of biosecurity published in *Nature* points out: 'The 2007 Foot and Mouth disease outbreak in the UK was attributed to a leaking pipe at Institute of Animal Health at Pirbright. The last known cases of smallpox and SARS were both caused by laboratory exposures, and involved secondary transmission from infected researchers to individuals outside of the laboratory. The 1977 influenza pandemic was caused by a strain closely related to those isolated in the 1950s, suggesting an anthropogenic origin.'[23]

Genome engineering opens up a fundamentally unstable frontier in which technoscience's ability to manipulate genomes surpasses its capacity to correctly predict the cascading effects of those very manipulations. The genetic integral accident is perpetually warded off, contained, predicted through simulations, and limited in its destabilising effects. Its haunting potentiality is carefully weighted against the potential benefits of further scientific advances. Yet, it continuously threatens the illusion of complete control over interwoven biological processes. Today, many lucidly recognise the new dangers – and the productive potentialities – inherent to emerging genomic biotechnologies. The literature on the multiple risks associated with genome editing is ponderous: off-target effects, unexpected mutations, genetic drifts, ecological unexpected consequences, bioaccidents, acts of bioterrorism, acts of genetic warfare. Even so, powerful structural tendencies are at play, pushing the search for ever-more effective means of 'genetic command and control'. Sovereign states and transnational corporations are locked in a competitive environment that fuels fears of falling behind in the technoscientific arms race for genetic control. The ecological crisis is portrayed as an existential threat of such a global magnitude that the speedy development of new genomic biotechnologies is no longer a political option, but an urgent necessity.

Expressing his social role as 'capital personified and endowed with consciousness and a will', Bill Gates remarks in 'Gene Editing for Good' that re-engineering the genomic fabric of life represents the only way to avoid a future scarred by hunger, scarcity and ecological collapse.[24] We are at a world-historical juncture: genomic capitalism, or the apocalypse. For capital, the genomic frontier – at once vast and submicroscopic – represents a pressing biological limit beyond which lies open pastures of expansion. While industrial production is undermining its very ecological conditions of possibility, genomics forms part of a swelling deluge of technological fixes which aims to save capital, and the industrial acceleration it continuously fuels, from itself. Subverting the neoliberal dogma that 'there is no alternative' to the large-scale re-engineering of Earth's biological inhabitants, we have argued that the contemporary ecological crisis is exposing capital's proliferating contradictions and totalising tendencies. Grandiose claims of techno-genomic salvation, ultimately, are little more than a veneer for a new wave of enclosures, and a powerful justification for the further expansion of the social system that endlessly generates hunger, scarcity and ecological crisis. Genome editing, synthetic biology, biogeoengineering – at least as they are currently declined – do not represent an alternative to the industrial Great Acceleration. They constitute, rather, an *intensification* and *biologisation* of that historical tendency.

This is why this new wave of accumulation-by-genomic-modification is resisted at every corner. If the large-scale territorialisation of capital social relations at the level of the genome constitutes one of the many proliferating instances of 'new imperialism', this violent expansion is not occurring without contestation.[25] Although we see capital's tendency towards the real subsumption of life as an expression of its internal relations – and therefore integral to its historical and geographical development – we equally contest the inevitability of this process. In multifarious ways, humans and non-humans are refusing and contesting this movement. Global struggles against the subsumption of life under capital – indigenous struggles against biopiracy and genetic drifts; peasant revolts against the Gene Revolution; juridical struggles against the enclosure of the human genome, the patenting of genetically modified life-forms and the enclosure of common knowledge; demands for free and equal access to genetic vaccines; resistance against new forms of genomic surveillance and discrimination; opposition to the geneticisation and medicalisation of human difference; actions against concentrated animal feeding operations and animal testing; pushbacks against projects of genetic pest control and planetary biogeoengineering – are proliferating.

Some commentators would have us believe that these struggles can be overcome by 'loving our monsters', embracing the boundary-transgressions of GM fish, humanised pigs and cancerous mice. Take for instance Bruno Latour's idealist reading of Mary Shelley's 1818 *Frankenstein; or, the Modern Prometheus*. A treasure trove of monstrous metaphors, Shelley's *Frankenstein* remains a prescient literary allegory of our present condition. It has become a popular myth whose differing interpretations filter global debates surrounding genetic engineering and genomic biotechnologies. In the novel, much like Prometheus created life from clay, Dr Frankenstein hopes to build 'a new species [that] would bless me as its creator and source'. He envisions a future in which 'many happy and excellent natures would owe their being to me'.[26] Shelley captures the outlines of a quintessentially modern quest to establish control and ownership over living processes. Yet, Latour insists that this allegory of the present should be read as an incitement to double-up on the modernist stance: celebrating the death of nature without worrying about 'unwanted consequences' and 'precautionary principles'. There is no point in slowing down the Great Acceleration in industrial impact, we should rather push through towards an imminent future in which 'we can fold ourselves into the molecular machinery of soil bacteria' and 'run robots on Mars'.[27]

Our reading of *Frankenstein* as a novel, of modernity as a historical process, and of the present as a political crisis is antagonistic to the one offered by Bruno Latour and his disciples at Silicon Valley's Breakthrough Institute. Latour imagines that modernists dreamed of a humanity freed from nature and that going beyond modernity simply entails embracing our entanglement in nature. Yet, the main thrust of modernity was not to expel nature from the *polis*, but rather to establish ever-new forms of control and domination over nature, new ways of extracting value from it, of enclosing its 'free gifts', and profiting from its subordinated existence within the *polis*. Descartes, much like Francis Bacon, dreamed of a world in which technoscientific knowledge of 'the powers and the effectiveness of fire, water, air, the stars, and all the other bodies that surround us' matched the knowledge 'of the various trades of our craftsmen', thereby making 'ourselves the masters and the possessors of nature'.[28] Frankenstein embodies this modern tendency. His lack of love for the monster he created, which Latour presents as the only issue in an otherwise rather admirable plan, was not an unexplainable accident but a direct consequence of his modernist approach to nature, of his vision of living organisms as machines to be manufactured, owned and traded. Today, the taste for technoscientific mastery over nature has not waned, and it is rather likely that the many forms-of-life currently designed as commodities and living means of production will be functionally employed, rather than loved. One cannot, as Latour wishes, replicate Frankenstein's industry without also replicating his alienation from life (in general) and from the living results of his labour (in particular). Techno-utopianism cannot be at once Promethean and ecological, functionalist and loving, instrumentalist and caring.

So, the centennial question resurfaces, what is to be done? If modern technoscience – with its quest to industrialise nature – delivered us into the Anthropocene, is it capable of fostering a liveable world? This book has formulated a critique of increasingly recurrent calls to engineer Earth's life-worlds; a vision relying on technocracy and industrialism and rooted in a rigid 'capitalist realism' which finds it easier to imagine mutating the many threads that constitute the web of life than changing the social organisation of production.[29] Against these 'monocultures of the mind', which are fettered to the artificial scarcity imposed by capital, we might ask what heterotopias of communal abundance can be imagined in the present.[30]

This question of heterotopias, however, quickly mitoses into two. First, one is confronted by a rather despairing question: 'Does it even matter what I think?' Many would ask themselves that, since they rightly feel disempowered by the fact that these debates occur in lavish hotels and conference halls with strict guest lists. Well-paid lobbyists have tried to keep these questions out of democratic debate by rendering them technical or juridical – by presenting them as unapproachable puzzles that can only be discussed by approved experts. Yet, technoscience is not a separate republic sanitised from political projects and social struggles. Regardless of one's positions on current developments and deployments of genetic biotechnologies, many would probably agree on the necessity of radically democratising any decision concerning what kinds of genome-edited lives – if any – should be created, under which conditions, and for what reasons. This question cannot be satisfactorily answered by corporations, scientists and technocrats, for the simple reason that those life-forms are not theirs to be decided upon. Precisely because every genetically engineered body is connected through the web of life to all other life-forms, there is a need for democratic deliberation on how these can shape the biosphere.

For all its intuitive appeal, 'democracy' can quickly collapse into an empty signifier. After all, who is against democracy? Few people, at least in rhetoric. Molecular biologists and corporations hail CRISPR as a democratic tool because it is cheaper and easier to use than previous technologies. Does CRISPR being cheaper make it 'democratic'? Is it enough to say that we need to 'democratise' decision-making, when the questions have already been asked according to principles inherited from the past? What would a more radical democratisation entail? Then, the demands of democracy become impossible to contain within the limited technical considerations peddled in scientific debate. Rather, it spills into other areas: into the need to construct a society in which every person has the time, education and interest to participate in collective decision-making processes about what natures should be collectively produced; in which self-governance is not restricted to answering ready-made questions; in which people can engage in meaningful collective decision-making about organising our being-in-nature.

The second question which arises is: 'if there *was* such a democratisation of decisions, what would be my position?' Our own position is rooted in the recognition that there is a fundamental tension at the heart of molecular biology; a troubling and persisting tension between the realisation of the complexity of even the simplest living organisms and the temptation to engineer them. Drew Endy, one of the synthetic biologists who developed the BioBrick concept, started off on his path to redesign 'better' and more 'efficient' forms of life exactly because he was not able to predict and control the replication patterns of even the most elementary life-form. Molecular biology can be an awe-inspiring discipline. Max Delbrück, the influential physicist who helped establish molecular biology as a discipline, once said that the 'closer one looks at these performances of matter in living organisms, the more impressive the show becomes. The meanest living cell becomes a magic puzzle box full of elaborate and changing molecules'.[31] The development of more and more systems of measurement generates ever-accumulating mountains of data about the irreducible complexity of even the simplest living organism. Models consistently fail, cutting straight to the heart of the central contradiction of the mechanistic conception of life. Yet, the same discipline has been driven by unrelenting socio-economic pressures towards intervening into this universe of irreducible complexity, which escapes each and every attempt at abstraction. Our position is that there should be cautious awareness of the limitations of our understanding of biological, ecological and social formations and the multifarious ways in which they interact.

More fundamentally, we argue that these decisions cannot be isolated from broader questions concerning economic structures, cultural tendencies, political struggles and ecological shifts. Capital is based on the separation and estrangement of people from the means of life. The world-ecology emerging from capitalist development will always reproduce this essential rift through dizzying cycles of crises and fixes; and no (bio) technological fix can ever resolve the underlying contradiction. The past two centuries of capital accumulation have profoundly transformed the biogeochemistry of the planet, while engendering ever-new means of adapting life-forms to capitalogenic environments. The structural tendencies characterising the Great Acceleration are clear. But structural tendencies are not all there is in history. Animals – human and non-human – continue to be, as Marx once said, the worst means of production 'because they have a head of [their] own'.[32] Rebellions and subversions are always possible; accelerations can be resisted and slowed down; trajectories can be shifted and reversed. Walter Benjamin once remarked that 'Marx says that revolutions are the locomotive of world history. But perhaps it is quite otherwise. Perhaps revolutions are an attempt by the passengers on this train – namely, the human race – to activate the emergency brake.'[33] Appealing as Benjamin's metaphor is, history is never confined to a train set on a track. Like life itself, history is organic and utterly unruly. After pulling the emergency break, we can get off the train altogether and begin nurturing radically different ecologies and socio-natural histories.

Notes

All websites were last accessed on 20 March 2022.

Introduction

1. These expressions are widely used. See, for instance: Doudna, J., and Sternberg, S. (2017) *A Crack in Creation: The New Power to Control Evolution*. Houghton Mifflin; Feng, Z., et al. (2013). Efficient genome editing in plants using a CRISPR/Cas system. *Cell Research*, *23*(10): 1229–1232; Dance, A. (2015). Core concept: CRISPR gene editing. *Proceedings of the National Academy of Sciences*, *112*(20): 6245–6246; Rosner, H. (2016). Tweaking genes to save species. *New York Times*, 16 April [www.nytimes.com/2016/04/17/opinion/sunday/tweaking-genes-to-save-species.html]; Phelan, M. (2015). Science Selects CRISPR Genome-Editing Tool as 2015 Breakthrough of the Year. [www.aaas.org/news/science-selects-crispr-genome-editing-tool-2015-breakthrough-year]; Kuchler, H. (2021). The code breaker and Crispr people. *Financial Times*, 18 February; Royal Society of Biology (2016). RSB at the British Science Festival: Can gene editing save the world? [www.rsb.org.uk/news/rsb-at-the-british-science-festival-can-gene-editing-save-the-world].
2. All quotes from the Nobel Prize Committee are available at www.nobelprize.org/prizes/chemistry/2020/press-release.
3. Ibid.
4. Marx, K. (1956). *Capital: A Critique of Political Economy, Volume II*. Progress Publishers, p. 142.
5. Ibid., pp. 142–144.
6. The study of biopolitical economy has already generated several critical accounts that inform our work. Although these authors have not presented their work as participating in a collective critique of biopolitical economy, we propose this concept as an epistemological bridge between our separate research efforts, and those of many others. See, for instance: Sunder Rajan's *Biocapital*, which focuses on the biomedical industry in relation to capital accumulation; Melinda Cooper's *Life as Surplus*, which draws on Foucault to offer a genealogy of biopolitical strategies since the beginning of the neoliberal era; Eugene Thacker's *The Global Genome*, which offers a lucid account of the complex relation between the digital and the material dimension of contemporary genomics; and Kenneth Fish's *Living Factories* that examines the living factory metaphor in contemporary accounts of biotechnology. Rajan, K. (2006). *Biocapital: The Constitution of Postgenomic Life*. Duke University Press; Cooper, M. (2008). *Life as Surplus: Biotechnology and Capitalism in the Neoliberal Era*. University of Washington Press; Thacker, E.

(2006). *The Global Genome: Biotechnology, Politics, and Culture*. MIT Press; Fish, K. (2013) *Living Factories*, McGill-Queen's University Press.

7. Marx, K. (1976). *Capital: A Critique of Political Economy. Volume I*. Penguin Classics, p. 493, n. 4.
8. Foucault, M. (1978). *The History of Sexuality: Volume I*. Pantheon Books, p. 138.
9. Marx, K. (1963). *The Eighteenth Brumaire of Louis Bonaparte*. International Publishers.
10. Clark, A. (2001). On stage with the supermodel sheep. *The Guardian*, 6 January.
11. Steffen, W., et al. (2015). The trajectory of the Anthropocene: the great acceleration. *The Anthropocene Review*, *2*(1): 81–98.
12. This very concept violently obscures the social context in which this 'anthropos' has operated, and the profoundly unequal outcomes of these processes. Upon closer inspection, the fingerprints observed by scientists in Earth's stratigraphic record seem not to read 'anthropos' but rather 'capital'. Indeed, the Anthropocene is said to have been inaugurated with the industrial revolution of the nineteenth century, which has since introduced dramatic changes to Earth's biogeochemistry. Malm, A. and Hornborg, A. (2014). The geology of mankind? *The Anthropocene Review*, *1*(1): 62–69.
13. Gates, B. (2018). Gene editing for good: how CRISPR could transform global development. *Foreign Affairs*, 97: 166–170.
14. Doudna, J. and Sternberg, S. (2017). *A Crack in Creation*. Houghton Mifflin, pp. xiii, 100.
15. Crutzen, P. (2002). Geology of mankind. *Nature*, *415*(3): 23.
16. This is an extremely common expression in the history of biology. See, for instance: Stern, C. (1953). The geneticist's analysis of the material and the means of evolution. *The Scientific Monthly*, *77*(4): 194; Paigen, K. and Petkov, P. (2010). Mammalian recombination hot spots. *Nature Reviews Genetics*, *11*(3): 221–233; Sun, J. X., et al. (2012). A direct characterization of human mutation based on microsatellites. *Nature Genetics*, *44*(10): 1161–1165.
17. Hershberg, R. (2015). Mutation – the engine of evolution. *Cold Spring Harbor Perspectives in Biology*, *7*(9): a018077.
18. Lucretius, T. (1942). *De Rerum Natura*. University of Wisconsin Press, pp. 334–343. For an interpretation of the swerve, see: Prigogine, I. and Stengers, I. (2018). *Order out of Chaos*. Verso; Parisi, L. and Terranova, T. (2000). Heat death: Emergence and control in genetic engineering and artificial life. *Critical Theory* [https://journals.uvic.ca/index.php/ctheory/article/view/14604/5455]. For a provocative account of the role of mutations in evolutionary patterns, see: Nei, M. (2013). *Mutation-Driven Evolution*. Oxford University Press.
19. A representative example of these early bioengineering sketches is published in: Cohen, S. (1975). The manipulation of genes. *Scientific American*, *233*(1): 24–33.
20. Joseph Masco has used the term 'mutant ecologies' to describe the nuclear waste sites of Los Alamos. Starting from the 1930s, radiation breeding and atomic gardening have been instrumental to generate mutant ecologies.

Today, however, many other tools for creating such ecologies are available. Masco, J. (2004). Mutant ecologies: radioactive life in post-Cold War New Mexico. *Cultural Anthropology*, *19*(4): 517–550.

Chapter 1 Life's Inner Workings: Cracking Codes, Mutant Flies and Recombinant Lives

1. For a well-documented history of Mendel's works and its lasting influence on modern molecular genetics see: Carlson, E. (2004). *Mendel's Legacy*. Cold Spring Harbor Press.
2. Poczai, P., Bell, N. and Hyvönen, J. (2014). Imre Festetics and the Sheep Breeders' Society of Moravia: Mendel's forgotten research network. *PLoS biology*,*12*(1): e1001772.
3. Wood, R. J. and Orel, V. (2005). Scientific breeding in central Europe during the early nineteenth century: background to Mendel's later work. *Journal of the History of Biology*, *38*(2): 239–272.
4. Orel, V. (2009). The 'useful questions of heredity' before Mendel. *Journal of Heredity*, *100*(4): 421–423.
5. Johannsen, W. (1911). The genotype conception of heredity. *The American Naturalist*, *45*(531): 129–159.
6. Loeb, J. (1912). *The Mechanistic Conception of Life*. University of Chicago Press, p. 3.
7. Loeb, J. (1922). *Proteins and the Theory of Colloidal Behavior.* McGraw-Hill. For an analysis of the importance of Loeb for biology, see Fruton, J. (1972). *Molecules and Life*. Wiley & Sons, pp.131–144; Pauly, P. (1987). *Controlling Life: Jacques Loeb and the Engineering Ideal in Biology.* Oxford University Press, pp. 151–153; Kay, L. (1992). *The Molecular Vision of Life*. Oxford University Press, pp. 104–120.
8. Shine, I. and Wrobel, S. (2014). *Thomas Hunt Morgan: Pioneer of Genetics*. University Press of Kentucky.
9. Muller, H. (1927). The problem of genetic modification. *Verhandlungen des V. Internationalen Kongress fur Veresbungswissenschaft*, *1927*: 234–260. See also Carlson, E. (1971). An unacknowledged founding of molecular biology. *Journal of the History of Biology*, *4*: 149–170.
10. Muller's quotes in this paragraph are taken from the following texts (in this order): Muller, H. (1927). Artificial Transmutation of the Gene. *Science*, *66*: 84–87. Muller, H. (1929). The Gene as the Basis of Life. *Proceedings of the International Congress of Plant Science*, *1*: 897–921. Muller, H. (1935). *Out of the Night: A Biologist's View of the Future*, Vanguard Press, p. 37. Muller, H. (1946). The production of mutations. In Nobel Foundation (1964). *Nobel Lectures. Physiology or Medicine 1942–1962*. Elsevier.
11. Kaempffer, W. (1928). The Superman: eugenics sifted. *New York Times*, 27 May, p. 2. Despite this bold beginning, and the sensational title, the article went on stressing that many scientists remained sceptical and that 'superman is still in the distance'.

12. Kaempffert, W. (1936). A biologist's view of man's future. *New York Times*, 15 March, p. 4.
13. Gray. G. (1937). *The Advancing Front of Science*. McGraw-Hill, p. 214.
14. Smith, H. (1958). Radiation in the production of useful mutations. *The Botanical Review*, *24*(1): 1–24. See also: Singleton, W. R. (1955). The Contribution of Radiation Genetics to Agriculture. *Agronomy Journal*, *47*(3): 113–117.
15. Hamblin, J. (2009). Let there be light … and bread: the United Nations, the developing world, and atomic energy's Green Revolution. *History and Technology*, *25*(1): 25–48.
16. Sanz Lafuente, G. (2021). Atoms for feeding. *Scandinavian Economic History Review*, *69*(3): 301–323; Kawai, T. (1988). Radiation breeding – 25 years and further on. *Gamma Field Symposia*, *35*(1): 1–36.
17. Howorth, M. (1960). *Atomic Gardening for the Layman*. KBP Publications.
18. Ma, L., et al. (2021). From classical radiation to modern radiation. *Frontiers in Public Health*, *9*: 10.3389.
19. The interview is reported in: Broad, W. (2007). Useful mutants, bred with radiation. *New York Times*, 28 August.
20. Muller, H. (1936). Physics in the attack on the fundamental problems of genetics. *Scientific Monthly*, *44*(3): 214.
21. Delbrück, M. (1949). A physicist looks at biology. *Transactions of the Connecticut Academy of Arts and Sciences*, *38*(1): 173–190.
22. Summers, W. (1993). How bacteriophage came to be used by the Phage Group. *Journal of the History of Biology*, *26*(2): 255–267.
23. Friedmann, H. (2004). From butyribacterium to *E. coli*. *Perspectives in Biology and Medicine*, *47*(1): 47–66.
24. Weaver, W. (1938). Molecular biology. In *Rockefeller Foundation Annual Report 1938*, pp. 203–204 [www.rockefellerfoundation.org/wp-content/uploads/Annual-Report-1938-1.pdf]. See also Weaver, W. (1970). Molecular biology: the origin of the term. *Science*, *170*(2): 581–582.
25. Kay, L. (1992). *The Molecular Vision of Life*. Oxford University Press; Morange, M. (2000). *A History of Molecular Biology*, Harvard University Press; Vettel, E. (2013). *Biotech: The Countercultural Origins of an Industry*, University of Pennsylvania Press.
26. Kay, *The Molecular Vision of Life*, pp. 9, 27–29.
27. Ibid., p. 44.
28. Quoted in: Kohler, R (1976). The management of science. *Minerva*, *14*(3): 296.
29. Quoted in: Kay, L. (1992). *The Molecular Vision of Life*, p. 44.
30. Ibid., p. 7.
31. Hessen, B. (2009). The social and economic roots of Newton's *Principia*. In Freudenthal, G. and McLaughin, P. (eds) *The Social and Economic Roots of The Scientific Revolution*. Springer, pp. 41–101.
32. For an analysis of molecular biology's role in the political economy of the 1940s and 1950s, see: Yoxen, E. (1984). *The Gene Business*. HarperCollins.
33. Schrödinger, E. (2012). *What Is Life?*, Cambridge University Press, pp. 21–22. For an assessment of Schrödinger's role in the history of genetics and molec-

ular biology see: Yoxen, E. (1979). Where does Schroedinger's 'What is Life?' belong in the history of molecular biology? *History of Science*, *17*(1): 17–52; and Kay, L. (2000). *Who Wrote the Book of Life?* Stanford University Press, pp. 59–72.

34. See, for instance, Morange, M. (2020). *The Black Box of Biology*. Harvard University Press.
35. Avery, O., MacLeod, C. and McCarty, M. (1944). Studies on the chemical nature of the substance inducing transformation of pneumococcal types. *Journal of Experimental Medicine*, *79*(1): 155.
36. All quotes by Boivin: Boivin, A. (1947). Directed mutation in colon bacilli, by an inducing principle of desoxyribonucleic nature. *Cold Spring Harbor Symposia on Quantitative Biology*, *12*, pp. 7–17.
37. Watson, J. and Crick, F., (1953). A structure for deoxyribose nucleic acid. *Nature*, *171*(4356): 737–738. The article was accompanied by two papers authored by Maurice Wilkins and Rosalind Franklin, displaying the experimental data on which Watson and Crick had based their theoretical speculations. A number of historians have emphasised the insufficient recognition of Franklin's role in the discovery of DNA structure. See: Maddox, B. (2002). *Rosalind Franklin: The Dark Lady of DNA*. HarperCollins.
38. Watson, J. and Crick, F. (1953). Genetical implications of the structure of deoxyribose nucleic acid. *Nature*, *171*: 964–967. See also: Watson, J. and Crick, F., (1953). The structure of DNA. *Cold Spring Harbor Symposia on Quantitative Biology*, *18*: 123–133.
39. Yoxen, E. (1981). Life as a productive force. In Levidow, L. and Young, B. (eds) *Science, Technology and the Labour Process*. CSE Books, pp. 66–122.
40. Wiener, N. (1965). *Cybernetics: Or, Control and Communication in the Animal and the Machine*. MIT Press.
41. Wiener, N. (1950). *The Human Use of Human Beings: Cybernetics and Society*. Houghton Mifflin, p. 103. See also: Cobb, M. (2013). 1953: when genes became 'information'. *Cell*, *153*(3): 503–506.
42. Crick, F. (1958). On protein synthesis. *Symposia of the Society for Experimental Biology*, *12*(1): 138–153. See also: Crick, F. (1970) Central dogma of molecular biology. *Nature*, *227*(5258): 561–563. Crick's model was further refined in the 1960s, when molecular biologists detailed the fundamental role played by RNA as a mediator between DNA and protein synthesis. This simple scheme was later complicated by the discovery of a number of circumstances in which the flow information appears to be modified or, even, reversed. Morange, M. (2009). The central dogma of molecular biology. *Resonance*, *14*(3): 236–247.
43. Gamow, G. (1955). Information transfer in the living cell. *Scientific American*, *193*(4): 70–79.
44. For a definition of Fordism and Taylorism: Jessop, B. (1994). The transition to post-Fordism and the Schumpeterian workfare state. In Burrows, R. and Loader, B. (eds). *Towards a Post-Fordist Welfare State?* Routledge, p. 15.
45. The expression was introduced by Nirenberg in an unpublished draft of his Nobel acceptance speech, suggesting that 'a new area of research will emerge during the next twenty-five years, that of molecular evolution, in which the

effects of synthetic genes upon the economy of the cell will be explored in a systematic fashion'. Nirenberg, M. (1968). Introduction to the Nobel Speech. Digitalised document available at the US National Library of Medicine, Marshall Nirenberg Papers [https://profiles.nlm.nih.gov].

46. Nirenberg, M. and Matthaei, J. (1961). The dependence of cell-free protein synthesis in E. coli upon naturally occurring or synthetic polyribonucleotides. *PNAS*, *47*(10): 1588–1602.
47. Lawrence, W. (1962). Structure of life. *New York Times*, 14 January, p. 11.
48. Ubell, E. (1961). The code of life finally cracked. *New York Herald Tribune*, 24 December, p. 1.
49. Cloud, W. (1962). DNA: The signals of life. *San Francisco Chronicle*, 20 October, p. 11.
50. Marx, K. (2010). Letter to Friedrich Engels, 18 June 1862. *Marx & Engels Collected Works Vol. 41: Letters 1860–64*. Lawrence & Wishart, p. 381.
51. Quoted in: Editor (1961) Cracking of the genetic code might lead to mutations. *The High Point Enterprise*, 30 December, p. 1.
52. Tiselius, A. (1964). Discours Solennel par le president de la Fondation Nobel. In *Les Prix Nobel En 1964*. Norstedt & Soner, pp. 15–17.
53. Huxley, J. (1964). Eugenics in evolutionary perspective. In *Essays of a Humanist*. Chatto, p. 255.
54. Nirenberg, M. (2004). Historical review: deciphering the genetic code. *Trends in biochemical sciences*, *29*(1): 46–54.
55. Nirenberg, M. (1967). Will society be prepared? *Science*, *157*(1): 633.
56. 'We are coming to the end of an era in molecular biology. If the DNA structure was the end of the beginning, the discovery of Nirenberg and Matthaei is the beginning of the end.' In: Crick, F. (1962). Towards the genetic code. *Scientific American*, *207*(3): 16.
57. Roberts, R. (2005). How restriction enzymes became the workhorses of molecular biology. *PNAS*, *102*(17): 5905–5908.
58. Jackson, D., Symons, R. and Berg, P. (1972). Biochemical method for inserting new genetic information into DNA of Simian Virus 40. *PNAS*, *69*(10): 2904–2909.
59. Cohen, S., et al. (1973). Construction of biologically functional bacterial plasmids in vitro. *PNAS*, *70*(11): 3240–3244.
60. The interview is reported in: Hughes, S. (2011). *Genentech: The Beginnings of Biotech*. University of Chicago Press, p. 1.
61. For a historical overview, see: Cohen, S. (2013). DNA cloning: a personal view after 40 years. *PNAS*, *110*(39): 15,521–15,529.
62. Interview reported in: Hughes, *Genentech*, p. 16.
63. Morrow, J., et al. (1974). Replication and transcription of eukaryotic DNA in *Escherichia coli*. *PNAS*, *71*(5): 1743–1747.
64. Wright, S. (1994). *Molecular Politics*. University of Chicago Press, p. 72.
65. Friedberg E. (2010). *Sydney Brenner*. Cold Spring Harbor Press, p. 185.
66. The quote is from one of the many recent books that have pushed the paradoxical idea that genetic engineering technologies are at once nothing new and something inherently revolutionary. This is, incidentally, the vision

popularised by biotech corporations' marketing strategists. Bastani, A. (2019). *Fully Automated Luxury Communism*. Verso.

67. Cohen, S. N. (1975). The manipulation of genes. *Scientific American*, *233*(1): 24–33.
68. Bylinsky, G. (1974). Industry is finding more jobs for microbes. *Fortune*, 9 February, pp. 96–102; McElheny, R. (1974). Animal gene shifted to bacteria: aid seen to medicine and farm. *New York Times*, 20 May, p. 61. Lewin, R. (1974). The future of genetic engineering. *New Scientist*, 17 October, p. 166.
69. Rose, J. (1972) Report to NIAID, 13 December. *Recombinant DNA History Collection*, MC100, MIT Libraries. As quoted in: Wright, *Molecular Politics*, p. 127.
70. Berg, P. and Singer, M. (1995). The recombinant DNA controversy: twenty years later. *PNAS*, *92*(20): 9011.
71. Berg, P. (2008). Asilomar 1975: DNA modification secured. *Nature*, *455*(7211): 290–291.
72. Chargaff, E. (1975). Profitable wonders. *The Sciences*, *15*(6): 21–26.
73. Sinsheimer, R. (1975). Troubled Dawn for Genetic Engineering. *New Scientist*, *68*(971): 148–151.
74. On the political and policy debates sparked by recombinant DNA, see: Wright, *Molecular Politics*, pp. 113–160.
75. Parliamentary Assembly of the Council of Europe (1982) Recommendation 934 – Genetic Engineering [https://assembly.coe.int/nw/xml/XRef/Xref-XML2HTML-en.asp?fileid=14968&lang=en].
76. Foucault, M. (1970). Croître et multiplier. In Foucault, M. (2001). *Dits and Écrits I*. Gallimard, pp. 967–972.
77. Foucault, M. (1978). *The History of Sexuality: Vol. I*. Pantheon Books, pp. 136–139.
78. Foucault, M. (2003). *'Society Must Be Defended': Lectures at the Collège de France, 1975–1976*. Picador, pp. 253–255.
79. Chargaff, E. (1976). On the dangers of genetic meddling. *Science*, *192*(4243): 937–938. 'The technology of genetic engineering', he went on, 'poses a greater threat to the world than the advent of nuclear technology. [...] What I see coming is a gigantic slaughterhouse, a molecular Auschwitz, in which valuable enzymes, hormones and so on will be extracted instead of gold teeth.' Chargaff, E. (1987). Engineering a molecular nightmare. *Nature*, *327*(6119): 199–200.
80. Bernardi, G. (1992). The Scientific Committee on Genetic Experimentation (COGENE). Library of Congress, *Maxine Singer Papers*, 43/20 [https://profiles.nlm.nih.gov/spotlight/dj/catalog/nlm:nlmuid-101584644X184-doc].
81. Genentech (1976). Outline for discussion, Kleiner & Perkins, April 1. Document reprinted in: Hughes, *Genentech*, p. 65.
82. Quoted in: Petit, C. (1977). A triumph in genetic engineering. *San Francisco Chronicle*, 3 December.
83. 'Wherever poverty and lack of sanitation drive families to low cost-per-calorie foods and packaged drinks, type 2 diabetes thrives.' Tabish, S. (2007). Is diabetes becoming the biggest epidemic of the twenty-first century? *International Journal of health sciences*, *1*(2): 5–23.

84. Goeddel, D., et al. (1979). Expression in Escherichia coli of chemically synthesized genes for human insulin. *PNAS, 76*(1): 106–110.
85. Valle, A. and Bolívar, J. (2021). *Escherichia coli*, the workhorse cell factory for the production of chemicals. In Singh, V. (ed.). *Microbial Cell Factories Engineering for Production of Biomolecules*. Academic Press, pp. 115–137.
86. Science News. (1978). Seizing the golden plasmid. *Science News, 114*(12): 195–196; Cohn, V. (1977). An artificial gene makes exact copy of brain hormone. *Washington Post*, 3 November; Editors (1977). Human gene in *E. coli*: it works! *Chemical Engineering News*, 55(45): 4.
87. Handler and Berg quoted in an untitled announcement published by the UCSF News Services on 2 December 1977. Discussed in: Hughes, *Genentech*, pp. 49–98.
88. Duménil, G. and Lévy, D. (2005). The neoliberal (counter-)revolution. In Saad-Filho, A. and Johnston, D. (eds). *Neoliberalism: A Critical Reader*. Pluto Press, pp. 9–19.
89. Lewontin, R. (2001). *It Ain't Necessarily So: The Dream of the Human Genome and Other Illusions*. New York Review of Books, p. xxii.
90. RAC (1933) Annual report, RG3, 915, Box 1.1, April 18. As quoted in: Kohler, R (1976). The management of science: 296.

Chapter 2 Manufacturing Lives: Corporate Genes, Genomic Rents and Living Assets

1. Marx, K. (1976). *Capital, Volume I*. Penguin, pp. 873–942.
2. For a history of the first enclosure movement, see: Linebaugh, P. (2014). *Stop, Thief!* PM Press. For an extensive discussion of present-day enclosures, see: May, C. (2015). *The Global Political Economy of Intellectual Property Rights*. Routledge.
3. Friello, D., Mylroie, J. and Chakrabarty, A. (2001). Use of genetically engineered multi-plasmid microorganisms for rapid degradation of fuel hydrocarbons. *International Biodeterioration & Biodegradation, 48*(1): 233–242. See also: Hughes, S. (2001). Making dollars out of DNA. *Isis, 92*(3): 541–575.
4. Rubenstein, L. (1979). Brief on behalf of the People's Business Commission, Amicus Curiae. [https://ipmall.law.unh.edu/content/diamond-v-chakrabarty-peoples-business-commission]. For an overview of the juridical discussion: Kevles, D. (2002) Of mice & money. *Daedalus, 131*(2): 78–88.
5. US Supreme Court. (1980) *Diamond v. Chakrabarty*, 447US303.
6. Brief Amicus Curiae of the Regents of the University of California, *US Court of Customs and Patent Appeals Dockets*, 76–712: 14–20. See also: Kevles, D. (1994). Ananda Chakrabarty Wins a Patent. *Historical Studies in the Physical and Biological Sciences, 25*(1): 111–135.
7. US Supreme Court. (1980) *Diamond v. Chakrabarty*, 447US303.
8. Powledge, F. (1995). Who owns rice and beans. *BioScience, 45*(7): 442.
9. Genentech. (1980). Supreme Court decision will spur genetics industry. Press release, 16 June.

10. For a different yet complementary analysis of the role of financial assets in the biotech industry, see: Birch, K. (2017). Rethinking value in the bioeconomy. *Science, Technology & Human Values*, 42(3): 460–490.
11. Hughes, S. (2011). *Genentech: The Beginnings of Biotech*. University of Chicago Press, pp. 158–164.
12. As convincingly shown by David Harvey, this period was characterised by 'surpluses of labour and capital side by side with seemingly no way to put them together in productive, i.e. "profitable" as opposed to socially useful ways'. There was thus a strong tendency towards 'a massive devaluation of both capital and labour'. See: Harvey, D. (2001). Globalization and the spatial Fix. *Geographische Revue*, 3(2): 23–30.
13. Fine, B. (2013). Financialization from a Marxist perspective. *International Journal of Political Economy*, 42(4): 47–66.
14. This tendency continues to this day. As of 2021, the global biotech market is estimated at a staggering $295 billion. See: Martin, D., et al. (2021). A brief overview of global biotechnology. *Biotechnology&Biotechnological Equipment*, 35(1): S5-S14; Hopkins, M. (2013). Buying big into biotech. *Industrial and Corporate Change*, 22(4): 903–952.
15. Marex, M. (2021). The Nasdaq Biotechnology Index. *Nasdaq Investment Intelligence*, 6 May.
16. USPT (1987). Animals – patentability. *USPT Official Gazette*, *18*(24): 1077.
17. A single exception was made by the Commissioner of Patents and Trademarks: genetically engineered humans – if ever produced – would not be considered patentable objects. Patenting humans would, in fact, constitute a new form of slavery in violation of the Thirteenth Amendment of the Constitution. Sharp, J. (1988). Of Transgenic Mice and Men. *Western State University Law Review*, *16*: 737–758.
18. Dismukes, K. (1980). Life is patently not human-made. *Hastings Center Report*, *10*(5): 11–12.
19. Editors (1988) Products of the year. *Fortune Magazine*, 5 December.
20. Outside its native land, the OncoMouse proved to be controversial. The Supreme Court of Canada rejected the patent on the ground that higher-life-forms could not be registered as 'products of manufacture'. The European Court of Appeals granted the patent in 1993, sparking intense political debate. See: Robins, R. (2008) Inventing Oncomice. *Genomics, Society and Policy*, 4(2): 1–15.
21. Haraway, D. (1997). *Modest_Witness@Second_Millenium. FemaleMan©_Meets_OncoMouse*™. Routledge, pp. 95, 52.
22. Ibid., p. 255.
23. The direct quotes from Francis Bacon are derived from the following writing: Bacon, F. (1858) [1625]. Of studies. In *Bacon's Essays*. Parker, p. 474; Bacon, F. (1999) [1620]. The new organon. In *Selected Philosophical Works*. Hackett, p. 148.
24. Bacon, F. (1999) [1626]. *New Atlantis*. Hackett, p. 263.
25. USPT (1987). Utility examination guidelines. In *Consolidated Listing of Official Gazette Notices*. USPT Publishing.

26. Sherkow, J. and Greely, H. (2015). The history of patenting genetic material. *Annual Review of Genetics, 49*: 161–182.
27. Dalton, R. (1999). Charges fly in $1 bn hormone patent battle. *Nature, 399*(6734): 289.
28. Holt, R. and Ho, K. (2019). The use and abuse of growth hormone in sports. *Endocrine Reviews, 40*(4): 1163–1185.
29. Feder, B. (1999). Genentech Agrees to Settle Patent Dispute. *New York Times*, 17 November, p. 11.
30. Rosenfeld, J. and Mason, C. (2013). Pervasive sequence patents cover the entire human genome. *Genome Medicine*, 5(3), 1–8.
31. King, M. (2014). The race to clone BRCA1. *Science, 343*(6178): 1462–1465.
32. Gold, E. and Carbone, J. (2010). Myriad Genetics: In the eye of the policy storm. *Genetics in Medicine, 12*(1): 39–70.
33. Burk, D. (2013). Are human genes patentable? *IIC, 44*(7): 747–749.
34. Rinehart, A. (2015). Myriad lessons learned. *UC Irvine Law Review*, 5(1): 1147.
35. Park, S. (2013). Gene patents and the public interest. *NC Jolt, 15*(1): 519–534.
36. *Association for Molecular Pathology et al. v. Myriad Genetics, Inc., et al.* 133S. Ct.2107 (June–13, 2013).
37. Ibid. See also: Aboy, M., et al. (2016). Myriad's impact on gene patents. *Nature Biotechnology, 34*(11): 1119–1123.
38. Sherkow, J. and Greely, H. (2015). The History of Patenting Genetic Material. *Annual Review of Genetics, 49*: 161–182.
39. Conley, J., Cook-Deegan, R. and Lázaro-Muñoz, G. (2014). Myriad after myriad. *NC Jolt, 15*(4): 597–614.
40. Jepson, P. and Canney, S. (2001). Biodiversity hotspots: hot for what? *Global Ecology and Biogeography, 10*(3): 225–227.
41. Crosby, A. (2004). *Ecological Imperialism: The Biological Expansion of Europe, 900–1900*. Cambridge University Press; Brockway, L. (1979). *Science and Colonial Expansion: The Role of the British Royal Botanical Gardens*, Academic Press.
42. Mgbeoji, I. (2005) *Global Biopiracy*. UBC Press.
43. Robinson, D. (2010). *Confronting Biopiracy*. Routledge.
44. Merson, J. (2000). Bio-prospecting or bio-piracy: intellectual property rights and biodiversity in a colonial and postcolonial context. *Osiris, 15*: 282–296. On unequal exchange: Bieler, A. and Morton, A. (2014). Uneven and combined development and unequal exchange: the second wind of neoliberal 'free trade'? *Globalizations, 11*(1): 35–45.
45. Narlikar, A. (2004). *International Trade and Developing Countries*. Routledge.
46. As quoted in: Bratspies, R. (2006). The new discovery doctrine. *American Indian Law Review, 31*(2): 315–340.
47. Williams, O. (2000). Life Patents, TRIPs and the International Political Economy of Biotechnology. In Russell, A. and Vogler, J. (eds) *The International Politics of Biotechnology*, Manchester University Press, pp. 67–84.
48. Shiva, V. (2016). *Biopiracy*. North Atlantic Books, pp. 2–6.
49. Parry, B. (2004) *Trading the Genome*. Columbia University Press.

50. Ibid., p. 227. For a more recent assessment, see: Blakeney, M. (2019). Remedying the misappropriation of genetic resources. In Singh, H., Keswani, C. and Singh, S. (eds). *Intellectual Property Issues in Microbiology*. Springer, pp. 147–161.
51. Blasiak, R., et al. (2018). Corporate control and global governance of marine genetic resources. *Science Advances, 4*(6): 312–329; Kintisch, E. (2018). UN tackles gene prospecting on the high seas. *Science Magazine, 361*(6406): 956–957.
52. Blasiak, R., et al. (2018). Corporate control and global governance of marine genetic resources: 312.
53. Parry's study, nearly two decades old, remains the most comprehensive survey to this date. The fact that comprehensive and transparent data is hardly traceable further illustrates the issues with this approach. Parry, *Trading the Genome*, p. 198.
54. Harvey, D. (2003) *The New Imperialism*. Oxford University Press, pp. 147–148.
55. Interview published in Reynolds, P. (2004) Nga Puni Whakapiri: Indigenous struggle and genetic engineering. Doctoral dissertation, Simon Fraser University, pp. 276–277.
56. Tauli-Corpuz, V. (2001). Biotechnology and indigenous peoples. In Tokar, B. (ed.). *Redesigning Life*. Wits University Press, pp. 232–253. See also: Nagan, W., et al. (2010). Misappropriation of Shuar traditional knowledge and trade secrets. *Journal of Technological Law & Policy*, *15*: 9.

Chapter 3 Genomic Infrastructures: Banking the Biosphere and Genomic Big Data

1. Mayr, E. (1961). Cause and effect in biology. *Science, 134*: 1503–1504.
2. Peluffo, A. (2015). The genetic program: behind the genesis of an influential metaphor. *Genetics, 200*(3): 685–696.
3. Dawkins, R. (1988). *The Blind Watchmaker*. Penguin, p. 112.
4. Chaitin, G. (2012). Life as evolving software. In Zenil, H. (ed.). *A Computable Universe*. World Scientific.
5. DeLisi, C. (2008). Santa Fe 1986: human genome baby-steps. *Nature, 455*(7215): 876–877; See also: Cook-Deegan, R. (1991). Human Genome Project interview with Charles DeLisi, *The Robert Cook-Deegan Human Genome Archive*. Bioethics Research Library of the Kennedy Institute of Ethics. [http://hdl.handle.net/10822/559556].
6. Cook-Deegan, R. (1996). Genes and the bomb. In *The Gene Wars*. Norton & Company.
7. Website of the National Human Genome Institute [www.genome.gov/human-genome-project].
8. See for instance: Pennisi, E. (2000). Finally, the book of life and instructions for navigating it. *Science, 288*(5475): 2304–2307; Stipp, D. (2000). Blessings from the book of life. *Fortune Magazine, 141*(5): F–17. For a critical analysis:

Kay, L (1998). A book of life? *Perspectives in Biology and Medicine, 41*(4): 504–528.

9. Watson, J. (1990). The human genome project: past, present, and future. *Science, 248*(4951): 44–49.
10. Gilbert, W. (1992) A vision of the grail. In Kevles, D. and Hood, L. (eds). *The Code of Codes,* Harvard University Press, p. 94. Similar assumptions are made in works of other prominent scientists and administrators, e.g., Collins, F., et al. (2003). A vision for the future of genomics research. *Nature, 422*(6934): 835–847.
11. David Botstein quoted in: Cook-Deegan, R. (1994). *The Gene Wars*, p. 98.
12. Keller, E. (2000). *The Century of the Gene*, Harvard University Press, p. 5, 69. Allen, G. (1994). The genetic fix: the social origins of genetic determinism. In Tobach, E. and Rosoff, B. (eds) *Challenging Racism and Sexism.* Feminist Press, pp.163–187; Lewontin, R. and Levins, R. (2007). *Biology Under the Influence.* Monthly Review Press, pp. 235–66. Lewontin, R., (1993). *The Doctrine of DNA*. Penguin; Rose, S., Lewontin, R., Kamin, L. (1984). *Not in Our Genes.* Penguin.
13. Tauber, A. and Sarkar, S. (1992). The Human Genome Project: has blind reductionism gone too far? *Perspectives in Biology and Medicine, 35*(2): 220–235.
14. Keller, E. (1994) Master molecules. In Cranor, C. (ed.) *Are Genes Us? The Social Consequences of the New Eugenics*, Rutgers University Press, pp. 89–98; Allen, The genetic fix, p. 164.
15. See for instance: Rechsteiner, M. C. (1991). The human genome project: misguided science policy. *Trends in Biochemical Sciences, 16*(1): 455–461; Murphy, T., (1994). The Genome Project and the meaning of difference. In *Justice and the Human Genome Project.* University of California Press, pp.1–13.
16. Quoted in Keller, E. (1992). Nature, nurture, and the human genome project. In Kevles and Hood, *The Code of Codes*, p. 282.
17. Gilbert, A vision of the grail, p. 211.
18. Ibid., p. 96.
19. Friedland, S. (1997). The criminal law implications of the Human Genome Project. *Kentucky Law Journal, 86*: 303–366; Jones, M. (2002). Overcoming the myth of free will in criminal law: the true impact of the genetic revolution. *Duke Law Journal, 52*: 1031–1053.
20. Emmert-Streib, F., Dehmer, M. and Yli-Harja, O. (2017). Lessons from the human genome project. *Frontiers in Genetics, 8*(1): 184–207.
21. Harvey, D. (2001). Globalization and the 'spatial fix': 24–25. See also: Harvey, D. (2003). The fetish of technology. *Macalester International, 13*(1): 3–30.
22. Cook-Deegan, R. (1991). Human Genome Project interview with Charles DeLisi. [http://hdl.handle.net/10822/559556].
23. US Senate (1988). *Department of Energy National Laboratory Cooperative Research Initiatives Act*, 15–17 September. [https://catalog.hathitrust.org/Record/007606746].
24. Quoted in Kolata, G. (1992). Biologist's speedy gene method scares peers but gains backer. *New York Times*, 28 July, pp. 11–12.

25. Marx, K. (1993). *Grundrisse*, Penguin, pp. 526–531; and: Marx, K. (1976). *Capital, Volume 1*. Penguin, pp. 474–475, 578–580.
26. Marx, *Grundrisse*, p. 531.
27. Office of Technology Assessment (1988). *Mapping Our Genes*. Johns Hopkins University Press.
28. Watson, J. (1990). The human genome project: past, present, and future. *Science, 248*(4951): 45.
29. Marx, *Capital, Volume I*, pp. 589–590, 645, 1019–1038.
30. Holley, R. W., *et al.* (1965). Structure of a ribonucleic acid. *Science, 147*(3664): 1462–1465.
31. Watson, J. (1990). The human genome project: 44.
32. *Science* named PCR the technological breakthrough of 1989, and awarded 'Molecule of the Year' to the polymerase molecule which drives the chain reaction. Koshland, D. (1992). The molecule of the year. *Science, 258*(5090): 1861–1861.
33. US Senate (1987). *Workshop on Human Gene Mapping*. Committee on Energy and Natural Resources, p. 12. [https://hdl.handle.net/2027/pst.000013681524].
34. Philip Mirowski, for instance, offers an insightful critique of the hallucinatory neoliberal hype surrounding Celera. He rightly questions Celera's contribution to the scientific endeavour and its ultimate commercial failure. While we subscribe to Mirowski's analysis, it is important to stress how this failure paradoxically contributed to the profitability of its mother company. From this point of view, the hype represents more than a misguided and naïve representation of the world; rather, it is reflective of a political ideology with material effects. See: Mirowski, P. (2011). *Science-Mart*. Harvard University Press.
35. See: Fortun, M. (2006). Celera genomics: the race for the human genome sequence. In Clarke, A. and Ticehurst, F. (eds) *Living with the Genome*. Palgrave, pp. 27–32.
36. Gorner, P. and Van, J. (2000). Now, Celera begins job of selling itself. *Chicago Tribune*, 28 June.
37. Collins, F., et al. (1998). New goals for the US human genome project: 1998–2003. *Science, 282*(5389): 682–689. Collins, F. S., Morgan, M. and Patrinos, A. (2003). The Human Genome Project: lessons from large-scale biology. *Science, 300*(5617): 286–290.
38. Venter, C. (2007). *A Life Decoded: My Genome, My Life*. Penguin, p. 259.
39. Cook-Deegan, R. and Heaney, C. (2010). Patents in genomics and human genetics. *Annual Review of Genomics and Human Genetics, 11*: 383–425.
40. Collins, F., et al. (1998). New goals for the US human genome project: 1998–2003. *Science*, 282(5389): 682–689.
41. Editors (2000). Rules of genome access. *Nature, 404*: 317.
42. Berenson, A. and Wade, N. (2000). A call for sharing of research causes gene stocks to plunge. *New York Times*, 15 March.
43. Pollack, A. (2005). Celera to quit selling genome information. *New York Times*, 27 April.

44. Nurk, S., et al. (2021). The complete sequence of a human genome. *bioRxiv*, 27 May. See also: Reardon, S. (2021). A complete human genome sequence is close. *Nature*, *594*(7862): 158–159.
45. White House (2000). President Clinton announces the completion of the first survey of the entire human genome. *Human Genome Information Archive* [https://web.ornl.gov/sci/techresources/Human_Genome/project/clinton1.shtml].
46. Ibid.
47. 'The real fruits of the HGP lie in the contrast between the primitive state of digital biology in the late 1980s and the current ease with which all scholars can access, harness and analyse biological data.' Gibbs, R. (2020). The human genome project changed everything. *Nature Reviews Genetics*, *21*(10): 575–576.
48. Barranco, C. (2021). The Human Genome Project. *Nature News*, 10 February: 55.
49. Dawkins, R. (1996). *The Selfish Gene*. Oxford University Press; Pinker, S. (2002). *The Blank Slate*. Viking. For a critique, see: McKinnon, S. (2005). *Neoliberal Genetics*. Prickly Paradigm.
50. On the postgenomic turn, see: Richardson, S. and Stevens, H. (eds). (2015). *Postgenomics: Perspectives on Biology after the Genome*. Duke University Press. On the epigenetics revolution, see: Guthman, J. and Mansfield, B. (2013). The implications of environmental epigenetics. *Progress in Human Geography*, *37*(4), 486–504; and Meloni, M. and Testa, G. (2014). Scrutinizing the epigenetics revolution. *BioSocieties*, *9*(4), 431–456.
51. The slogan was first launched in 2016 by Airtel as part of a campaign to promote internet access. Since then, it has become a viral hashtag on social media and won several marketing prizes around the world. The same slogan has also been used to emphasize and promote how the collection of biological data – on track heart rate, sleep quality, time, blood pressure – can guide customers in shaping their lifestyle.
52. See, e.g., Fuchs, C. (2010). Labor in informational capitalism and on the internet. *The Information Society*, *26*(3): 179–196; Srnicek, N. (2017). *Platform Capitalism*. Wiley & Sons; Moulier-Boutang, Y. (2011). *Cognitive capitalism*. Polity; Dean, J. (2005). Communicative capitalism. *Cultural Politics*, *1*(1): 51–74; Zuboff, S. (2019). *The Age of Surveillance Capitalism*. Profile.
53. IBM. (2010). A smarter planet relies on data analysis. [www.youtube.com/watch?v=x81hb8nT4yQ].
54. Turkheimer, E. (2000). Three laws of behavior genetics and what they mean. *Current directions in psychological science*, *9*(5): 160–164.
55. Philips, A. (2016). Only a click away. *Applied & Translational Genomics*, *8*: 16–22.
56. Regalado, A. (2019). More than 26 million people have taken an at-home ancestry test. *MIT Technology Review*, *11*(2): 60–64.
57. Calico (2021). Mission and values. [https://calicolabs.com/mission-and-values].

58. Newcomer, E. (2015). Your genome isn't really secret, says Google Venture's Bill Maris. Bloomberg, 20 October.
59. Seife, C. (2013). 23andMe is terrifying, but not for the reasons the FDA thinks. *Scientific American*, 27 November.
60. Abbasi, J. (2020). 23andMe develops first drug compound using consumer data. *JAMA*, *323*(10): 916–916.
61. Morange, M. (2006). Post-genomics, between reduction and emergence. *Synthese*, *151*(3): 355–360.
62. See Daviet, R., Nave, G. and Wind, J. (2020). Genetic data. *Journal of Marketing*, *13*(2): 12.
63. Paul, K. (2021). Fears over DNA privacy as 23andMe plans to go public in deal with Richard Branson. *The Guardian*, 9 February. See also: Hayden, E. (2017). The rise and fall and rise again of 23andMe. *Nature News*, *550*(7675): 174.
64. Nordgren, A. and Juengst, E. (2009). Can genomics tell me who I am? Essentialistic rhetoric in direct-to-consumer DNA testing. *New Genetics and Society*, *28*(2): 157–172.
65. Merz, S. (2016). Health and ancestry start here. *Ephemera*, *16*(3).
66. Phillips, C. (2018). The Golden State Killer investigation and the nascent field of forensic genealogy. *Forensic Science International*, *36*: 186–188.
67. Erlich, Y., et al. (2018). Identity inference of genomic data using long-range familial searches. *Science*,362(6415): 690–694.
68. Ancestry (2021). Transparency report 2021. [www.ancestry.com/cs/transparency].
69. Lewontin and Levins, *Biology under the Influence*, pp. 254–256. For a recent, high-profile case see: Koerner, B. (2015). Your relative's DNA could turn you into a suspect. *Wired*, 13 October.
70. Roberts, D. (2011). Collateral consequences, genetic surveillance, and the new biopolitics of race. *Howard Law Journal*, *54*(3): 12–54.
71. Khan, S., Can, N. and Machado, H. (eds). (2021). *Racism and Racial Surveillance*. Routledge.
72. Zaretsky, A. (2021). DNA collection in immigration custody and the threat of genetic surveillance. *California Law Review*, *109*: 317.
73. Rockefeller Foundation (2021). Accelerating national genomic surveillance. [www.rockefellerfoundation.org/wp-content/uploads/2021/03/The-Rockefeller-Foundation_Accelerating-National-Genomic-Surveillance.pdf].
74. Moreau, Y. (2019). Crack down on genomic surveillance. *Nature*, *576:* 36–38.
75. Kumar, P. (2019). NEC enters the race to lead Malaysia's digital ID program. [https://asia.nikkei.com/Business/Companies/NEC-enters-the-race-to-lead-Malaysia-s-digital-ID-project].
76. Joly, Y., Marrocco, G. and Dupras, C. (2019). Risks of compulsory genetic databases. *Science*, *363*(6430): 938–940.
77. Skinner, D. and Wienroth, M. (2019). Was this an ending? The destruction of samples and deletion of records from the UK police national DNA database. *BJHS Themes*, *4*(1): 99–121.

78. Cyranoski, D. (2010). Chinese bioscience: the sequence factory. *Nature News*, *464*(7285): 22–24; Cyranoski, D. (2016). The sequencing superpower. *Nature*, *534*(7608): 462–464.
79. Dorfman, R. (2013). Falling prices and unfair competition in consumer genomics. *Nature Biotechnology*, *31*(9), 785–786.
80. Chen, F., et al. (2020). CNGBdb: China National Genebank Database. *Yi Chuan*, *42*(8): 799–809.
81. Needham, K. (2020). Covid opens new doors for China's gene giant. Reuters, 5 August.
82. BGI Shenzhen (2011). BGI unveils significant new global research collaborations at ICG–6. *EurekAlert*, 16 November.
83. Yang, H., Wang, X. and Tian, J. (2019). Beautiful genes, beautiful plants. *Plants, People, Planet*, *1*(1): 27–31.
84. Stevens, H. (2018). Starting up biology in China. *Osiris*, *33*(1): 85–106.
85. Wang, K., Shen, X. and Williams, R. (2021). Sequencing BGI. *New Genetics and Society*, *40*(3): 305–330.
86. Wong, W. (2017). Speculative authorship in the city of fakes. *Current Anthropology*, *58*(S15): S103–S112.
87. Zhu, J. (2012). A year of great leaps in genome research. *Genome Medicine*, *4*(1): 4
88. Exposito-Alonso, M., et al. (2019). The Earth BioGenome project: opportunities and challenges for plant genomics and conservation. *The Plant Journal*, *102*(2): 222–229.
89. Lewin, H., et al. (2018). Earth BioGenome Project: sequencing life for the future of life. *PNAS*, *115*(17): 4327.
90. Ibid., p. 4326.
91. EBP (2019). Earth BioGenome Project builds foundation. [https://research.ucdavis.edu/earth-biogenome-project-builds-foundation-for-10-year-mission-to-sequence-life/].
92. Nobre, C., et al. (2016). Land-use and climate change risks in the Amazon and the need of a novel sustainable development paradigm. *PNAS*, *113*(39): 10765.
93. McAfee, K., (1999) Selling nature to save it? *Environment and Planning D*, *17*(2): 133.
94. WEF. (2018). Harnessing the fourth industrial revolution for life on land. [www3.weforum.org/docs/WEF_Harnessing_4IR_Life_on_Land.pdf].
95. Hayden, C., (2003). *When Nature Goes Public*. Princeton University Press, pp. 58–59. See also: Reid, W., (1993). Bioprospecting. *Environmental Science and Technology*, *27*(9): 1730–2.
96. Zeller, C., (2007). From the gene to the globe. *Review of International Political Economy*, *15*(1): 86–115. See also: Mazzucato, M., (2015) *The Entrepreneurial State*. Public Affairs.
97. Paddon, C. and Keasling, J. (2014). Semi-synthetic artemisinin. *Nature Reviews Microbiology*, *12*(5): 355–367; Cankar, K. (2014). Valencene oxidase CYP706M1 from Alaska cedar. *FEBS*, *588*(6): 1001–1007.
98. EBP, Earth BioGenome Project builds foundation.

99. All quotes in this section are from: WEF, Harnessing the fourth industrial revolution for life on land, pp. 5, 16; and Lewin et al., Earth BioGenome Project, p. 4326. See also: www.earthbankofcodes.org/
100. Editor (2018). Sequencing the world. *Economist*, 27 January.
101. All quotes found in: WEF (2018). *Harnessing the Fourth Industrial Revolution for Life on Land*, pp. 5, 16.
102. Lewin et al., Earth BioGenome Project, pp. 4325–4333.
103. WEF, Harnessing the fourth industrial revolution for life on land, p. 9.
104. Ibid., pp. 5, 15. See also: WEF (2019). Building block(chains) for a better planet. [www3.weforum.org/docs/WEF_Building-Blockchains.pdf].

Chapter 4 CRISPR Assembly Lines: Speeding up the Molecular Factory

1. Jansen, R., et al. (2002). Identification of genes that are associated with DNA repeats in prokaryotes. *Molecular Microbiology*, *43*(6): 1565–1575.
2. Mojica, F., et al. (2005). Intervening sequences of regularly spaced prokaryotic repeats derive from foreign genetic elements. *Journal of Molecular Evolution*, *60*(2): 174–182.
3. Koonin, E. and Makarova, K. (2013). CRISPR-Cas: evolution of an RNA-based adaptive immunity system in prokaryotes. *RNA Biology*, *10*(5): 679–686. See also: Makarova, K. (2011). Evolution and classification of the CRISPR–Cas systems. *Nature Reviews Microbiology*, *9*(6): 467–477.
4. The production of 'herd immunity' is a complex labour process. For a review of traditional techniques in comparison to new CRISPR-based techniques: McDonnell, B., et al. (2018). Generation of bacteriophage-insensitive mutants of *Streptococcus thermophilus*. *Applied and Environmental Microbiology*, *84*(4): e01733–17.
5. Du Pont (2012). Du Pont launches cultures for pizza cheese. *New Hope Magazine*, 3 April.
6. Horvath, P. and Barrangou, R. (2010). CRISPR/Cas, the immune system of bacteria and archaea. *Science*, *327*(5962): 167–170.
7. The CRISPR system could be extracted from one organism and moved into another one without losing its properties; and the Cas9 enzyme could be tricked into slashing through any DNA sequence chosen in advance by simply replacing or tweaking the RNA it uses as a guiding compass. Gasiunas, G., et al. (2012). Cas9–crRNA ribonucleoprotein complex mediates specific DNA cleavage for adaptive immunity in bacteria. *PNAS*, *109*(39): E2579–E2586.
8. Jinek, M., et al. (2012). A programmable dual-RNA–guided DNA endonuclease in adaptive bacterial immunity. *Science*, *337*(6096): 816–821.
9. Gasiunas et al., Cas9–crRNA ribonucleoprotein complex, pp. E2579–E2586.
10. The term began to be used around 2005; for example, see: Urnov, F., et al. (2005). Highly efficient endogenous human gene correction using designed zinc-finger nucleases. *Nature*, *435*(7042): 646–651.
11. The term 'genome editing' is increasingly used in order to describe genetic interventions performed with tools such as CRISPR-Cas9, ZFNs

and TALENs. It is not the only one. Many molecular biologists often use a number of alternative terms, such as 'gene editing', 'genome engineering', 'genomic design' and 'targeted mutagenesis'. While we prefer this last term, we consider all these concepts to be roughly synonyms and we sometimes use them as such.

12. The quoted interviews by Church and Charpentier are reported in: Rogers, A. (2015) A CRISPR cut. *Pomona College Magazine*, 23 February, pp. 14–15.
13. Specter, M. (2015) The gene hackers. *New Yorker*, 15 November, pp. 52–61.
14. Doudna, J. (2019). On the future of genome editing. [https://news.berkeley.edu/2019/04/10/berkeley-talks-transcript-jennifer-doudna-future-of-gene-editing]; Bayer (2020). Facts about CRISPR. [https://release.ace.bayer.com/sites/default/files/2020-04/hereare-the-facts-about-agriculture-and-nutrition-brochure.pdf].
15. Loria, K. (2015). The genetic technology that's going to change everything is at a critical turning point. *Business Insider*, 12 August; Doudna, J. and Sternberg, S. (2017). *A Crack in Creation*. Houghton Mifflin; Regalado, A. (2014) EmTech. *MIT Technology Review*, 24 September; Kaul, T., et al. (2020). How Crisp is CRISPR? In Tuteja, N., et al. (eds). *Advancement in Crop Improvement Techniques*. Woodhead.
16. Royal Swedish Academy of Sciences (2020). Press release: the Nobel Prize in Chemistry 2020. [www.nobelprize.org/prizes/chemistry/2020/press-release/].
17. Maron, L. (2019). Quilting plant chromosomes with CRISPR/Cas9. *The Plant Journal*, *98*(4): 575–576.
18. Marx, K. (1992). *Capital, Volume I*. Penguin, p. 601.
19. Today, the acronym CRISPR is widely used to refer to a whole panoply of different gene-editing systems, each having specific characteristics. Following this linguistic tendency, we often use Crispr as a shorthand to indicate Crispr-Cas genome editing technologies. See: Makarova, K. and Koonin, E. (2015). Annotation and classification of CRISPR-Cas systems. *CRISPR*: 47–75.
20. An indicative list of gene edited plants is offered in: Liu, X., et al. (2017). Application of CRISPR/Cas9 in plant biology. *Acta Pharmaceutica Sinica B*, *7*(3): 292–302. A list of gene edited animals is offered in: Cohen, J. (2019). The CRISPR Animal Kingdom. *Science*, *365*(6452): 426–429.
21. For instance, in 2014 a research team at the University of Texas designed CRISPR to target a twenty-letter DNA sequence in the TYR gene and injected into fertilised mice eggs. The function of the gene was suggested in a visible way: many of the resulting gene-edited mice were albino. Mizuno, S. (2014). Simple generation of albino C57BL/6J mice with G291T mutation in the tyrosinase gene by the CRISPR/Cas9 system. *Mammalian Genome*, *25*(7): 327–334.
22. The resulting 'knock-in organisms' are characterised by the presence of exogenous genetic material, which may be synthetically produced, or extracted from a different organism. For an example, see: Platt, R., et al. (2014). CRISPR-Cas9 knockin mice for genome editing and cancer modeling. *Cell*, *159*(2): 440–455.

23. For a discussion of epigenome editing applications, see: Nakamura, M., et al. (2021). CRISPR technologies for precise epigenome editing. *Nature Cell Biology*, *23*(1): 11–22.
24. Ibid.
25. A useful archive of CRISPR covers in major magazine and newspapers is offered by the Centre for Genetics & Society; see: www.geneticsandsociety.org/internal-content/human-gene-editing-timeline-crispr-cover-stories#chronological.
26. For an overview of CRISPR companies, see: Cohen, J. (2017). The birth of CRISPR Inc. *Science*, *355*(6326): 680–684; and Brinegar, K., et al. (2017). The commercialization of genome-editing technologies. *Critical Reviews in Biotechnology*, *37*(7): 924–932.
27. For an overview of the distribution of CRISPR patents see: Ghosh, P. (2020). Patent Landscape of CRISPR/Cas. In *CRISPR/Cas Genome Editing*. Springer, pp. 213–220.
28. Lander, E. S. (2016). The heroes of CRISPR. *Cell*, *164*(1–2): 18–28. For a review of the debate: Ledford, H. (2016). The unsung heroes of CRISPR. *Nature News*, *535*(7612): 342–347.
29. Carlyle, T. (1927). *Sartor Resartus*. Macmillan.
30. This form of social cooperation that engenders science and knowledge could also be conceptualised in Marxian terms as the 'General Intellect'. Marx, K. (1993). *Grundrisse*. Penguin, p. 706. For a discussion of the role of this concept in Marx's critique, see: Vercellone, C. (2007). From Formal Subsumption to General Intellect. *Historical Materialism*, *15*(1): 13–36.
31. Baltimore, D., et al. (2015). A prudent path forward for genomic engineering and germline gene modification. *Science*, *348*(6230): 36–38.
32. Lander, E., et al. (2019). Adopt a moratorium on heritable genome editing. *Nature*, *567*(7747): 165–168.
33. Mullin, E. (2016). Obama advisers urge action against CRISPR bioterror threat. *MIT Technology Review*, 17 November.
34. Tumpey, T., et al. (2005). Characterization of the reconstructed 1918 Spanish influenza pandemic virus. *Science*, *310*(5745): 77–80.
35. Noyce, R. and Evans, D. (2018). Synthetic horsepox viruses and the continuing debate about dual use research. *PLoS Pathogens*, *14*(10): e1007025. See also: Lewis, G., et al. (2020). The biosecurity benefits of genetic engineering attribution. *Nature Communications*, *11*(1): 1–4.
36. Fu, Y., et al. (2013). High-frequency off-target mutagenesis induced by CRISPR-Cas nucleases in human cells. *Nature Biotechnology*, *31*(9): 822–826.
37. Davies, K. and Church, G. (2019). Radical technology meets radical application. *CRISPR Journal*, *2*(6): 346–351.
38. Virilio, P. (1999). *Politics of the Very Worst*. Semiotext(e), p. 89.
39. The introduction of exogenous genes in complex genomic structures represents only a first layer of unpredictability. The resulting gene-edited organisms are subsequently inserted into delicate ecological systems, composed of countless interacting elements in motion. Ecological complexity thus creates further unpredictability.
40. Vermeulen, N. (2010). *Supersizing Science*. Universal Publishers, pp. 18–25.

41. In the following years, the study of 'junk DNA' emerged as a new frontier in genomics studies, which has slowly shed light on the many different functions that this apparently meaningless genomic regions may play. Palazzo, A. and Gregory, T. (2014). The case for junk DNA. *PLoS Genetics*, *10*(5).
42. Project presentation available at https://engineeringbiologycenter.org. See also: Schindler, D., Dai, J. and Cai, Y. (2018). Synthetic genomics. *Current Opinion in Chemical Biology*, *46*(1): 56–62.
43. Specter, M. (2015). The gene hackers. *New Yorker*, 15 November, pp. 52–61.
44. The interview is reported in: Duncan, D. (2018). Is the world ready for synthetic people? *Neo-Life*, 5 April. See also: Endy, D. (2005). Foundations for engineering biology. *Nature*, *38*(7067): 449–453.
45. The statement appears in a public lecture presented by Endy at the US National Academies, archived at: [https://sites.nationalacademies.org/cs/groups/pgasite/documents/webpage/pga_052088.pdf].
46. Knight, T. (2014). Biosimplicity: engineering simple life. Unpublished lecture notes. [https://slidetodoc.com/biosimplicity-engineering-simple-life-tom-knight-mit-computer/].
47. Horkheimer, M. (1974). *Eclipse of Reason*. Seabury Press, p. 97.
48. Horkheimer argues that living beings are increasingly reduced to functional machines, mere tools and 'objects of total exploitation'. Synthetic biology goes one step further: it not only treats animal bodies as tools, it also aims at transforming those very bodies according to the functions they are meant to perform in a given process of production. Horkheimer, M. (2013). *Critique of Instrumental Reason*. Verso.
49. JCVI (2010) First self-replicating synthetic bacterial cell. [www.jcvi.org/sites/default/files/assets/projects/first-self-replicating-synthetic-bacterial-cell/press-release-final.pdf].
50. Callaway, E. (2016). Race to design life heats up. *Nature*, *531*(7596): 557–559.
51. Ironically in piecing together its components, Venter and colleagues realised that the function of over one third of the genes in the stripped down 'minimal cell' remains unknown. See: Bedau, M., et al. (2010). Life after the synthetic cell. *Nature*, *465*(7297): 422–424.
52. The quote is from Synthetic Genomics' website (now Viridos); see www.viridos.com/technology.
53. Malm, A. (2016). *Fossil capital*. Verso.
54. Church, G. and Regis, E. (2014). *Regenesis*. Basic Books, p. 19.
55. Endy, D. (2009) Synthetic biology: overview. Unpublished Lecture at the National Academies of Science. [https://sites.nationalacademies.org/cs/groups/pgasite/documents/webpage/pga_052088.pdf].
56. Scott, J. (2020) *Seeing Like a State*. Yale University Press, p. 21.
57. Shetty, R., Endy, D. and Knight, T. (2008) Engineering BioBrick vectors from BioBrick parts. *Journal of Biological Engineering*, *2*(1): 1–12.
58. Hill, P., et al. (2020). Clean manufacturing powered by biology. *JIMB*, *47*(11): 965–975.
59. Ro, D., et al. (2006). Production of the antimalarial drug precursor artemisinic acid in engineered yeast. *Nature*, *440*(7086): 940–943.

60. Weathers, P., et al. (2014). Dried-leaf Artemisia annua. *World Journal of Pharmacology*, *3*(4): 39.
61. A full list of synthetic biology products is available on Amyris' website. [https://amyris.com/ingredients].
62. Eunjung, A. (2013). Scientists now creating app-style life-forms. *Japan Times*, 27 October, p. 16.
63. Regalado, A. (2021). Is Ginkgo's synthetic-biology story worth $15 billion? *MIT Technology Review*, 24 August.
64. Weber, J. (2019). Ginkgo Bioworks CEO wants biology to manufacture physical goods. *Bloomberg Businessweek*, 6 November.
65. Witze, A. (2012). Factory of life. *ScienceNews*, 27 December.
66. Feldman, A. (2019). The life factory. *Forbes*, 5 August.
67. Regalado, A. (2021). Is Ginkgo's synthetic-biology story worth $15 billion? *MIT Technology Review*, 24 August.
68. Hill et al., Clean manufacturing powered by biology, pp. 965–975.
69. de Leon, R. (2021) Bill Gates-backed Ginkgo Bioworks going public via $15 billion SPAC. CNBC, May–11.
70. Frist, B. (2021) Could this be the next Apple or Microsoft? *Forbes*, 7 September.
71. Lipschutz, B. (2021) Wall Street darling Ginkgo caught in SPAC. *Bloomberg*, 6 October. [www.bloomberg.com/news/articles/2021-10-06/wall-street-darling-ginkgo-caught-in-spac-short-seller-selloff]. The complete report is available at https://scorpioncapital.s3.us-east-2.amazonaws.com/reports/DNA1.pdf.
72. Ibid.
73. A description of the magazine is available at www.growbyginkgo.com/about.
74. Tsing, A. (2000). Inside the economy of appearances. *Public Culture*, *12*(1): 115–144.
75. Feldman, A. (2019). The life factory. *Forbes*, 5 August.
76. Roosth, S. (2017) *Synthetic: How Life Got Made*. University of Chicago Press, p. 11.
77. Ginsberg, A. (2017). Design as the machines come to life. In Ginsberg, A., et al. (eds) *Synthetic Aesthetics*. MIT Press, pp. 39–70.
78. For a critique of 'contemporary design practices', which may be extended to contemporary practices of genetic design: Wizinsky, M. (2022). *Design after Capitalism*. MIT Press.
79. For Marx, the generalisation of commodity exchange engenders a 'fetishism' by which social relations are misrecognised and reified into *things*. This perspective was expanded in Horkheimer, M. and Adorno, T. (1972) *Dialectic of Enlightenment*. Stanford University Press, pp. 94–136.

Chapter 5 Molecular Factory Farms: Engineering Living Means of Production

1. We consider genome-edited organisms as being a historically specific form of genetically modified organisms. This classification follows the one

established by the European Court of Justice in its landmark 2018 ruling. ECJ. (2018). Press release 111/18. *CURIA Press*, 25 July.

2. Marx, K. (1976). *Capital, Volume I.* Penguin, p. 637.
3. Halliday, S. (2001). *The Great Stink of London.* History Press.
4. Foster, J. B. (1999). Marx's theory of metabolic rift. *American Journal of Sociology, 105*(2): 366–405.
5. Liebig, J. (1843). *Organic Chemistry and its Application to Agriculture and Physiology.* Taylor & Walton, p. 47.
6. Ibid., p. 54.
7. Crookes, W. (1899). *The Wheat Problem.* John Murray, p. 3.
8. Today, chemical factories produce roughly the same amount of nitrogen as that produced by natural processes (circa 160 million metric tons per year). This means that the input of fixed nitrogens in the biosphere is doubled. Kirchman, D. (2021). *Dead Zones.* Oxford University Press, pp. 94–95.
9. Smith, C., Hill, A. and Torrente-Murciano, L. (2020). Current and future role of Haber–Bosch ammonia in a carbon-free energy landscape. *Energy & Environmental Science, 13*(2): 331–344.
10. Szöllösi-Janze, M. (2001). Pesticides and war: the case of Fritz Haber. *European Review, 9*(1): 97–108.
11. Russell, E. (2001). *War and nature.* Cambridge University Press. Also: Krimsky, S. (2019). *GMOs Decoded.* MIT Press.
12. Rani, L., et al. (2021). An extensive review on the consequences of chemical pesticides on human health and environment. *Journal of Cleaner Production, 283*(1): 124657.
13. Gaud, W. (1968). *The Green Revolution: Accomplishments and Apprehensions.* The Society for International Development, Washington, DC, March 8, 1968. See: Patel, R. (2013). The long green revolution. *The Journal of Peasant Studies, 40*(1): 1–63.
14. Kloppenburg, J. (2005). *First the Seed.* University of Wisconsin Press, pp. 157–162.
15. Ibid., pp. 91–129.
16. Yapa, L. (1993). What are improved seeds? An epistemology of the Green Revolution. *Economic Geography, 69*(3): 254–273. See also: Kloppenburg, *First the Seed*, pp. 9–11, 128–129; Dahlberg, K. (ed.). (2012). *Beyond the Green Revolution.* Springer.
17. Pimentel, D. (1996). Green revolution agriculture and chemical hazards. *Science of the total environment, 188*(1): S86–S98.
18. Mahapatra, S. K. (2003). Taming the waters: the political economy of large dams in India. *Indian Journal of Agricultural Economics, 58*(4): 856.
19. Fischer, K. (2016). Why new crop technology is not scale-neutral. *Research Policy, 45*(6): 1185–1194.
20. Friedman, H. and McMichael, P. (1989), Agriculture and the state system. *Sociologia ruralis, 29*(2): 105–108.
21. Clapp, J. (2018). Mega-mergers on the menu. *Global Environmental Politics, 18*(2): 12–33; Howard, P. (2015). Intellectual property and consolidation in the seed industry. *Crop Science, 55*(6): 2489–2495.

22. Clapp, J. and Purugganan, J. (2020). Contextualizing corporate control in the agrifood and extractive sectors. *Globalizations, 17*(7): 1265–1275.
23. Clapp, Mega-mergers on the menu, p. 16.
24. This steady pace of concentration continues to this day. See: Clapp, J. (2021). The problem with growing corporate concentration and power in the global food system. *Nature Food, 2*(6): 404–408.
25. Holden, C. (1994). Tomato of tomorrow. *Science, 264*(5158): 512–513.
26. Martineau, B. (2001). *First Fruit: The Creation of the Flavr Savr Tomato and the Birth of Biotech Foods.* McGraw-Hill.
27. Pline-Srnic, W. (2006). Physiological mechanisms of glyphosate resistance. *Weed technology, 20*(2): 290–300.
28. Duke, S. (2018). The history and current status of glyphosate. *Pest Management Science, 74*(5): 1027–1034.
29. James, C. (2011) *Global Status of Commercialized Biotech/GM Crops.* Brief no. 43. ISAAA.
30. Sharkey, S., Williams, B. and Parker, K. (2021). Herbicide drift from genetically engineered herbicide-tolerant crops. *Environmental Science & Technology, 55*(23): 15,559–15,568.
31. Clapp, J. (2021). Explaining growing glyphosate use. *Global Environmental Change, 67*(1): 102239.
32. Benbrook, C. (2016). Trends in glyphosate herbicide use in the United States and globally. *Environmental Sciences Europe, 28*(1): 6.
33. Bonny, S. (2016). Genetically modified herbicide-tolerant crops, weeds, and herbicides. *Environmental Management, 57*(1): 31–48.
34. Schütte, G., et al. (2017). Herbicide resistance and biodiversity. *Environmental Sciences Europe, 29*(1): 1–12.
35. Clay, S. (2021). Near-term challenges for global agriculture: herbicide-resistant weeds. *Agronomy Journal, 113*(6): 4463–4472.
36. Neuman, W. and Pollack, A. (2010). US farmers cope with Roundup-resistant weeds. *New York Times*, 3 May.
37. Green, J. and Siehl, D. (2021). History and outlook for glyphosate-resistant crops. *Reviews of Environmental Contamination and Toxicology, 255*: 67–91.
38. Schütte, G. (2017). Herbicide resistance and biodiversity. *Environmental Sciences Europe, 29*(1): 1–12.
39. Ibrahim, M., et al. (2010). *Bacillus thuringiensis*: a genomics and proteomics perspective. *Bioengineered Bugs, 1*(1): 31–50.
40. Sanahuja, G., et al. (2011). *Bacillus thuringiensis*: a century of research. *Plant Biotechnology Journal, 9*(3): 283–300.
41. Whiteley, H. and Schnepf, H. (1985). *Bacillus thuringiensis* crystal protein gene toxin segment. International Patent Application: WO1986001536A.
42. Yamamoto, T. (2001). One hundred years of Bacillus thuringiensis research and development. *Journal of Insect Biotechnology and Sericology, 70*(1): 1–23.
43. Baum, J., Kakefuda, M. and Gawron-Burke, C. (1996). Engineering *Bacillus thuringiensis* bioinsecticides. *Applied and Environmental Microbiology, 62*(12): 4367–4373.
44. Federici, B. (2005). Insecticidal bacteria. *Journal of Invertebrate Pathology, 89*(1): 30–38; Baum, J. (1998). Transgenic *Bacillus thuringiensis. Phytoprotec-*

tion, *79*(4): 127–130; Lucena, W. (2014). Molecular approaches to improve the insecticidal activity of *Bacillus thuringiensis* Cry toxins. *Toxins*, *6*(8): 2393–2423.

45. Whiteley and Schnepf, *Bacillus thuringiensis* crystal protein gene toxin segment.
46. Abbas, M. (2018). Genetically engineered (modified) crops and the world controversy on their safety. *Egyptian Journal of Biological Pest Control*, *28*(1): 1–12.
47. Lu, Y. (2022) Bt cotton area contraction drives regional pest resurgence, crop loss, and pesticide use. *Plant Biotechnology Journal*, *20*(2): 390–398.
48. The data presented in this paragraph is publicly available through the institutional database of the International Service for the Acquisition of Agri-Biotech Applications at www.isaaa.org/gmapprovaldatabase.
49. Koch, M., et al. (2015). The food and environmental safety of Bt crops. *Frontiers in Plant Science*, *6*(1): 283.
50. Gatehouse, A., et al. (2011). Insect-resistant biotech crops and their impacts on beneficial arthropods. *Philosophical Transactions of the Royal Society B*, *366*(1569): 1438–1452. A comprehensive review of the sprawling scientific literature on the environmental effects of GM crops is well beyond the scope of this work. For an accessible review, see: Krimsky, *GMOs Decoded*.
51. Benbrook, C. (2012). Impacts of genetically engineered crops on pesticide use in the US. *Environmental Sciences Europe*, *24*(1): 7.
52. Tabashnik, B., Brévault, T. and Carrière, Y. (2013). Insect resistance to Bt crops: lessons from the first billion acres. *Nature Biotechnology*, *31*(6): 510–521.
53. Carrière, Y., Crickmore, N. and Tabashnik, B. (2015). Optimizing pyramided transgenic Bt crops for sustainable pest management. *Nature Biotechnology*, *33*(2): 161–168.
54. Ives, A., et al. (2011). The evolution of resistance to two-toxin pyramid transgenic crops. *Ecological Applications*, *21*(2): 503–515.
55. Gould, F., Brown, Z. and Kuzma, J. (2018). Wicked evolution. *Science*, *360*(6390): 728–732.
56. Gatehouse, A., et al. (2011). Insect-resistant biotech crops and their impacts on beneficial arthropods. *Philosophical Transactions of the Royal Society B*, *366*(1569): 1438–1452.
57. Ma, C., et al. (2021). Climate warming promotes pesticide resistance. *Nature Communications*, *12*(1): 1–10.
58. Randerson, J. (2008) Consequences of GM crop contamination are set to worsen. *The Guardian*, 18 February.
59. Bucchini, L. and Goldman, L. (2002). Starlink corn. *Environmental Health Perspectives*, *110*(1): 5–13.
60. Quist, D. and Chapela, I. (2001). Transgenic DNA introgressed into traditional maize landraces in Oaxaca, Mexico. *Nature*, *414*(6863): 541–543.
61. McAfee, K. (2008). Beyond techno-science. *Geoforum*, *39*(1): 148–160.
62. Aguila-Way, T. (2014). The Zapatista 'mother seeds in resistance' project. *Social Text*, *32*(1): 67–92; Brandt, M. (2014). Zapatista corn. *Social Studies of*

Science, 44(6): 874–900; Curry, H. (2022). *Endangered Maize*. University of California Press.

63. E.g. Piñeyro-Nelson, A., et al. (2009). Transgenes in Mexican maize. *Molecular Ecology, 18*(4): 750–761; Agapito-Tenfen, S., et al. (2017). Transgene flow in Mexican maize revisited. *Ecology and Evolution, 7*(22): 9461–9472.
64. Price, B. and Cotter, J. (2014) The GM Contamination Register. *International Journal of Food Contamination, 1*(1): 1–13.
65. Paredes, A. (2021). Weedy activism: women, plants, and the genetic pollution of urban Japan. *Journal of Political Ecology, 28*(1): 70–90.
66. Chakraborty, T. (2014). Release of Bt-Brinjal in Bangladesh. *Economic and Political Weekly, 11*(2): 24–26.
67. Repp, R. (1999). Biotech pollution. *Idaho Law Review, 36*: 585–628.
68. For another case of 'gene piracy', see: Saldanha, L. and Rao, B. (2011) Monsanto's Brinjal biopiracy. *India Law News*, 1 September. [https://indialawnews.org/2011/09/01/monsantos-brinjal-biopiracy-a-shocking-eexpose-of-callous-disregard-for-biodiversity-laws-in-india/].
69. Robin, M. (2014). *The world according to Monsanto*. The New Press, pp. 206–208.
70. Ibid., p. 206.
71. Burrell, R. and Hubicki, S. (2005). Patent liability and genetic drift. *Environmental Law Review, 7*(4), 278–286.
72. See for instance the use of the concept of 'seed piracy' in: *Monsanto Company v. Omega Farm Supply, Inc.*, Case No. 4:14-cv-00870 (2015). For a history of the concept of piracy in relation to the expansion of the world market, see: Policante, A. (2016). *The Pirate Myth*, Routledge.
73. Bayer (n.d.) Follow the law. Promotional leaflet. [https://traits.bayer.com/stewardship/Documents/follow-the-law-stewardship.pdf].
74. Marx, *Capital, Volume 1*, p. 638.
75. Patel, R. (2012). *Stuffed and starved*. Melville House.
76. Corteva (2019). Ag at a crossroads: CRISPR. [www.corteva.com/our-impact/innovation/crispr.html]. Since June 2019, Corteva spun off as an independent public company.
77. Brodwin, E. (2019). We'll be eating the first CRISPR'd foods within 5 years. *Business Insider*, 20 April.
78. Doudna, J. and Sternberg, S. (2017). *A Crack in Creation*. Houghton-Mifflin.
79. Shan Q et al. (2013). Targeted genome modification of crop plants using a CRISPR-Cas system. *Nature Biotechnology, 31*(8): 686–688.
80. Woo, J., et al. (2015). DNA-free genome editing in plants with preassembled CRISPR-Cas9 ribonucleoproteins. *Nature Biotechnology, 33*(11):1162–1164. Dalla Costa, L., et al. (2020). Strategies to produce T-DNA free CRISPRed fruit trees via *Agrobacterium tumefaciens* stable gene transfer. *Scientific Reports, 10*(1): 1–14.
81. Regalado, A. (2015). DuPont predicts CRISPR plants on dinner plates in five years. *MIT Technology Review*, 8 October.
82. Yin, X., et al. (2019). Editing a stomatal developmental gene in rice with CRISPR/Cpf1. In *Plant Genome Editing with CRISPR Systems*. Humana Press, pp. 357–268.

83. Ordonez, N., et al. (2015). Worse comes to worst. *PLoS Pathogens, 11*(11): e1005197.
84. Maxmen, A. (2019). CRISPR might be the banana's only hope against a deadly fungus. *Nature, 574*(15): 10–1038.
85. Hofvander, P., Andreasson, E. and Andersson, M. (2022). Potato trait development going fast-forward with genome editing. *Trends in Genetics, 38*(3): 218–221.
86. Donnelly, J. (2002). *The Great Irish Potato Famine*. Sutton.
87. Braa, D. (1997) The great potato famine and the transformation of Irish peasant society. *Science & Society, 61*(2): 193–215.
88. Fry, W. and Goodwin, S. (1997). Resurgence of the Irish potato famine fungus. *Bioscience, 47*(6): 363–371.
89. Mann, C. (2011). How the potato changed the world. *Smithsonian Magazine*, 8 November.
90. Kieu, N., et al. (2021). Mutations introduced in susceptibility genes through CRISPR/Cas9 genome editing confer increased late blight resistance in potatoes. *Scientific reports, 11*(1): 1–12.
91. Finnegan, G. (2017). Can CRISPR feed the world? *Horizon*, 18 May.
92. Zafar, S., et al. (2020). Engineering abiotic stress tolerance via CRISPR/Cas-mediated genome editing. *Journal of Experimental Botany, 71*(2): 470–479.
93. DARPA's Biological Technology Office applies the 'the unique properties of biology' in order to develop 'technology-driven capabilities to detect novel threats and protect US force readiness, deploy physiological interventions to maintain operational advantage, support warfighter performance, and focus on operational biotechnology for mission success'. [www.darpa.mil/about-us/offices/bto].
94. Darpa Outreach Office. (2016). DARPA enlists insects to protect agricultural food supply. [www.darpa.mil/news-events/2016-10-19].
95. A 'horizontal gene transfer' is any event in which genetic material between organisms that are not in a parent–offspring relationship. For a review of horizontal gene transfers and their role in ecology and evolution, see: Soucy, S., Huang, J. and Gogarten, J. (2015). Horizontal gene transfer. *Nature Reviews Genetics, 16*(8), 472–482.
96. Reeves, R., et al. (2018). Agricultural research, or a new bioweapon system? *Science, 362*(6410): 35–37.
97. Ibid.
98. Bextine's interview was published in: Achenbach, J. (2018) The Pentagon is studying an insect army to defend crops: critics fear a bioweapon. *Washington Post*, 4 October.
99. USDA. (2015). Request for confirmation that transgene-free, CRISPR-edited mushroom is not a regulated article. [www.aphis.usda.gov/biotechnology/downloads/reg_loi/15-321-01_air_response_signed.pdf].
100. Wang, T., Zhang, H. and Zhu, H. (2019). CRISPR technology is revolutionizing the improvement of tomato and other fruit crops. *Horticulture Research, 6*(1): 1–13.

101. Zaidi, S., et al. (2020). Engineering crops of the future. *Genome Biology*, *21*(1): 1–19.
102. Patel, R. (2012). *Stuffed and Starved*. Melville House.
103. Potrykus, I. (2001). Golden rice and beyond. *Plant Physiology*, *125*(3): 1157–1161.
104. For a history of Golden Rice: Nestle, M. (2010). *Safe Food*. University of California Press, pp. 145–166.
105. Dong, O., et al. (2020). Marker-free carotenoid-enriched rice generated through targeted gene insertion using CRISPR-Cas9. *Nature Communications*,11(1): 1–10; Sakellariou, M. and Mylona, P. (2020). New Uses for Traditional Crops. *Agronomy*, *10*(12): 1964–1982; Jha, A. and Warkentin, T. (2020). Biofortification of pulse crops. *Plants*, *9*(1): 73–81.
106. Masipag (2013). Masipag upholds farmers' action. [https://masipag.org/2013/09/masipag-upholds-farmers-action-against-the-golden-rice-field-trials].
107. Stop Golden Rice Network. (2020). Golden Rice is 'Trojan horse'. *The Ecologist*, 19 August.
108. Ibid.
109. See, for instance: Rosset, P. (2003). Food sovereignty. *Food First Backgrounder*, *9*(4): 1–4; Pellegrini, P. (2009). Knowledge, Identity and Ideology in Stances on GMOs. *Science Studies*, *22*(1): 50; Wittman, H. (2009). Reworking the metabolic rift. *The Journal of Peasant Studies*, *36*(4): 805–826; Masucci, M. (2021). The GMO revival. Navdanya International, 11 April. [https://navdanyainternational.org/de/the-gmo-revival/].
110. Sanatech. (2020). Launch of the genome-edited tomato with increased GABA in Japan. [https://sanatech-seed.com/en/20201211-2-2/].
111. Eurofruit. (2021). Sanatech Seed launches world's first GE tomato. [www.fruitnet.com/eurofruit/article/184662/sanatech-seed-launches-worlds-first-ge-tomato].
112. Gates, B. (2018). Gene editing for good: how CRISPR could transform global development. *Foreign Affairs*, 97: 166.
113. Montenegro de Wit, M., Kapuscinski, A. and Fitting, E. (2020). Democratizing CRISPR? *Elementa*, *8*: 9.
114. For instance, DuPont has been criticised for pushing a 'patent land grab', accumulating a significant number of CRISPR-related patents: Grushkin, D. (2016). DuPont in CRISPR-Cas patent land grab. *Nature Biotechnology*, *34*(1): 13–14.
115. Menz, J., et al. (2020). Genome edited crops touch the market. *Frontiers in Plant Science*, *11*(1): 6–12; see also: Stokstad, E. (2021). UK set to loosen rules for gene-edited crops and animals. *Science Insider*, 26 May.
116. ECJ. (2018). Press Release 111/18. *CURIA*, 25 July. See also: Gelinsky, E. and Hilbeck, A. (2018). European Court of Justice ruling regarding new genetic engineering methods scientifically justified. *Environmental Sciences Europe*, *30*(1): 1–9.
117. The Greens/EFA. (2021). GM crops and consumer rights. [https://extranet.greens-efa.eu/public/media/file/1/6910].

118. Mutant crop varieties produced by x-ray mutagenesis *are* classified as GMOs. Nevertheless, they are *exempted* from the obligations of the GMO Directive due to the fact that, according to the EJ, they 'have conventionally been used in a number of applications and have a long safety record' (mutagenesis exemption in Dir. 2001/18/EC, art. 3-1B). See: Van der Meer, P., et al. (2020). The status under EU law of organisms developed through novel genomic techniques. *European Journal of Risk Regulation* online ahead of print. [https://doi.org/10.1017/err.2020.105].
119. Ahmad, A., et al. (2021). An outlook on global regulatory landscape for genome-edited crops. *International Journal of Molecular Sciences*, *22*(21): 11753.
120. Urnov, F., Ronald, P. and Carroll, D. (2018). A call for science-based review of the European court's decision on gene-edited crops. *Nature Biotechnology*, *36*(9): 800–802.
121. For a lucid discussion of the debate, see: Gelinsky, E. and Hilbeck, A. (2018). European Court of Justice ruling regarding new genetic engineering methods scientifically justified. *Environmental Sciences Europe*, *30*(1): 1–9.
122. Corporate Europe Observatory (2021). CRISPR-Files expose lobbying tactics to deregulate new GMOs. [https://corporateeurope.org/en/2021/03/derailing-eu-rules-new-gmos].
123. Mitchell, T. (2002). *Rule of Experts*. University of California Press, pp. 19–53.
124. Marx, *Capital, Volume I*, p. 285.
125. Ibid., p. 497.
126. Ibid., pp. 517, 499, 503. See also: Malm, A. (2016). *Fossil Capital*. Verso.
127. Marx, K. (1956). *Capital, Volume II*. Progress Publishers, p. 142.
128. Ibid., p. 144.
129. Marx, *Capital, Volume I*, p. 446.
130. Marx, *Capital, Volume II*, p. 144.
131. Ibid., p. 72.
132. Ibid., p. 142.
133. 'Formerly English sheep, like the French as late as 1855, were not fit for the butcher until four or five years old. According to the Bakewell system, sheep may be fattened when only one year old and in every case have reached their full growth before the end of the second year.' In: Marx, *Capital, Volume II*, p. 142. See also: Derry, M. (2003). *Bred for Perfection*. JHU Press.
134. Regulatory approval came after several years of back and forth with US governmental agencies. See: Waltz, E. (2016). GM salmon declared fit for dinner plates. *Nature Biotechnology*, *34*(1): 7.; Van Eenennaam, A. and Muir, W. (2011). Transgenic salmon. *Nature Biotechnology*, *29*(8): 706–710.
135. Smith, C. (2021). Genetically modified salmon head to US dinner plates. *Associated Press News*, 28 May.
136. Clausen, R. and Longo, S. (2012). The tragedy of the commodity and the farce of AquAdvantage salmon. *Development and Change*, *43*(1): 229–251.
137. Du, S., et al. (1992). Growth enhancement in transgenic Atlantic salmon by the use of an 'all fish' chimeric growth hormone gene construct. *Bio/technology*, *10*(2): 176–179.
138. Ibid., pp. 178–181.

139. Cook, J., et al. (2000). Growth rate, body composition and feed digestibility/conversion of growth-enhanced transgenic Atlantic salmon. *Aquaculture, 188*(1–2): 15–32.
140. Aquabounty. (2016). Annual report. [https://investors.aquabounty.com/static-files/9be72322-7eaf-4ea0-a617-0c76ac00ab84].
141. Genetic Literacy Project (2013). The salmon dialogue. [https://geneticliteracyproject.org].
142. Editor (2022). Japan embraces CRISPR-edited fish. *Nature Biotechnology*, 4(1): 10; Jinguji, M. and Segawa, S. (2022). Fleshier sea bream. *Asahi Shimbun*, 30 September.
143. Business partners are listed on Regional Fish Institute's website: https://regional.fish/en/#team/. See also: Forbes. (2021). Asia-Pacific's small companies and startups on the rise. *Forbes Asia*, 31 August.
144. RFI. (2022). Fish for future. [https://regional.fish/en/].
145. ISAAA. (2021). Japan begins sale of genome-edited 'Madai' red sea bream, October–20. [www.isaaa.org/kc/cropbiotechupdate/article/default.asp?ID=19061].
146. RFI. (2021) RFI has launched an e-commerce website to sell genome-edited seafood. [https://regionalfish.online].
147. RFI. (2022). Creating a next-generation aquaculture model. [https://regional.fish/en/farming/].
148. Forbes. (2021). Asia-Pacific's small companies and startups on the rise. *Forbes Asia*, 31 August.
149. Tait-Burkard, C., et al. (2018). Livestock 2.0. *Genome Biology, 19*(1): 1–11.
150. See for instance: Proudfoot, C., et al. (2015). Genome edited sheep and cattle. *Transgenic Research, 24*(1): 147–153; Crispo, M., et al. (2015). Efficient generation of myostatin knock-out sheep. *PLoS One, 10*(8): e0136690; Wang, X., et al. (2018). CRISPR/Cas9-mediated MSTN-disruption and heritable mutagenesis in goats causes increased body mass. *Animal Genetics, 49*(1): 43–51; Moro, L. (2020). Generation of myostatin-edited horse embryos using CRISPR/Cas9 technology. *Scientific Reports, 10*(1): 1–10; Bastón, J., et al. (2021). 94 Myostatin gene editing by CRISPR/Cas9 technology of Brangus fetal fibroblasts to produce edited embryos by cloning. *Reproduction, Fertility & Development, 33*(2): 154–168.
151. Matika, O., et al. (2019). Balancing selection at a premature stop mutation in the myostatin gene underlies a recessive leg weakness syndrome in pigs. *PLoS Genetics, 15*(1): e1007759.
152. ISAAA. (2020). UC Davis scientists use CRISPR technology to develop bull that produces 75% male offspring. [www.isaaa.org/kc/cropbiotechupdate/article/default.asp?ID=18234].
153. Quinton, A. (2020). Meet Cosmo, a bull calf designed to produce 75% male offspring. *UC Davis News*. [www.ucdavis.edu/news/placeholder-cow-story].
154. Owen, J. (2021). One-step generation of a targeted knock-in calf using the CRISPR-Cas9 system in bovine zygotes. *BMC Genomics, 22*(1): 1–11.
155. See EggXYt's website at www.eggxyt.com; see also: Cremer, J. (2021). How CRISPR can create more ethical eggs. *Alliance For Science*, 2 March.

156. Espinosa, R., Tago, D. and Treich, N. (2020). Infectious diseases and meat production. *Environmental & Resource Economics*, *76*(4): 1019–1044.
157. Tait-Burkard, C., et al. (2018). Livestock 2.0 – genome editing for fitter, healthier, and more productive farmed animals. *Genome Biology*, *19*(1): 1–11.
158. Silpa, M. V., et al. (2021). Climate resilient dairy cattle production. *Frontiers in Veterinary Science*, *8*: 327.
159. Laible, G., et al. (2020). Holstein Friesian dairy cattle edited for diluted coat color as adaptation to climate change. *bioRxiv*, 15 September.
160. PR Newswire. (2020). Acceligen launches program for precision crossbreeding of African dairy production systems. *PR Newswire*, 24 September. Bill & Melinda Gates Foundation (2020). Committed grants: Acceligen Inc. [www.gatesfoundation.org/about/committed-grants/2020/08/inv004986].
161. Dikmen, S., et al. (2014). The SLICK hair locus derived from Senepol cattle confers thermotolerance to intensively managed lactating Holstein cows. *Journal of Dairy Science*, *97*(9): 5508–5520.
162. Foundation for Food & Agriculture Research (2020). FFAR Grant Evaluates Gene Editing to Improve Heat Resistance in Cattle. *Foundation for Food & Agriculture Research*, 31 August.
163. US FDA (2022). Press release: FDA makes low-risk determination for marketing of products from genome-edited beef cattle after safety review. *US FDA Press*, 7 March.
164. The 'Naturally Cool' trademark has been filed in 2018 by Recombinetics. [https://trademarks.justia.com/owners/recombinetics-inc-3147679/].
165. Roy, E. (2020). From red seaweed to climate-smart cows. *The Guardian*, 1 January.
166. Giddings, V., Rozansky, R., and Hart, D. (2020). Gene editing for the climate. [www2.itif.org/2020-gene-edited-climate-solutions.pdf].
167. Barber, G. (2019). A more humane livestock industry brought to you by CRISPR. *Wired*, 12 February.
168. Young, A., et al. (2020). Genomic and phenotypic analyses of six offspring of a genome-edited hornless bull. *Nature Biotechnology*, *38*(2): 225–232.
169. Marx, K. (1993). *Grundrisse*. Penguin, p. 506.
170. Targeted mutagenesis represents a powerful way of mastering biological functions and steering metabolic processes in desired directions. It must be stressed that this form of subsumption relies on the particular relation between genotype and metabolic processes that characterises the living cell. This is why we prefer the expression 'real subsumption of life' to the most often used 'real subsumption of nature'. It is the metabolic capacities of the living cell and not 'nature' that is mobilised as an increasingly adaptable platform of production. This conceptualization of the 'real subsumption of life' relies on, but slightly diverges from, half a century of debates on 'real subsumption'. The distinction between formal and real subsumption came to occupy a central place in Marxist theory in the 1970s, partly as a result of the rediscovery of a series of Marxian texts – such as 'Results of the Immediate Process of Production' – in which these concepts plays a prominent role. The distinction became central to a number of thinkers – including Jacques

Camatte and Antonio Negri – who emphasise an historical trend towards the intensification of capitalist relations of production. In the late 1990s, ecological Marxists began theorising the 'subsumption of nature' as a direct product of capital's accumulation. In a series of important contributions, Paul Burkett proposes that capital's subsumption of the labour process is an inherently ecological process, co-constituted by the 'real subsumption of nature' under capital: Burkett, P. (1999). *Marx and Nature.* Springer, pp. 11, 64–67. Boyd, Prudham and Schurman have revisited Burkett's conceptualization of the 'real subsumption of nature' in order to 'refer to systematic increases in or intensification of biological productivity': Boyd, W., Prudham, W. and Schurman, R. (2001). Industrial dynamics and the problem of nature. *Society & Natural Resources, 14*(7): 555–570. Most recently, Troy Vettesse has advanced a *Marxist Theory of Extinction,* contributing to rethinking how 'real subsumption has allowed the expansion of animal industry', and 'propels the Sixth Extinction': Vettesse, T. (2020). A Marxist theory of extinction. *Salvage, 7*: 62–74.

Chapter 6 Engineering Extinction Ecologies: Resurrection, Annihilation and Genetic Biocontrol

1. Lorimer, J. and Driessen, C. (2016). From 'Nazi cows' to cosmopolitan 'ecological engineers'. *Annals of the American Association of Geographers, 106*(3): 631–652.
2. Shapiro, B. (2017). Pathways to de-extinction. *Functional Ecology, 31*(5): 996–1002.
3. Orlando, L. (2015). The first aurochs genome reveals the breeding history of British and European cattle. *Genome Biology, 16*(1): 1–3. Also: Lorimer, J. and Driessen, C. (2013). Bovine biopolitics and the promise of monsters in the rewilding of Heck cattle. *Geoforum, 48*: 249–259.
4. Stokstad, E. (2015). Bringing back the aurochs. *Science, 350*(6265): 1144–1147; Gordon, I. J., Pérez-Barbería, F. J. and Manning, A. D. (2021). Rewilding lite. *Sustainability, 13*(6): 3347.
5. Pina-Aguilar, R., et al. (2009) Revival of extinct species using nuclear transfer. *Cloning & Stem Cells, 11*(3): 341–346. See also: Adams, W. (2017). Geographies of conservation I. *Progress in Human Geography, 41*(4): 534–545.
6. Siipi, H. and Finkelman, L. (2017). The extinction and de-extinction of species. *Philosophy & Technology, 30*(4): 427–441.
7. Brondizio, E., et al. (2019). Global assessment report on biodiversity and ecosystem services. [https://doi.org/10.5281/zenodo.3831673].
8. Herper, M. and Molteni, M. (2021). Return of the mammoth. *STAT News*, 13 September.
9. Betuel, E. (2022). Colossal grabs $60 million Series A for moonshot mammoth project. *Tech Crunch*, 9 March.
10. Shapiro, Pathways to de-extinction, pp. 998–999; Shapiro, B. (2015). Mammoth 2.0. *Genome Biology, 16*(1): 1–3.

11. The description of the project and its aims can be consulted directly on the company's website at https://colossal.com/mammoth.
12. TEDxDeExtinction. (2013). Hybridizing with extinct species: George Church at TEDxDeExtinction. [www.youtube.com/watch?v=oTH_fmQo3Ok].
13. Shapiro, B. (2015). *How To Clone a Mammoth*. Princeton University Press.
14. Line, B. (2013). Behind the cover: meet the extinct animals. *National Geographic*, 15 April.
15. Most of this fundamental research was conducted by public researchers with no connection to Colossal. Miller, W., et al. (2008). Sequencing the nuclear genome of the extinct woolly mammoth. *Nature*, *456*(7220): 387–390.
16. Dalén, L., et al. (2015) Complete genomes reveal signatures of demographic and genetic declines in the woolly mammoth, *Current Biology*, *25*(10): 1395–1400; Dalén, L., et al. (2021) Million-year-old DNA sheds light on the genomic history of mammoths. *Nature*, *591*(7849): 265–269.
17. Miller, W., et al. (2008). Sequencing the nuclear genome of the extinct woolly mammoth. *Nature*, *456*(7220): 387–390.
18. Ibid.
19. De Luce, I. (2021). Colossal co-founder Ben Lamm explains why resurrecting the woolly mammoth will save the planet. *B2 Magazine*, 21 September.
20. Chen, L., et al. (2014). Advances in genome editing technology and its promising application in evolutionary and ecological studies. *Gigascience*, *3*(1): 2047–2172.
21. Partridge, E., et al. (2017). An extra-uterine system to physiologically support the extreme premature lamb. *Nature Communications*, *8*(1): 1–16.
22. Zimov, S. (2005). Pleistocene park: return of the mammoth's ecosystem. *Science*, *308*(5723): 796–798. See also https://pleistocenepark.ru.
23. Robbins, P. (2012). *Political Ecology: A Critical Introduction*. Wiley & Sons.
24. Van Dooren, T. and Rose, D. (2017). Keeping faith with the dead. *Australian Zoologist*, *38*(3): 375–378.
25. De Luce, Colossal co-founder Ben Lamm explains why resurrecting the woolly mammoth will save the planet.
26. Betuel, E. (2022). Colossal grabs $60 million Series A for moonshot mammoth project, *Tech Crunch*, 9 March.
27. Clifford, C. (2021). Lab-grown woolly mammoths could walk the Earth in six years if geneticist's new start-up succeeds. *CNBC News*, 13 September.
28. At One Ventures. (2021). Our portfolio. [www.atoneventures.com/portfolio].
29. Capelj, R. (2021). Animal Capital, alongside partner Josh Richards, spurs Thomas Tull's Colossal stake. *Yahoo Finance*, 13 September.
30. Shapiro, Pathways to de-extinction, p. 997. See also: IUCN (2016). *Guiding Principles on Creating Proxies of Extinct Species for Conservation Benefit*. IUCN Press.
31. Colossal. (n.d.). Woolly mammoth de-extinction: Earth's old friend and new hero. [https://colossal.com/mammoth/].
32. Zimmer, C. (2021). A new company with a wild mission. *New York Times*, 30 September.
33. Shapiro, Pathways to de-extinction, pp. 996–1002.

34. De Luce, Colossal co-founder Ben Lamm explains why resurrecting the woolly mammoth will save the planet.
35. For a discussion of the concept in the mediaeval period, see: Fort, G. (2017). Penitents and their proxies: penance for others in early medieval Europe. *Church History*, *86*(1): 1–32.
36. On niche-making, see: Royle, C. (2017). Complexity, dynamism, and agency. *Antipode*, *49*(5), 1427–1445.
37. Revive & Restore. (2021). Press release. Woolly mammoths will walk the Arctic tundra again. [https://reviverestore.org/press-release/colossal-press-release-091321].
38. Browning, H. (2018). Won't somebody please think of the mammoths? *Journal of Agricultural & Environmental Ethics*, *31*(6): 785–803.
39. McCauley, D., et al. (2017). A mammoth undertaking. *Functional Ecology*, *31*(5): 1003–1011.
40. On the sociotechnical imaginary: Jasanoff, S. and Kim, S. (2015). *Dreamscapes of Modernity.* University of Chicago Press.
41. Vettese, T. (2019). A Marxist theory of extinction. *Salvage*, *7*: 251–262.
42. This teleological view of history is implicit in much of the literature on the Anthropocene, where the present ecological crisis is often presented as the inevitable outcome of a chain of events stretching back to the early Stone Age. For instance, Steffen and Crutzen in their seminal theorisation of the Anthropocene posit that: 'The mastery of fire by our ancestors provided humankind with a powerful monopolistic tool unavailable to other species, that put us firmly on the long path towards the Anthropocene.' Steffen, W., Crutzen, P. and McNeill, J. (2007). The Anthropocene. *Ambio-Journal of Human Environment Research & Management*, *36*(8): 614.
43. Haraway, D. J. (1996). *Modest_Witness@Second_Millennium. FemaleMan©_Meets_OncoMouse™. Feminism and Technoscience*. Routledge, p. 10.
44. Conceived in this way, wilderness retains colonial conceptual undertones. We use it here to indicate the operative imaginary driving these projects. Cronon, W. (1996). The trouble with wilderness. *Environmental History*, *1*(1): 7–28.
45. Sayer, K. (2017). The 'modern' management of rats. *BJHS Themes*, 2: 235–263.
46. IPBES (2019). Global assessment report on biodiversity and ecosystem services. IPBES Publication.
47. Thomas, C. (2020). The development of Anthropocene biotas. *Philosophical Transactions of the Royal Society B*, *375*(1794): 20190113. See also: Doubleday, Z. and Connell, S. (2018). Weedy futures. *Frontiers in Ecology and the Environment*, *16*(10): 599–604.
48. Cameron, G. and Scheel, D. (2001). Getting warmer: effect of global climate change on distribution of rodents in Texas. *Journal of Mammalogy*, *82*(3): 652–680.
49. Taitingfong, R. I. (2021). Editing islands. Doctoral dissertation, UC San Diego, p. 27.
50. Taitingfong, R. I. (2020). Islands as laboratories. *Human Biology*, *91*(3): 179–188.

51. Russell, J. and Kueffer, C. (2019). Island biodiversity in the Anthropocene. *Annual Review of Environment & Resources*, *44*(1): 31–60.
52. Salinas-de-León, P., et al. (2020). Evolution of the Galapagos in the Anthropocene. *Nature Climate Change*, *10*(5): 380–382.
53. Marris, E. (2018). WIRED features GBIRd and Island Conservation. [www.islandconservation.org].
54. Marris, E. (2021). *Wild souls*. Bloomsbury, p. 181.
55. Hall, S. (2017). Could genetic engineering save the Galápagos? *Scientific American*, *317*(6): 55.
56. E.g. Siers, S., et al. (2020). Brodifacoum residues in fish three years after an island-wide rat eradication attempt in the tropical Pacific. *Management of Biological Invasions*, *11*(1): 105–121; Pitt, W., et al. (2015). Non-target species mortality and the measurement of brodifacoum rodenticide residues after a rat eradication on Palmyra Atoll. *Biological Conservation*, *185*: 36–46.
57. Veitch, C., et al. (eds) (2019). *Island invasives*. IUCN.
58. Atkinson, U. (1973). Spread of the ship rat in New Zealand. *Journal of the Royal Society of New Zealand*, *3*(3): 457–472.
59. Russell, J., et al. (2015). Predator-free New Zealand. *BioScience*, *65*(5): 520–541.
60. Ibid., p. 522.
61. Morton, J. (2021). NZ's predator-free dream on path to failure without new tech. *New Zealand Herald*, 31 January.
62. Russell, Predator-free New Zealand, p. 523.
63. Goals and tactics of the project are detailed on the NZ Department of Conservation's website: www.doc.govt.nz/nature/pests-and-threats/predator-free-2050/goal-tactics-and-new-technology.
64. Teem, J., et al. (2020) Genetic biocontrol for invasive species. *Frontiers in Bioengineering and Biotechnology*, *8*: 452–466.
65. The first successful gene-drive was conducted in 2011 using homing endonuclease genes. Windbichler, N., et al. (2011). A synthetic homing endonuclease-based gene drive system in the human malaria mosquito. *Nature*, *473*(7346): 212–215.
66. Del Amo, V., et al. (2020). Small-molecule control of super-Mendelian inheritance in gene drives. *Cell Reports*, *31*(13): 107841.
67. Champer, J., Buchman, A. and Akbari, O. (2016). Cheating evolution: engineering gene drives to manipulate the fate of wild populations. *Nature Reviews Genetics*, *17*(3): 146–159.
68. DiCarlo, J., et al. (2015). Safeguarding CRISPR-Cas9 gene drives in yeast. *Nature Biotechnology*, *33*(12): 1250–1255; Gantz, V. and Bier, E. (2015). The mutagenic chain reaction. *Science*, *348*(6233): 442–444.
69. Grunwald, H., et al. (2019). Super-Mendelian inheritance mediated by CRISPR–Cas9 in the female mouse germline. *Nature*, *566*(7742): 105–109.
70. Andra Crisanti as interviewed in: Coffey, D. (2020). What is a gene drive? *Live Science*, 17 April.
71. Simon, S., Otto, M., and Engelhard M. (2018). Synthetic gene drive. *EMBO Reports*, *19*(5): e45760.

72. Scudellari, M. (2019). Self-destructing mosquitoes and sterilized rodents: the promise of gene drives. *Nature, 571*(7764): 160–163.
73. Ibid., p. 162.
74. Godwin, J., et al. (2019). Rodent gene drives for conservation. *Proceedings of the Royal Society B, 286*(1914): 20191606.
75. On island laboratories, see: Taitingfong, R. (2020). Islands as laboratories. *Human Biology, 91*(3): 179–188.
76. Godwin, Rodent gene drives for conservation.
77. Faber, N., et al. (2021). Novel combination of CRISPR-based gene drives eliminates resistance and localises spread. *Scientific Reports, 11*(1): 1–15.
78. Kaiser, S. (2021). Gene Drive: a game-changing technology for grey squirrel control. *Forestry & Timber News*, October: 58–60.
79. Ibid., p. 58.
80. See, for example, the perspective presented by the Centre for Scientific Research on Squirrels: http://i-csrs.com/effect-squirrels-trees.
81. Redford, K., et al. (2019). *Genetic Frontiers for Conservation*. IUCN, p. 122.
82. ETC Group. (2019). Driving under the influence. [www.etcgroup.org/files/files/etc-iucn-driving_under_influence.pdf].
83. National Academies of Sciences, Engineering, and Medicine. (2016). *Gene Drives on the Horizon*, The National Academies Press. While the NAS report supports continued development of gene drive technologies, it does recognise the emergence of pressing questions 'about who would be affected by the benefits and harms, who will be able to conduct research into gene drive technologies and study the release of gene-drive modified organisms, and who will make the decisions about whether to pursue the benefits and risk the potential harms'.
84. Noble, C., et al. (2018) Current CRISPR gene drive systems are likely to be highly invasive in wild populations. *eLife, 7*: e33423.
85. Le Page, M. (2016). 'Daisy-chain' gene drive vanishes after only a few generations. *New Scientist*, 16 June.
86. Latour, B. (2007). 'It's development, Stupid!' or: how to modernize modernization. In Nordhaus, T. and Shellenberger, M. (eds). *Break-through: From the Death of Environmentalism to the Politics of Possibility*. Houghton-Mifflin.
87. Latour, B. (2011). Love your monsters. *Breakthrough Journal, 2*(11): 19–26.
88. Latour, B. (2004). *Politics of Nature*. Harvard University Press, p. 26.
89. Simoni, A., et al. (2020). A male-biased sex-distorter gene drive for the human malaria vector Anopheles gambiae. *Nature Biotechnology, 38*(9): 1054–1060.
90. Galizi, R., Doyle, et al. (2014). A synthetic sex ratio distortion system for the control of the human malaria mosquito. *Nature Communications*, 5(1): 1.
91. WHO. (2017). *World Malaria Report 2017*. WHO Publications.
92. Snodgrass, M. (2017). *World Epidemics: A Cultural Chronology of Disease from Prehistory to the Era of Sars*. McFarland & Co. See also: e.g., Whitfield, J. (2002). Portrait of a serial killer. *Nature News*, 3 October.
93. WHO. (2015). *World Malaria Report 2015*. WHO Publications.
94. WHO. (2020). *World Malaria Report 2020*. WHO Publications.

95. Regalado, A. (2016). Bill Gates doubles his bet on wiping out mosquitoes with gene editing. *MIT Technology Review*, 6 September.
96. WHO. (2020). Evaluation of genetically modified mosquitoes for the control of vector-borne diseases: position Statement. [www.who.int/publications/i/item/9789240013155].
97. Moloo, Z. (2018). Cutting corners on consent. *Project Syndicate*, 19 December.
98. Achenbach, J. (2018). Gene drive research to fight diseases can proceed cautiously. *Washington Post*, 30 November.
99. Matthews, D. (2018). Gene drives could end malaria, and they just escaped a UN ban. *Vox*, 7 December.
100. Watts, J. (2018). GM mosquito trial sparks 'sorcerer's apprentice' lab fears. *The Guardian*, 25 November.
101. African Centre for Biodiversity. (2019). Civil society denounces the release of GM mosquitoes in Burkina Faso. *Grain*, 3 July.
102. Science and Technology Committee of the House of Lords. (2015) *Genetically Modified Insects, 1st Report of Session 2015–2016*. Stationery Office, p. 26.
103. African Centre for Biodiversity (2018). *Critique of African Union and NEPAD's Positions on Gene Drive Mosquitoes for Malaria Elimination*. The African Centre for Biodiversity.
104. Servick, K. (2019). Study on DNA spread by genetically modified mosquitoes prompts backlash. *Science*, 17 September: 365.
105. Waltz, E. (2021). First genetically modified mosquitoes released in the United States. *Nature*, *593*(7858): 175–176.
106. Kimbrell, A. (2021). Genetically engineered mosquitoes released in Florida Keys. *Center for Food & Safety News*, 26 April.
107. Waltz, First genetically modified mosquitoes released in the United States.
108. Oxitec (2021). Oxitec announces ground-breaking commercial launch of its Friendly™ *Aedes aegypti* solution. *Oxitec News Releases*, 3 November. [www.oxitec.com/en/news].
109. Ibid.
110. Ibid.
111. Scott, M. and Benedict, M. (2015). Concept and history of genetic control. In: Adelman, Z. (ed.). *Genetic Control of Malaria and Dengue*. Academic Press, p. 31.
112. Gutierrez, A., Ponti, L. and Arias, P. (2019). Deconstructing the eradication of new world screwworm in North America. *Medical & Veterinary Entomology*, *33*(2): 282–295.
113. Vargas-Terán, M., et al. (2021). Impact of screwworm eradication programmes using the sterile insect technique. In Dyck, V., Hendrichs, J. and Robinson, A. (2021). *Sterile Insect Technique*. Springer, pp. 629–650.
114. The worldwide directory of SIT Facilities (DIRSIT) collects information on sites dedicated to the mass-rearing of sterile insects for operations of genetic control: https://nucleus.iaea.org/sites/naipc/dirsit.
115. Alphey, L., et al. (2013). Genetic control of *Aedes* mosquitoes. *Pathogens & Global Health*, *107*(4): 170–179.

116. Knipling, E. (1959). Sterile-male method of population control. *Science, 130*(3380): 902–904.
117. Scott and Benedict, Concept and history of genetic control, p. 43.
118. Alphey, L. (2014). Genetic control of mosquitoes. *Annual Review of Entomology, 59*(1): 205–224.
119. Windbichler, N., et al. (2011) A synthetic homing endonuclease-based gene drive system in the human malaria mosquito. *Nature, 473*(7346): 212–215.
120. Gantz, V., et al. (2015). Highly efficient Cas9-mediated gene drive for population modification of the malaria vector mosquito *Anopheles stephensi. PNAS, 112*(49): E6736-E6743; Hammond, A., et al. (2016). A CRISPR-Cas9 gene drive system targeting female reproduction in the malaria mosquito vector *Anopheles gambiae. Nature Biotechnology, 34*(1): 78–83.
121. Kyrou, K., et al. (2018). A CRISPR–Cas9 gene drive targeting doublesex causes complete population suppression in caged *Anopheles gambiae* mosquitoes. *Nature Biotechnology, 36*(11): 1062.
122. Bier, E. (2022). Gene drives gaining speed. *Nature Reviews Genetics, 23*(1): 5–22.
123. Kyrou, A CRISPR–Cas9 gene drive targeting doublesex causes complete population suppression in caged *Anopheles gambiae* mosquitoes.
124. Associated Press (2015) Gene editing research seeks to balance ethics with potential for fighting disease. *National Post*, 9 October.
125. Wang, G. (2021). Combating mosquito-borne diseases using genetic control technologies. *Nature Communications, 12*(1): 1–12.
126. Patterson, G. (2009). *The Mosquito Crusades*. Rutgers University Press, p. 9.
127. Snowden, F., (2008). *La conquista della malaria.* Einaudi, pp.116–130.
128. Russell, E. P. (1999). The strange career of DDT. *Technology & Culture, 40*(4): 770–796; Stapleton, D. (1998). The dawn of DDT and its experimental use by the Rockefeller Foundation in Mexico, 1943–1952. *Parassitologia, 40*(1–2): 149–158.
129. Carson, R. (1962). *Silent Spring*. Houghton Mifflin.
130. Sánchez-Bayo, F. and Wyckhuys, K. (2019). Worldwide decline of the entomofauna. *Biological Conservation, 232*: 8.
131. Ibid., p. 17.
132. McNeill, J. R. (2010). *Mosquito Empires*. Cambridge University Press, p. 3.
133. Mitchell, T. (2002). *Rule of Experts*. University of California Press, pp. 19–53.
134. Samanta, A. (2001). Crop, climate and malaria. *Economic & Political Weekly*: 4887–4890.
135. Swyngedouw, E. (2006). Circulations and metabolisms. *Science as Culture, 15*(2): 105.
136. Chaves, L., et al. (2020). Global consumption and international trade in deforestation-associated commodities could influence malaria risk. *Nature Communications, 11*(1): 1–10; Caminade, C., et al. (2014). Impact of climate change on global malaria distribution. *PNAS, 111*(9): 3286–3291.
137. National Academies of Science, *Gene Drives on the Horizon*, p. 41.
138. Poulin, B., Lefebvre, G. and Paz, L. (2010). Red flag for green spray. *Journal of Applied Ecology, 47*(4): 884–889.

139. Walker, E., Kaufman, M. and Merritt, R. (2010). An acute trophic cascade among microorganisms in the tree hole ecosystem following removal of omnivorous mosquito larvae. *Community Ecology*, *11*(2): 171–178.
140. Thien, L. and Utech, F. (1970). The mode of pollination in *Habenaria obtusata*. *American Journal of Botany*, *57*(9): 1031–1035; Barredo, E. and DeGennaro, M. (2020). Not just from blood: Mosquito nutrient acquisition from nectar sources. *Trends in Parasitology*, *36*(5): 473–484.
141. Clapper, J. (2016). Worldwide threat assessment of the US intelligence community. [http://dni.gov/files/documents/SASC_Unclassified_2016_ATA_SFR_FINAL.pdf].
142. Darpa (2016) Setting a safe course for genome editing research. [www.darpa.mil/news-events/2016-09-07].
143. Cheever, A. (n.d.) Safe genes. Defense Advanced Research Projects Agency. [www.darpa.mil/program/safe-genes].
144. Darpa, Setting a safe course for genome editing research.
145. Cheever, Safe genes.
146. Gene Drive Files. (2017). Gene drive files. Synbio-Watch. Obtained by Edward Hammond by North Carolina Public Records Law request. [http://genedrivefiles.synbiowatch.org/2017/12/01/us-military-gene-drive-development].
147. Darpa (2019) Safe Genes tool kit takes shape. [www.darpa.mil/news-events/2019-10-15].
148. Ibid.
149. National Academies of Science, *Gene Drives on the Horizon*, p. 34.
150. Akbari, O. S., et al. (2015). Safeguarding gene drive experiments in the laboratory. *Science*, *349*(6251): 927–929.
151. Evans, B., et al. (2019). Transgenic *Aedes aegypti* mosquitoes transfer genes into a natural population. *Scientific Reports*, *9*(1): 1–6.
152. Servick, K. (2019). Study on DNA spread by genetically modified mosquitoes prompts backlash. *Science*, 17 September: 365.
153. Power, M. (2021). Synthetic threads through the web of life. *PNAS*, *118*(22): e2004833118.
154. Kahn, J. (2020). The Gene Drive Dilemma. *New York Times Magazine*, 8 January. See also: ETC Group. (2019). Gene drive organisms. [www.etcgroup.org/sites/www.etcgroup.org/files/files/etc_gene_drive_organisms-web_en.pdf].
155. Montenegro de Wit, M. (2019). Gene driving the farm. *Agroecology&Sustainable Food Systems*, *43*(9): 1054–1074.
156. Oxitec's marketing statements are publicized on the company's public website. [www.oxitec.com/en/our-technology]. See also: Simmons, G., et al. (2011) Field performance of a genetically engineered strain of pink bollworm. *PloS*, *6*(9): e24110.
157. On Marx conceptualization of metabolism, see: Marx, K. (1976). *Capital, Volume 1*. Penguin, pp. 283, 290; Marx, K. (1988). *Capital, Volume II*. Progress Publishers, pp. 40, 63. On the 'metabolic rift', see: Marx, *Capital, Volume 1*, pp. 637–638. On the concept of 'metabolic shift', see: Moore, J. (2017). Metabolic rift or metabolic shift? *Theory & Society*, *46*(4): 285–318.

Chapter 7 Pharmaceutical Lives: Humanised Mice, Pharma-Pigs, and the Molecularisation of Production

1. Lexchin, J. (2018). The Pharmaceutical Industry in Contemporary Capitalism. *Monthly Review*, *69*(10): 37–50.
2. Ibid., p. 38–39.
3. Lexchin, J. (2018). *Private Profits versus Public Policy*. University of Toronto Press.
4. Marx, K. (1976). *Capital, Volume I*. Penguin, p. 134.
5. Löwy, I. (2000). Experimental bodies. In Cooter, R. and Pickstone, J. (eds). *Medicine in the Twentieth Century*, Harwood: pp. 435–449.
6. It must be remarked that the distinction between valuable subjects and expendable experimental bodies has always been political, rather than ontological. The history of colonial medicine shows that human bodies, once de-humanised, may be swiftly turned into expendable experimental bodies. See: Graboyes, M. (2014). Incorporating medical research into the history of medicine in East Africa. *The International Journal of African Historical Studies*, *47*(3): 379–398; Washington, H. (2006) *Medical Apartheid*. Doubleday.
7. Löwy, I Experimental bodies, pp. 435–445. See also: Feinstein, Alvan R. (1987). The intellectual crisis in clinical science. *Perspectives in Biology and Medicine*, *30*: 215–229.
8. Steensma, D., Kyle, R. and Shampo, M. (2010). Abbie Lathrop, the 'mouse woman of Granby'. *Mayo Clinic Proceedings*, *85*(11): e83.
9. Rader, K. (2004) *Making Mice: Standardizing Animals for American Biomedical Research, 1900–1955*. Princeton University Press, p. 217.
10. Black6, in other words, enables the global circulation of experimental data just as money enables the exchange of commodities on the world market. In Marxian terms it plays 'within the world of commodities the part of the universal equivalent'. Marx, *Capital, Volume 1*, p. 162.
11. Rader, *Making Mice*, p. 86.
12. Rockefeller Foundation. (1936). *Annual Report 1935*, pp. 126, 159–160. [rockefellerfoundation.org/wp-content/uploads/Annual-Report–1935.pdf].
13. Little, C. (1933). Not dead but sleeping. *Journal of Heredity*, *24*(4): 149–150.
14. Little, C. (1928). Opportunities for research in mammalian genetics. *The Scientific Monthly*, *26*(6): 521–534.
15. US Committee on Commerce (1937) *Cancer Research: Hearings Before A Subcommittee of the Committee on Commerce, Seventy-Fifth Congress, First Session*. US Government Printing Office, p. 54.
16. Holstein, J. (1979) *The First Fifty Years at the Jackson Laboratory, 1929–1979*. Jackson Laboratory.
17. Little, C. (1935). A new deal for mice. *Scientific American*, *152*(1), pp. 16–18.
18. Ibid., p. 18.
19. Rader, *Making Mice*, p. 5.
20. Jackson Laboratory. (2020). 2020 report. [www.jax.org/about-us/legal-information/financials].

21. Woolston, C. (2020). The tale of a mouse-lab mastermind. *Nature*, *586*(7831): 818–819.
22. Taft, R. A., Davisson, M. and Wiles, M. V. (2006). Know thy mouse. *TRENDS in Genetics*, *22*(12): 649–653.
23. Zeldovich, L. (2017). Genetic drift: the ghost in the genome. *Lab-animal*, *46*(6): 255–257.
24. Wiles, M., and Taft, R. (2010). The sophisticated mouse: protecting a precious reagent. In: Proetzel, G. and Wiles, M. (ed.). *Mouse Models for Drug Discovery*. Humana Press, pp. 23–36.
25. Parker, M. and Mulder, G. (2021). The history of Black 6 mice. *Charles River Laboratory Eureka Magazine*, 18 August.
26. Chandra, K., Datta, A. and Mondal, D. (2013). Development of mouse models for cancer research. In: Verma, A. and Singh, A. (eds). *Animal Biotechnology*. Academic Press, p. 80.
27. Waterston, R. and Pachter, L. (2002). Initial sequencing and comparative analysis of the mouse genome. *Nature*, *420*(6915): 520–562.
28. Botting, J. and Morrison, A. (2015). *Animals and Medicine*. Open Book, p. 244.
29. Ali, I., et al. (2012). Thalidomide. *Current Drug Therapy*, *7*(1): 13–23.
30. Shah, R. R. (2001). Thalidomide, drug safety and early drug regulation in the UK. *Adverse Drug Reactions and Toxicological Reviews*, *20*(4): 199–255. See also: Botting, J. (2002). The history of thalidomide. *Drug News and Perspectives*, *15*(9): 604–11.
31. Vargesson, N. (2015). Thalidomide-induced teratogenesis. *Embryo-Today*, *105*(2): 140–156.
32. Canguilhem, G. (1991) *The Normal and the Pathological*. Zone Books, pp. 148–149.
33. Hackam, D. and Redelmeier, D. (2006). Translation of research evidence from animals to humans. *Jama*, *296*(14), 1727–1732.
34. Akhtar, A. (2015). The flaws and human harms of animal experimentation. *Cambridge Quarterly of Healthcare Ethics*, *24*(4): 407–419.
35. Engber, D. (2011). The mouse trap. *Slate Magazine*, 15 November.
36. McManus, R. (2013). Ex-director Zerhouni surveys value of NIH research. *NIH Record*, *65*(13): 6–7.
37. BCC Research. (2019). Global market for laboratory animal models to top $7 billion by 2023. BCC Research Press Releases, 20 March. [www.bccresearch.com/pressroom/phm/global-market-for-laboratory-animal-models-to-top-$7-billion-by-2023].
38. Hanahan, D., Wagner, E. and Palmiter, R. (2007). The origins of oncomice. *Genes & Development*, *21*(18): 2258–2270.
39. Leder, P. and Stewart, T. (1988). US Patent No. 4,736,866. All subsequent references in this paragraph are extracts from this patent.
40. Jaffe, S. (2004). Ongoing battle over transgenic mice. *The Scientist*, *18*(14): 46–48.
41. Kittredge, C. (2005). A question of chimeras. *The Scientist*, *19*(7): 54–56.
42. Grimm, D. (2006). A mouse for every gene. *Science*, *312*(5782): 1862–1866.

43. Yamada, A. (2012). Bigger muscles, stronger muscles? The strange case of myostatin. *Experimental physiology*, *97*(5): 562–563.
44. Lu, D., et al. (2018). Impairment of social behaviors in Arhgef10 knockout mice. *Molecular Autism*, *9*(1): 1–14.
45. Hisaoka, T., et al. (2018). Abnormal behaviours relevant to neurodevelopmental disorders in Kirrel3-knockout mice. *Scientific Reports*, *8*(1): 1–12.
46. Interview reported in: Viney, K. (2012). Inside the industry supplying millions of mutant mice. *New Scientist*, 16 July.
47. Thon, R., et al. (2002). Welfare evaluation of genetically modified mice. *Scandinavian Journal of Laboratory Animal Science*, *29*(1): 45–53. See also: Bailey, J. (2019). Genetic modification of animals: Scientific and ethical issues. in Herrmann, K. and Jayne, K. (ed.). *Animal Experimentation*. Brill, pp. 443–479.
48. Rosen, B., Schick, J. and Wurst, W. (2015). Beyond knockouts. *Mammalian Genome*, *26*(9): 456–466.
49. See International Mouse Phenotyping Consortium website at www.mousephenotype.org.
50. Clements, Peter J., et al. (2021). Animal models in toxicologic research: rodents. In Haschek, W., et al. (eds). *Haschek and Rousseaux's Handbook of Toxicologic Pathology, Volume I*. Academic Press, pp. 653–694.
51. Cohen, J. (2016). Mice made easy. *Science*, *354*(6312): 538–542.
52. Wang, H. (2013). One-step generation of mice carrying mutations in multiple genes by CRISPR/Cas-mediated genome engineering. *Cell*, *153*(4): 910–918.
53. Cohen, Mice made easy, pp. 540–542.
54. Cohen, J. (2016). Any idiot can do it: CRISPR could put mutant mice in everyone's reach. *Science News*, 3 November.
55. Ibid.
56. Yang, H., Wang, H. and Jaenisch, R. (2014). Generating genetically modified mice using CRISPR/Cas-mediated genome engineering. *Nature Protocols*, *9*(8): 1956–1968.
57. Akhtar, The flaws and human harms of animal experimentation, pp. 407–419.
58. In 2006, clinical trials for a new drug intended for patients affected by leukaemia ended in disaster. The drug had no toxic effects on mice, but it caused sudden and systematic organ failure in all six trial participants. Clair, E. (2008). The calm after the cytokine storm. *Journal of Clinical Investigation*, *118*(4): 1344–1347.
59. Zhu, F. (2019). Humanising the mouse genome piece by piece. *Nature Communications*, *10*(1): 1–13.
60. Macchiarini, F., et al. (2005) Humanized mice: are we there yet? *Journal of Experimental Medicine*, *202*(10): 1307–1311.
61. Lapidot, T., Fajerman, Y. and Kollet, O. (1997). Immune-deficient SCID and NOD/SCID mice models as functional assays for studying normal and malignant human hematopoiesis. *Journal of Molecular Medicine*, *75*(9): 664–673.
62. Morton, J., et al. (2016). Humanized mouse xenograft models. *Cancer Research*, *76*(21): 6153–6158.

63. Shultz, L., et al. (2012). Humanized mice for immune system investigation. *Nature Reviews Immunology*, *12*(11): 786–798.
64. Regalado, A. (2022) Going bald? Lab-grown hair cells could be on the way. *MIT Technology Review*, 18 January.
65. Raghavan, Shreya, et al. (2011) Successful implantation of bioengineered, intrinsically innervated, human internal anal sphincter. *Gastroenterology*, *141*(1): 310–319.
66. Himmel, L., et al. (2022). Genetically engineered animal models in toxicologic research. In Haschek, W., et al. (eds). *Haschek and Rousseaux's Handbook of Toxicologic Pathology, Volume I*. Academic Press, pp. 859–924.
67. Kim, Y. and Ko, J. (2018). Humanized model mice by genome editing and engraftment technologies. *Molecular & Cellular Toxicology*, *14*(3): 255–261.
68. Liu, E., et al. (2017). Of mice and CRISPR. *EMBO Reports*, *18*(2): 187–193.
69. Ibid., p. 189.
70. Naritomi, Y., Sanoh, S. and Ohta, S. (2019). Utility of chimeric mice with humanized liver for predicting human pharmacokinetics in drug discovery. *Biological & Pharmaceutical Bulletin*, *42*(3): 327–336.
71. Kazuki, Y., et al. (2016). Thalidomide-induced limb abnormalities in a humanized CYP3A mouse model. *Scientific Reports*, *6*(1): 1–6.
72. Macdonald, L., et al. (2014). Precise and in situ genetic humanization of 6Mb of mouse immunoglobulin genes. *PNAS*, *111*(14): 5147–5152.
73. Liu, Of mice and CRISPR, pp. 187–193.
74. The description appears on the company's website: www.regeneron.com/science/technology.
75. Wu, J., et al. (2016). Stem cells and interspecies chimaeras. *Nature*, *540*(7631): 51–59.
76. Rashid, T., Kobayashi, T. and Nakauchi, H. (2014). Revisiting the flight of Icarus. *Cell Stem Cell*, *15*(4): 406–409.
77. Hinterberger, A. (2018). Marked 'h' for human: Chimeric life and the politics of the human. *BioSocieties*, *13*(2): 453–469.
78. Academy of Medical Sciences. (2011). *Animals Containing Human Material*. AMS Press.
79. Pittius, C., et al. (1988). A milk protein gene promoter directs the expression of human tissue plasminogen activator cDNA to the mammary gland in transgenic mice. *PNAS*, *85*(16): 5874–5878.
80. Ibid: 5876–5878.
81. Pennisi, E. (1998). After Dolly, a pharming frenzy. *Science*, *279*(5351): 646–648. See also: Rehbinder, E., et al. (2008). *Pharming*. Springer.
82. Zangeneh, F. and Dolinar, R. (2016). Biosimilar drugs are not generics. *Endocrine Practice*, *22*(1): 6–21.
83. Melmer, G. (2006). Biopharmaceuticals and the industrial environment. In Gellissen, G. (ed.). *Production of recombinant proteins*. Wiley & Sons.
84. Pfeffer, N. (2017). *Insider Trading: How Mortuaries, Medicine and Money Have Built a Global Market in Human Cadaver Parts*. Yale University Press, pp. 34–47.
85. Spadiut, O., et al. (2014). Microbials for the production of monoclonal antibodies and antibody fragments. *Trends in Biotechnology*, *32*(1): 54–60.

86. Butler, M. (2005). Animal cell cultures, *Applied Microbiology and Biotechnology, 68*(3): 283–291.
87. Gadgil, M. (2017). Cell culture processes for biopharmaceutical manufacturing. *Current Science, 112*(7): 1478–1488.
88. Walsh, G. (2018). Biopharmaceutical benchmarks 2018. *Nature Biotechnology, 36*(12): 1136–1145.
89. Owczarek, B., Gerszberg, A. and Hnatuszko-Konka, K. (2019). A brief reminder of systems of production and chromatography-based recovery of recombinant protein biopharmaceuticals. *BioMed*: 4216060.
90. Noh, S., Sathyamurthy, M. and Lee, G. (2013). Development of recombinant Chinese hamster ovary cell lines for therapeutic protein production. *Current Opinion in Chemical Engineering*, 2(4): 391–397; Mukherjee, S. (2019) Protect at all costs. *Fortune*, 18 July.
91. Walsh, Biopharmaceutical benchmarks 2018, p. 1140.
92. Jozala, A., et al. (2016). Biopharmaceuticals from microorganisms. *Brazilian Journal of Microbiology, 47*: 51–63.
93. Stix, G. (2005). The land of milk & money. *Scientific American, 293*(5): 102–105.
94. Mahmuda, A., et al. (2017). Monoclonal antibodies. *Tropical Journal of Pharmaceutical Research, 16*(3): 713–722.
95. Holden, C. (1994). Raising a herd of therapeutic goats. *Science, 264*(5161): 902–902.
96. Meade, H., and Lonberg, N. (1989). Isolation of exogenous recombinant proteins from the milk of transgenic mammals. US Patent No. 4,873,316.
97. Ibid.
98. Echelard, Y., Meade, H. and Ziomek, C. (2005). The first biopharmaceutical from transgenic animals: ATryn. In Klablein, J. (ed.). *Modern Biopharmaceuticals*. Wiley, pp. 995–1020.
99. Suhashini, M. and Siridewa, K. (2021). Molecular pharming. *Current Trends in Biotechnology&Pharmacy, 15*(3): 315–324.
100. Editor. (2014). FDA approves second transgenic milk drug. *Nature Reviews Drug Discovery, 13*(1): 644.
101. Mullard, A. (2016). FDA approves drug from transgenic chicken. *Nature Reviews Drug-Discovery, 15*(1): 7–8.
102. Jenkins, K. (2021) Insect-based vaccine production platform validated by EMA. *Pharmafile*, 22 June.
103. Escribano, J., et al. (2020). Chrysalises as natural production units for recombinant subunit vaccines. *Journal of Biotechnology, 324*: 100019.
104. The interviews have been published in: Raper, P. (2019). Insect pupae may transform insect production. *Genetic Engineering & Biotechnology News*, 3 December; and: Alvarez, R. and Cortes, A. (2021). Impact of Covid-19 on biotech firms. *Asebio Reports*, 27 January.
105. United States Securities and Exchange Commission. (2014). *rEvo Biologics, Inc. Registration Statement*, 4 September.
106. Wilmut, I. (1998). Cloning for medicine. *Scientific American, 279*(6): 58–63.
107. Colman, A. (1999). Dolly, Polly and other 'ollys'. *Genetic Analysis, 15*(3–5): 167–173.

108. Himmel, Genetically engineered animal models in toxicologic research, p. 914.
109. Velander, W., Lubon, H. and Drohan, W. (1997). Transgenic livestock as drug factories. *Scientific American*, *276*(1): 70–74.
110. Ibid.
111. Ibid.
112. Transplant Observatory. (2021). *Transplant Report 2020*. Transplant Observatory. [www.transplant-observatory.org].
113. Vanholder, R., et al. (2021). Organ donation and transplantation. *Nature Reviews Nephrology*, *17*(8): 554–568.
114. Ahmad, M., et al. (2019). A systematic review of opt-out versus opt-in consent on deceased organ donation and transplantation. *World Journal of Surgery*, *43*(12): 3161–3171.
115. Editor. (2006) Psst, wanna buy a kidney? Governments should let people trade kidneys, not convict them. *The Economist*, 18 November, p. 15. See also: Clay, M. and Block, W. (2002). A free market for human organs. *The Journal of Social, Political, and Economic Studies*, *27*(2): 227–236.
116. Deschamps, J., et al. (2005). History of xenotransplantation. *Xenotransplantation*, *12*(2): 91–109.
117. Buehler, L., et al. (1999). Xenotransplantation. *Frontiers in Bioscience*, *4*: D416–32.
118. Cooper, D. (1993). Identification of α-galactosyl and other carbohydrate epitopes that are bound by human anti-pig antibodies. *Transplant Immunology*, *1*(3): 198–205.
119. Moore, S. (1997). Novartis expands partnership in xenotransplant competition. *Wall Street Journal*, 10 October.
120. Stolberg, S. G. (1999). Could this pig save your life? *New York Times*, 3 October.
121. Engel, M. E. and Hiebert, S. W. (2010). The enemy within: dormant retroviruses awaken. *Nature Medicine*, *16*(5): 517–518.
122. Denner, J. (2021). Porcine endogenous retroviruses and xenotransplantation, 2021. *Viruses*, *13*(11): 2156–2171. See also: Kimsa, M., et al. (2014). Porcine endogenous retroviruses in xenotransplantation. *Viruses*, *6*(5): 2062–2083.
123. Lopata, K., et al. (2018). Porcine endogenous retrovirus (PERV). *Frontiers in Microbiology*, *9*: 730.
124. Wadman, M. (2001). Novartis to pigs: keep your kidneys. *Fortune Magazine*, 16 April.
125. Reardon, S. (2015). New life for pig organs. *Nature*, *527*: 152–154.
126. Dolgin, E. (2021). First GM pigs for allergies: could xenotransplants be next? *Nature Biotechnology*, *39*(4): 397–400.
127. Meier, R., et al. (2018). Xenotransplantation: back to the future? *Transplant International*, *31*(5): 465–477. See also: Servick, K. (2022). Here's how scientists pulled off the first pig-to-human heart transplant. *Science Insider*, 12 January.
128. Cooper, D. (2021). Genetically engineered pig kidney transplantation in a brain-dead human subject. *Xenotransplantation*, *28*(6): e12718.

129. Weintraub, K. (2022) 'The new heart is still a rockstar'. *USA Today*, 12 January.
130. Cookson, C. (2022) Gene editing: pig hearts and the new era of organ transplants. *Financial Times*, 2 February.
131. Ayares, D. and Phelps, C. (2018). US Patent Application No.15/758,895.
132. University of Maryland School of Medicine (2022). Successful transplant of porcine heart into adult human with end-stage heart disease. *Science Daily*, 10 January. See also: Reardon, S. (2022). First pig-to-human heart transplant. *Nature*, *601*(1): 305–306.
133. The interview is reported in: Arianne, G. (2022). Pig-to-human transplant. *Business Times*, 21 January.
134. Shirk, C. (2022). Breakthrough transplants signal BIG opportunity. *Investor Place*, 13 January.
135. Ekblom, J. (2021). First ever pig-heart transplant boosts shares of Swedish Medtech. *Bloomberg News*, 11 January.
136. Regalado, A. (2022). The gene-edited pig heart given to a dying patient was infected with a pig virus, *MIT Technology Review*, 4 May.
137. Rabin, R. (2022) Signs of an animal virus discovered in man who received a pig's heart. *New York Times*, 5 May.
138. Niu D., et al (2017). Inactivation of porcine endogenous retrovirus in pigs using CRISPR-Cas9. *Science*, *357*(6357): 1303–1307.
139. eGenesis (2021). eGenesis announces $125 million series C financing. *eGenesis Press Releases*, 2 March.
140. Park, A. (2021). Miromatrix goes hog-wild with upsized $43M IPO to bioengineer pig organs into human transplants. *Fierce Biotech*, 24 June.
141. Reardon, First pig-to-human heart transplant, pp. 305–306.
142. Pellerin, C. (2016). DoD funds new tissue biofabrication manufacturing consortium. *Department of Defense News*, 22 December; Zhang, S. (2017). 'Big pork' wants to get in on organ transplants. *The Atlantic*, 1 May.
143. Scheper-Hughes, N. (2000). The global traffic in human organs. *Current Anthropology*, *41*(2): 191–224.
144. Goździak, E. and Vogel, K. (2020). Palermo at 20. *Journal of Human Trafficking*, *6*(2): 109–118.
145. Jafar, T. (2009). Organ trafficking. *American Journal of Kidney Diseases*, *54*(6): 1145–1157.
146. Callender, C. and Miles, P. (2010). Minority organ donation: the power of an educated community. *Journal of the American College of Surgeons*, *210*(5): 708–715.
147. Callender, C., et al. (2016). Organ donation in the United States. *Transplantation Proceedings*, *48*(7): 2392–2395.
148. Callender and Miles, Minority organ donation, p. 708.

Chapter 8 Bioengineering the Human: Human Genetic Capital in a Neoliberal Environment

1. Weiss, M., et al. (2018) The last universal common ancestor between ancient Earth chemistry and the onset of genetics. *PloS Genetics*, *14*(8): e1007518.

2. All quotes in this paragraph are from Marx, K. (1976). *Capital, Volume 1*, Penguin, pp. 283,290.
3. Marx, K. (1989). Notes on Adolph Wagner's Lehrbuch der politischen Ökonomie. *MECW*, *24*: 553.
4. Federici, S. (2020). *Beyond the Periphery of the Skin*. PM Press.
5. An article in *Nature* sums up well how viruses reproduce by hijacking cellular production: 'No matter whether their genomes are RNA or DNA, and regardless of their mRNA production method, the goal remains the same: to ensure that cellular ribosomes are recruited to viral mRNAs. The ensuing synthesis of viral proteins is required for viral genome replication and progeny virion production.' Walsh, D. and Mohr, I. (2011). Viral subversion of the host protein synthesis machinery. *Nature Reviews Microbiology*, *9*(12): 860–875.
6. See, for instance: Strand, M. and Burke, G. R. (2014). Polydnaviruses: nature's genetic engineers. *Annual review of virology*, *1*: 333–354; and Temin, M. (1992). Retroviruses: nature's genetic engineers. In: Kurth, R. and Schwerdtfeger, W. K. (eds). *Current Topics in Biomedical Research*. Springer, pp. 79–84.
7. Villarreal, L. (2005). *Viruses and the Evolution of Life*. ASM Press, p. ix. See also: Van Regenmortel, M. and Mahy, B. (2004). Emerging issues in virus taxonomy. *Emerging Infectious Diseases*, *10*(1): 8–29.
8. Fuhrman, J. A. (1999). Marine viruses and their biogeochemical and ecological effects. *Nature*, *399*(6736): 541–548.
9. Editorial Board. (2011). Microbiology by numbers. *Nature Review of Microbiology*, *9*(9): 628.
10. The quote appears in an interview with Frank Aylward in: Nelson, E. (2020). Viruses don't have a metabolism, but some have the building blocks for one. *Virginia Tech Magazine*, 6 April.
11. Danovaro, R., et al. (2011). Marine viruses and global climate change. *FEMS Microbiology Reviews*, *35*(6): 993–1034.
12. Editorial Board (2011). Microbiology by numbers: 628.
13. French, R. and Holmes, E. (2020). An ecosystems perspective on virus evolution and emergence. *Trends in Microbiology*, *28*(3), 165–175.
14. For a review of the data presented in this paragraph, including references to the relevant reports issued by the WHO and the IMF, please see: Sarkodie, S. and Owusu, P. (2021). Global assessment of environment, health and economic impact of the novel coronavirus. *Environment, Development & Sustainability*, *23*: 5005–5015.
15. UN Secretariat. (2020). Tackling the Inequality Pandemic. UN Secretary-General's lecture, 18 July.
16. Revollo, P. (2021). *The Inequality Virus: Methodology Note*. Oxfam Publications.
17. Kim, J. et al. (2021). Operation Warp Speed. *The Lancet Global Health*, *9*(7): e1017–e1021.
18. Foucault, M. (2007). *Security, Territory, Population*. Springer, p. 10.
19. Ibid., p. 57.

20. Depeuille, G. (1800). A milk maid shows her cowpoxed hand to a physician, while a farmer or surgeon offers to a dandy inoculation with cowpox that he has taken from a cow. Wellcome Library, no. 16139i.
21. Willis, N. (1997). Edward Jenner and the eradication of smallpox. *Scottish Medical Journal*, *42*(4): 118–121.
22. Pead, P. (2006). Benjamin Jesty: the first vaccinator revealed. *The Lancet*, *368*(9554): 2202.
23. For a detailed account of this history, see: Pead, P. (2020). *Benjamin Jesty*. Cambridge Scholars.
24. Esparza, J., et al. (2020). Early smallpox vaccine manufacturing in the United States. *Vaccine*, *38*(30): 4773–4779.
25. For a succinct history of vaccine production techniques, see: Plotkin, et al. (2018). *Vaccines*. Elsevier, pp. 1–15, 50–60.
26. Goodpasture, E., Woodruff, A. and Buddingh, G. (1931). The cultivation of vaccine and other viruses in the chorio-allantoic membrane of chick embryos. *Science*, *74*(1919).
27. Yeung, J. (2020). The US keeps millions of chickens in secret farms to make flu vaccines. *CNN Report*, 29 March.
28. Buckland, B. (2015). The development and manufacture of influenza vaccines. *Human Vaccines & Immunotherapeutics*, *11*(6): 1357–1360.
29. Salk, J. (1955). Considerations in the preparation and use of poliomyelitis virus vaccine. *Journal of the American Medical Association*, *158*(14), 1239–1248.
30. On vaccine production in cell cultures, see: Barrett, P., et al. (2009). Vero cell platform in vaccine production. *Expert Review of Vaccines*, *8*(5): 607–618.
31. Payette, P. and Davis, H. (2001). History of vaccines and positioning of current trends. *Current Drug Targets-Infectious Disorders*, *1*(3): 241–247.
32. On recombinant vaccines, see: Soler, E. and Houdebine, L. M. (2007). Preparation of recombinant vaccines. *Biotechnology Annual Review*, *13*: 65–94.
33. Tregoning, J. (2021). Progress of the COVID-19 vaccine effort. *Nature Reviews Immunology*, *21*(10): 626–636.
34. Dolgin, E. (2021). How protein-based COVID vaccines could change the pandemic. *Nature*, *599*(7885): 359–360; Yadav, T., et al. (2020). Recombinant vaccines for COVID-19. *Human Vaccines & Immunotherapeutics*, *16*(12): 2905–2912.
35. Kis, Z., et al. (2020). Rapid development and deployment of high-volume vaccines for pandemic response. *Journal of Advanced Manufacturing & Processing*, *2*(3): e10060.
36. Rinaldi, A. (2020). RNA to the rescue. *EMBO Reports*, *21*(7): e51013.
37. Gertner, J. (2021). Genome sequencing and Covid vaccines. *New York Times*, 25 March.
38. Malone, R., Felgner, P. and Verma, I. (1989). Cationic liposome-mediated RNA transfection. *PNAS*, *86*(16): 6077–6081. See also: Malone, R (1989). mRNA transfection of cultured eukaryotic cells and embryos using cationic liposomes. *Focus*, *11*(4): 61–66.
39. See: Verbeke, R., et al. (2021). The dawn of mRNA vaccines. *Journal of Controlled Release*, *333*: 511–520.

40. See: Kariko, K., et al. (1998) Phosphate-enhanced transfection of cationic lipid-complexed mRNA and plasmid DNA. *BBA-Biomembranes, 1369*(2): 320–334.
41. Tam, Y., Chen, S. and Cullis, P. (2013). Advances in lipid nanoparticles for siRNA delivery. *Pharmaceutics*, *5*(3): 498–507.
42. Allegretti, A. and Elgot, J. (2021). Covid: greed and capitalism behind vaccine success, Johnson tells MPs. *The Guardian*, 24 March.
43. Dolgin, E. (2021). The tangled history of mRNA vaccines. *Nature*, *597*(7876): 318–324.
44. Ibid.
45. Gaviria, M., and Burcu K. (2021) A network analysis of COVID-19 mRNA vaccine patents. *Nature Biotechnology*, *39*(1): 546–548.
46. Malpani, R. and Maitland, A. (2021). Dose of reality. *People's Vaccine Alliance*, 21 October.
47. Kis, Z. and Rizvi, Z. (2021). How to make enough vaccine for the world in one year. *Public Citizen Report*, 26 May. See also: Light, D. and Lexchin, J. (2021). The costs of coronavirus vaccines and their pricing. *Journal of the Royal Society of Medicine*, *114*(11): 502–504.
48. Maxmen, A. (2021). The fight to manufacture COVID vaccines in lower-income countries. *Nature*, *597*(7877): 455–457.
49. Erman, M. and Mishra, M. (2021). Pfizer sees robust COVID-19 vaccine demand for years. *Reuters*, 4 May.
50. Mazzucato, M. (2020). Capitalism after the pandemic. *Foreign Affairs*, *99*: 50–61.
51. Oxfam (2021). More than a million COVID deaths in 4 months since G7 leaders failed to break vaccine monopolies. *Oxfam Reports*, 3 June.
52. Maxmen, A. (2022). South African scientists copy Moderna's COVID vaccine. *Nature News*, 3 February.
53. Responding to a query if Moderna would enforce patents in Africa at the end of the pandemic, Chief Executive Stéphane Bancel replied: 'We have not decided yet for low- and middle-income countries.' Roelf, W. and Steenhuysen, J. (2022) Moderna patent application raises fears for Africa COVID vaccine-hub. *Reuters*, 17 February.
54. Marx, K. (1973). *Grundrisse*. Penguin, p. 705.
55. Ibid., p. 706.
56. Antonio Negri and Paolo Virno have interpreted this Marxian thesis as anticipating the emergence of a cognitive capitalism, which increasingly relies on the diffused production of knowledge. Their analysis has been largely based on the new strategies of accumulation emerging in the realm of the Internet, of media production, of finance, etc. Our analysis suggests that the prominence they grant to cognitive labour is not to be taken for granted. In the realm of molecular biology, and its industrial applications, it is the entire body with its metabolic potential that is mobilised as a source of value and as a site of production. See: Negri, A. and Vercellone, C. (2008). The capital/labor relationship in cognitive capitalism. *Multitudes*, *1*: 39–50; and Virno, P. (2007). General intellect. *Historical Materialism*, *15*(3): 3–8.

57. Porta, M. (ed.). (2014). *A Dictionary of Epidemiology*. Oxford University Press, pp.182–183.
58. Abraham, J. (2010). Pharmaceuticalization of society in context. *Sociology*, *44*(4): 603–622.
59. Williams, S., Martin, P. and Gabe, J. (2011). The pharmaceuticalisation of society? *Sociology of Health & Illness*, *33*(5): 710–725.
60. Pardi, et al. (2018). mRNA vaccines – a new era in vaccinology. *Nature Reviews Drug Discovery*, *17*(4): 261–279.
61. Verma, I. and Weitzman, M. (2005). Gene therapy. *Annual Review of Biochemistry*, *74*(1): 711–738.
62. Friedmann, T. and Roblin, R. (1972). Gene therapy for human genetic disease? *Science*, *175*(4025): 949–955.
63. Landhuis, E. (2021). The definition of gene therapy has changed. *Nature: Innovations in Gene Therapy*, 26 October.
64. Blaese, R. M., et al. (1995). T-lymphocyte-directed gene therapy for ADA – SCID. *Science*, *270*(5235): 475–480.
65. Landhuis, The definition of gene therapy has changed.
66. Wirth, T., Parker, N. and Ylä-Herttuala, S. (2013). History of gene therapy. *Gene*, *525*(2): 162–169.
67. Orkin, S. and Motulsky, A. (1995). Report and Recommendations of the Panel to Assess the NIH Investment in Research on Gene Therapy, *NIH Reports*.
68. Marshall, E. (1999). Gene therapy death prompts review of adenovirus vector. *Science*, *286*(5448): 2244–2245.
69. Hollon, T. (2000). Researchers and regulators reflect on first gene therapy death. *Nature Medicine*, *6*(1): 6–7.
70. Wilson, J. (2009). Lessons learned from the gene therapy trial for ornithine transcarbamylase deficiency. *Molecular Genetics and Metabolism*, *96*(4): 151–157.
71. Wilson's interview is published in: Wenner, M. (2009). Gene therapy: an interview with an unfortunate pioneer. *Scientific American*, 1 September, pp. 7–8.
72. Ibid. See also: Wilson, R. (2010). The death of Jesse Gelsinger: new evidence of the influence of money and prestige in human research. *American Journal of Law & Medicine*, *36*(2–3): 295–325.
73. Ylä-Herttuala, S. (2012). Endgame. *Molecular Therapy*, *20*(10): 1831–1832.
74. Crowe, K. (2018). The million-dollar drug. *CBC News*, 17 November.
75. Mullin, E. (2017). The world's most expensive medicine is being pulled from the market. *MIT Technology Review*, 21 April.
76. Pearlman, A. (2019). Biohackers are pirating a cheap version of a million-dollar gene therapy. *MIT Technology Review*, 30 August.
77. Ibid.
78. Crowe, K. (2019). Canadian breakthrough that became the world's most expensive drug, then vanished, gets second chance. *CBC News*, 17 October.
79. Kozubek, J. (2018). *Modern Prometheus*. Cambridge University Press, pp. 22–23.

80. Kleutghen, Paul, et al. (2018) Drugs don't work if people can't afford them. *Health Affairs*, 8 February.
81. Marselis, D. and Hordijk, L. (2020). From blockbuster to nichebuster. *BMJ*, *370*: m2983.
82. Alhakamy, N., Curiel, D. and Berkland, C. (2021). The era of gene therapy. *Drug Discovery Today*, *26*(7): 1602–1619.
83. Ma, C., et al. (2020). The approved gene therapy drugs worldwide. *Biotechnology Advances*, *40*: 107502.
84. Cohn, R. (2016). Gene Therapy 2.0. *Neuromuscular Disorders*, *26*: S208–S209.
85. Uddin, F., Rudin, C. and Sen, T. (2020). CRISPR gene therapy. *Frontiers in Oncology*, *10*: 1387.
86. Cyranoski, D. (2016). First trial of CRISPR in people. *Nature*, *535*(7613): 476–477.
87. Cyranoski, D. (2016). CRISPR gene-editing tested in a person for the first time. *Nature News*, *539*(7630): 479.
88. Alhakamy, N., Curiel, D. and Berkland, C. (2021). The era of gene therapy. *Drug Discovery Today*, *26*(7): 1602–1619.
89. Ledford, H. (2020). Quest to use CRISPR against disease gains ground. *Nature*, *577*(7789): 156–157.
90. Frangoul, H., et al. (2021). CRISPR-Cas9 gene editing for sickle cell disease and β-thalassemia. *New England Journal of Medicine*, *384*(3): 252–260.
91. Demirci, S., et al. (2021). CRISPR-Cas9 to induce fetal hemoglobin for the treatment of sickle cell disease. *Molecular Therapy: Methods & Clinical Development*, *23*: 276–285.
92. Philippidis, A. (2020). One small dose, one giant leap for CRISPR gene editing. *Human Gene Therapy*, *31*(7–8): 402–404.
93. Auwerter, T., et al. (2020). Biopharma portfolio strategy in the era of cell and gene therapy. *McKinsey Report*, 8 April.
94. Alliance for Regenerative Medicine. (2021). *2020 Annual Report*. ARM Publishing.
95. Cohen, J. (2017). How the battle lines over CRISPR were drawn. *Science News*, 15 February.
96. Kozubek, J. (2017). Who will pay for CRISPR? *STAT*, 26 June.
97. Ibid.
98. See, for instance: Fu, Y., et al. (2013). High-frequency off-target mutagenesis induced by CRISPR-Cas nucleases in human cells. *Nature Biotechnology*, *31*(9): 822–826; Cullot, G., et al. (2019). CRISPR-Cas9 genome editing induces megabase-scale chromosomal truncations. *Nature Communications*, *10*(1): 1–14.
99. Harper, A. (2021). Crispr genome editing may select for cells with cancer-associated mutations. *BioNews*, *1122*(1): 11.
100. Lane, D. (1992). p53, guardian of the genome. *Nature*, *358*(6381): 15–16.
101. Sanford Burnham Cancer Institute (2021). Study encourages cautious approach to CRISPR therapeutics. *Science Daily*, 11 November.
102. Azzazy, H., Mansour, M. and Christenson, R. (2009). Gene doping. *Clinical Biochemistry*, *42*(6): 435–441.

103. Cantelmo, R., et al. (2020). Gene doping. *European Journal of Sport Science, 20*(8): 1093–1101.
104. Gao, M. and Liu, D. (2014). Gene therapy for obesity. *Discovery Medicine, 17*(96): 319–328.
105. American Medical Association (2013). Is obesity a disease? *Report of the Council on Science and Public Health. CSAPH Report, 3A*(13): 1–14.
106. Ortiz, S., Kawachi, I. and Boyce, A. (2017). The medicalization of obesity. *Health, 21*(5): 498–518.
107. Wong, D., Sullivan, K. and Heap, G. (2012). The pharmaceutical market for obesity therapies. *Nature Reviews. Drug Discovery, 11*(9): 669–670.
108. Claussnitzer, M., et al. (2015). FTO obesity variant circuitry and adipocyte browning in humans. *NEJM, 373*(10), 895–907. The reported statement by Manolis Kellis was published in: Knight, H. (2015). A metabolic master switch underlying human obesity. *MIT News*, 19 August.
109. Cold Spring Harbor Laboratory (2019). Gene therapy reduces obesity and reverses type 2 diabetes in mice. *Science Daily*, 29 August.
110. Chung, J., et al. (2019). Targeted delivery of CRISPR interference system against Fabp4 to white adipocytes ameliorates obesity. *Genome Research, 29*(9): 1442–1452.
111. These expressions are used, for instance, in the following articles: Friedmann, T. (1992). A brief history of gene therapy. *Nature Genetics, 2*(2): 93–98; Gura, T. (1999). Repairing the genome's spelling mistakes. *Science, 285*(5246): 316–318.
112. Naeem, M., et al. (2020). Latest developed strategies to minimize the off-target effects in CRISPR-Cas-mediated genome editing. *Cells, 9*(7): 1608–1626.
113. Silver, L. M. (2000). Reprogenetics. *EMBO Reports, 1*(5): 375–378.
114. Sharma, A., et al. (2020). Germline gene editing for sickle cell disease. *The American Journal of Bioethics, 20*(8): 46–49.
115. Doudna, J. and Sternberg, S. (2017). *A Crack in Creation*. Houghton Mifflin, p. 159.
116. Church, G. (2015). Perspective: encourage the innovators. *Nature, 528*(7580): S7–S7.
117. Kevles, D. (1980). Genetics in the United States and Great Britain, 1890–1930. *Isis, 71*(3): 441–455.
118. Muller, H. (1950). Our load of mutations. *American Journal of Human Genetics, 2*(2): 111–176.
119. Fletcher, J. (1974). *The Ethics of Genetic Control: Ending Reproductive Roulette*. Anchor Press. See also: Shaw, M. (1977). Perspectives on today's genetics and tomorrow's progeny. *Journal of Heredity, 68*(5): 274–279.
120. Regalado, A. (2015). Engineering the perfect baby. *MIT Technology Review*, 5 March.
121. Doudna and Sternberg, *A Crack in Creation*, p. 187.
122. Specter, M. (2015). The gene hackers. *New Yorker*, 16 November.
123. Baltimore, D., et al. (2015). A prudent path forward for genomic engineering and germline gene modification. *Science, 348*(6230): 36–38 (emphasis added).

124. Liang, P., et al. (2015). CRISPR/Cas9-mediated gene editing in human tripronuclear zygotes. *Protein Cell*, *6*: 363–372.
125. Fessenden, M. (2015). Gene editing in human embryos ignites controversy. *Smithsonian Magazine*, 23 April.
126. US National Academy of Sciences. (n.d.). International Summit on Human Gene Editing. [www.nationalacademies.org/our-work/international-summit-on-human-gene-editing#sectionProjectScope].
127. LaBarbera, A. (2016). Proceedings of the international summit on human gene editing. *Journal of Assisted Reproduction and Genetics*, *33*(9): 1123–1127.
128. Ma, H., et al. (2017). Correction of a pathogenic gene mutation in human embryos. *Nature*, *548*(7668): 413–419.
129. See: Liu, N. and Loucas, L. (2018). Chinese scientist claims to have created genetically edited babies. *Financial Times*, 26 November.
130. Cyranoski, D. and Ledford, H. (2018). Genome-edited baby claim provokes international outcry. *Nature*, *563*(7731): 607–609.
131. Cyranoski, D. (2020). What CRISPR-baby prison sentences mean for research. *Nature*, *577*(7789): 154–156.
132. Davies, K. (2020). Guilty as charged: a Chinese court delivered a three-year prison sentence and hefty fine to He Jiankui. *Genetic Engineering and Biotechnology News*, *40*(2): 30–32.
133. Regalado, A. (2019). China's CRISPR babies: read exclusive excerpts from the unseen original research. *MIT Technology Review*, 3 December.
134. Qiu, J. (2019). Chinese government funding may have been used for 'CRISPR babies' project. *STAT News*, 25 February.
135. Ouagrham-Gormley, S. and Vogel, K. (2020). Follow the money: what the sources of Jiankui He's funding reveal about what Beijing authorities knew about illegal CRISPR babies, and when they knew it. *Bulletin of the Atomic Scientists*, *76*(4): 192–199.
136. Dyer, O. (2019). Creator of first gene edited babies faces stern punishment. *British Medical Journal*, *364*: i346.
137. Ouagrham-Gormley and Vogel, Follow the money, p. 194.
138. Nie, J. (2018). He Jiankui's genetic misadventure. *Hastings Center*, December–5. See also: Nie, J., et al. (2020). Conflict of interest in scientific research in China. *Journal of Bioethical Inquiry*, *17*(2): 191–201.
139. Cohen, J. (2019). The untold story of the 'circle of trust' behind the world's first gene-edited babies. *Science*, 1 August.
140. Ibid.
141. WHO. (2018). Evaluation of the safety and efficacy of gene editing with human embryo CCR5 gene. WHO International Clinical Trials Registry Platform. [www.chictr.org.cn/showprojen.aspx?proj=32758]. See also: Song, L. and Joly, Y. (2021). After He Jiankui. *Medical Law International*, *21*(2): 174–192.
142. Wilson, *Lessons Learned*, pp. 151–157.
143. Lander, E., et al. (2019). Adopt a moratorium on heritable genome editing. *Nature*, *567*: 165–168.
144. Begley, S. (2019). Fertility clinics around the world asked 'CRISPR babies' scientist for how-to help. *STAT News*, 28 May.

145. Scheufele, D., et al. (2017). US attitudes on human genome editing. *Science*, *357*(6351): 553–554.
146. President's Council on Bioethics. (2003). *Beyond Therapy*. Executive Office of the President.
147. Sufian, S. and Garland-Thomson, R. (2021). The Dark Side of CRISPR, *Scientific American*, 16 February.
148. Garland-Thomson, R. (2020). How we got to CRISPR. *Perspectives in Biology and Medicine*, *63*(1): 28–43.
149. Agar, N. (1998). Liberal eugenics. *Public Affairs Quarterly*, *12*(2): 137–155; Agar, N. (2004). *Liberal Eugenics: In Defence of Human Enhancement*. Wiley & Sons; Agar, N. (2008). How to defend genetic enhancement. In Gordijn, B. and Chadwick, R. (eds). *Medical Enhancement and Posthumanity*. Springer, pp. 55–67. Agar, N. (2019). Why we should defend gene editing as eugenics. *Cambridge Quarterly of Healthcare Ethics*, *28*(1): 9–19.
150. Agar, *Liberal Eugenics*, p. 135.
151. Ibid., p. 7.
152. Ibid., p. 5.
153. All quotes in this paragraph are from: Sinsheimer, R. (1969). The prospect for designed genetic change. *American Scientist*, *57*(1): 134–142.
154. Nozick, R. (1974). *Anarchy, State, and Utopia*. Basic Books, pp. 314–315.
155. Galton, F. (1883). *Inquiries into Human Faculty and its Development*. Macmillan. Also: Agar, N. (2013). Eugenics. In Lafollette, H. (ed.). *International Encyclopedia of Ethics*. Blackwell Publishing, pp. 1766–1771.
156. Savulescu, J. (2001). Procreative beneficence. *Bioethics*, *15*(5–6): 413–426. Savulescu expresses the view that genome engineering should be deployed as a means of 'screening out' a whole list of people he presumes of inferior value, writing that 'when it comes to screening out personality flaws, such as potential alcoholism, psychopathy and disposition to violence, you could argue that people have a moral obligation to select ethically better children'. For a detailed critique of Savulescu's positions see: Sparrow, R. (2011). A not-so-new eugenics. *Hastings Center Report*, *41*(1): 32–42.
157. Foucault, M. (2004). *The Birth of Biopolitics*. Palgrave, p. 225.
158. All quotes in this paragraph are from: Foucault, *The Birth of Biopolitics*, pp. 224–229.
159. The lecture is available at www.practicalethics.ox.ac.uk.

Conclusion

1. Marx, K. (1976). *Capital, Volume I*. Penguin, p. 313.
2. Kaempffert, W. (1936). A biologist's view of man's future. *New York Times*, 15 March, p. 4.
3. Muller, H. (1929). The gene as the basis of life. *Proceedings of the International Congress of Plant Science*, *1*: 897–921.
4. Chan, L., Kosuri, S. and Endy, D. (2005). Refactoring bacteriophage T7. *Molecular Systems Biology*, *1*(1): 2005–0018.
5. Doudna, J. and Sternberg, S. (2017). *A Crack in Creation*, p. 90.

6. Muller, H. (1946). The production of mutations. In Nobel Foundation (1964). *Nobel Lectures: Physiology or Medicine 1942–1962*. Elsevier, p. 1.
7. 'Estrangement appears not only in the fact that the means of my life belong to another and that my desire is the inaccessible possession of another, but also in the fact that all things are other than themselves, that my activity is other than itself, and that finally – and this goes for the capitalists too – an inhuman power rules over everything.' Marx, K. (1992). Economic and philosophical manuscripts of 1844. In Marx, K. (1992). *Early Writings*, Penguin, p. 366. See also: Dyer-Witheford, N., Kjøsen, A. and Steinhoff, J. (2019). *Inhuman Power*. Pluto Press.
8. Crispo, M., (2015). Efficient generation of myostatin knock-out sheep. *PloS One*, *10*(8), e0136690; Editors (2015). Brazil approves transgenic eucalyptus. *Nature Biotechnology*, *33*: 577; Devanthi, P. and Gkatzionis, K. (2019). Soy sauce fermentation. *Food Research International*, *120*: 364–374. See also: Kromdijk, J., et al. (2016). Improving photosynthesis and crop productivity by accelerating recovery from photoprotection. *Science*, *354*(6314): 857–861.; Parry, M., et al. (2003). Manipulation of Rubisco. *Journal of Experimental Botany*, *54*(386): 1321–1333.
9. Sonstergard, T. (2019). Stolen Kiss and Naturally Cool – two advanced breeding solutions for the animal welfare traits of castration and heat stress. *Proceedings of the 2019 RSPCA Australia Animal Welfare Seminar*. [www.rspca.org.au/sites/default/files/AWS-2019-Proceedings.pdf].
10. Walker, N., Patel, S. and Kalif, K. (2013). From Amazon pasture to the high street. *Tropical Conservation Science*, *6*(3): 446–467.
11. Biotechnology Innovation Organization (2018). Recombinetics' animal gene editing could transform the beef industry. [www.bio.org/blogs/recombinetics-animal-gene-editing-could-transform-beef-industry].
12. Grossi, G., et al. (2019). Livestock and climate change. *Animal Frontiers*, *9*(1): 69–76.
13. Karavolias, N., et al. (2021). Application of gene editing for climate change in agriculture. *Frontiers in Sustainable Food Systems*, *5*: 685801.
14. Bebber, D. (2015). Range-expanding pests and pathogens in a warming world. *Annual Review of Phytopathology*, *53*: 335–356; Bett, B., et al. (2017). Effects of climate change on the occurrence and distribution of livestock diseases. *Preventive Veterinary Medicine*, *137*: 119–129.
15. Langner, T., Kamoun, S. and Belhaj, K. (2018). CRISPR crops: plant genome editing toward disease resistance. *Annual Review of Phytopathology*, *56*: 479–512.
16. deLorenzo, V., Marliere, P. and Solé, R. (2016). Bioremediation at a global scale: from the test tube to planet Earth. *Microbial Biotechnology*, *9*(5): 618–625.
17. Solé, R. Montañez, R. and Duran-Nebreda, S. (2015). Synthetic circuit designs for earth terraformation. *Biology Direct*, *10*(1): 1–10.
18. Cyranoski, D. (2016). CRISPR gene-editing tested in a person for the first time. *Nature News*, *539*(7630): 479.
19. E.g., Charlet, K. (2018). The new killer pathogens: countering the coming bioweapons threat. *Foreign Affairs*, *97*: 178.

20. Marx, K. (1976). *Capital: Volume I*, p. 512.
21. Like CRISPR, retrons are part of a bacterial immune system; some scientists have speculated that they might constitute the next technological frontier. Brownell, L. (2022). Move over CRISPR, the retrons are coming. *Wyss Institute News*, 30 April.
22. Virilio, P. (1998) Surfing the accident. In V2 Organization (eds). *The Art of the Accident*. NAI, pp. 30–44. See also: Virilio, P. (1999). *Politics of the Very Worst*. Semiotext(e), p. 89.
23. Lewis, G., et al. (2020). The biosecurity benefits of genetic engineering attribution. *Nature Communications*, *11*(1): 1–4.
24. The reference to 'capital personified' is from: Marx, *Capital, Volume I*, p. 254. The second quote is from: Gates, B. (2018). Gene editing for good: how CRISPR could transform global development. *Foreign Affairs*, 97: 166.
25. Harvey, D. (2004) *The New Imperialism*. Oxford University Press.
26. Shelley, M. (2017) [1818]. *Frankenstein*. MIT Press, p. 72.
27. Latour, B. (2011). Love your monsters. *Breakthrough Journal*, 2(11): 21.
28. Descartes as quoted in: Marx, *Capital, Volume 1*, p. 513.
29. Fisher, M. (2009). *Capitalist Realism*. John Hunt.
30. Shiva, V. (1993). *Monocultures of the Mind*. Palgrave.
31. Delbrück, M. (1949). A physicist looks at biology. *Transactions of the Connecticut Academy of Arts and Sciences*, *38*: 173–190.
32. Marx, *Capital, Volume I*, p. 497.
33. Benjamin, W. (2003). Paralipomena to 'On the concept of history'. In *Selected Writings: 1938–1940, Vol. 4*. Harvard University Press, p. 402.

Index

Thanks to our Patreon subscribers:

Andrew Perry
Ciaran Kane

Who have shown generosity and comradeship in support of our publishing.

Check out the other perks you get by subscribing to our Patreon – visit patreon.com/plutopress.

Subscriptions start from £3 a month.